Tupolev Aircraft

*Impressive take-off from Farnborough by a Tu-22M-3 in 1992, the first visit to the West
by a Tupolev military aircraft.*

Tupolev Aircraft
since 1922

Bill Gunston

PUTNAM

© Bill Gunston 1995

First published in Great Britain in 1995 by
Putnam Aeronautical Books,
an imprint of Brassey's (UK) Ltd,
33 John Street,
London WC1N 2AT

British Library Cataloguing in Publication Data

Gunston, Bill
Tupolev Aircraft since 1922
I. Title
629.13309

ISBN 0 85177 866 6

Typesetting and page make-up by
Florencetype Ltd, Stoodleigh, Devon
Printed and bound by
Butler & Tanner Ltd, Frome

Contents

Appendices

Index of Aircraft

Index of Persons

Introduction and Acknowledgements

In September 1977, one of the world's biggest-selling aviation magazines, the New York-based *Flying*, produced a massive gold-covered 50th Anniversary issue. They asked me if I would contribute a brief feature on a famous aircraft manufacturer. I said 'Certainly, I'll do Tupolev'. The reply was, 'Who? ... no, we said, someone *famous*'. In the end I had my way, though the piece bore little similarity to what I wrote. Sad, because the name Tupolev – which in recent English-language books has sometimes appeared as 'Tupelov' and even 'Tulepov' and is often mispronounced 'Tyewpoleff' – deserves to be the *most* famous in the whole history of aircraft design.

This is for the simple reason that this name identifies the greatest and most diverse family of aircraft in the world. What makes this doubly remarkable is that, almost from the beginning, the name Tupolev has tended to mean big, heavy and powerful aircraft. In the 1930s one of these, with a span of 63m, was the biggest aeroplane in the world, and it was almost followed by a production bomber with the amazing span of 95m (nearly 312ft). Even today the heaviest and most powerful combat aircraft of all time is a Tupolev.

Whereas the early giants had severe, uncompromising angular outlines, today's monster, the Tu-160, is slender and graceful. Where they are alike is that every one of the one hundred-plus types featured in this book was a thoroughly sound design. In stark contrast to some of the most famous makes of Western aircraft, Tupolev aircraft have no history of structural failure in the sky except when induced aeroelastically. At the start, the structural technology was based on the

aluminium-alloy airframes of Junkers. Several Junkers F 13 transports broke up in the air, one of them over England, and young Tupolev was determined he would do better. Many Tupolev aircraft flying today have lasted several times as long as their Western counterparts, despite having to suffer the worst weather in the world.

Of course, the founder of the dynasty did not do it all himself. Like all great aircraft designers, he made it his business to gather round him talented draughtsmen and engineers, many of whom, under his tutelage, became famous in their own right. Even British and American aviation enthusiasts have heard of Arkhangel'-skii, Petlyakov, Sukhoi and Mya-sishchev. This book includes several aircraft, such as the Ar-2, Pe-8 and Su-2, which were later designated for Tupolev's aides.

Andrei Nikolayevich Tupolev was born on 10 November (old calendar 29 October) 1888, at Pustomazov, near Kalinin. He picked the right father,

because Nikolai Ivanovich Tupolev was a lawyer who was arrested for revolutionary activities. Such an attitude inevitably carried across to his son, and after the October (1917) Revolution this furnished him with the right credentials.

In 1909, at the age of 20, A N Tupolev gained admission to the Moscow Higher Technical School (MVTU). This was one of the best such establishments in the world. It even had a 1.5m by 0.3m (59in by 12in) wind tunnel, erected in 1902. Here he fell under the spell of the great Professor Nikolai Yegorovich Zhukovskii, who was primarily an aerodynamicist. Tupolev quickly realised that aircraft design was to be his lifetime's work. In 1910 he made many flights with a biplane hang-glider, towed by runners at Lefortovskii Park. He was also made the Supervisor of the MVTU tunnel, but in 1911 the Tsar's police interrupted his studies when Andrei Nikolayevich was arrested for inciting student unrest.

The Zhukovskii Museum, once the house of the fur merchant Mikhailov on Moscow's Voznesensky Street, where Tupolev began his career. Note Zhukovskii's bust at left. (ANTK Tupolev)

After the outbreak of war in 1914 Tupolev was able to get a job with the Duks factory at Khodinka, Moscow, where he gained his first practical experience of aircraft manufacture, modification and, to a small degree, design. Thus, when the Russian war effort crumbled from within in 1917–18 he was one of the few people who was simultaneously a theoretically qualified aerodynamicist, an engineer with shop-floor experience, and an ardent revolutionary.

On 28 June 1918 V I Lenin published a decree nationalising the aircraft factories. Many had been vandalised or even burned, and all were essentially at a standstill. The workers were either fighting with Red forces or scouring the countryside looking for food, and the entire vast country was in chaos. Tragically, the reaction against authority extended to anyone who was an officer, or in any position of authority, and such people quickly learned that, if they wished to survive, it would pay to put on rough 'proletarian' clothes and try and get their hands dirty. The alternative was to try to escape abroad, as did such designers as Sikorsky and de Seversky. Thus, the infant Soviet Union lost the services of about half its most highly qualified aviation engineers.

One of the few aircraft factories left intact was the Duks works, and as it was famous and in Moscow it was proclaimed GAZ No.1 (*Gosudarstvennyi aviatsionnyi zavod*, State Aircraft Factory, No. 1). Henceforth, everything in the Soviet State was to be organised and ruled with absolute authority from Moscow. The first thing to do was to set up organisations, and one of the first to be created was the NTO-VSNKh, in English the Scientific/Technical Department of the Supreme Council of the National Economy. As early as the summer of 1917 Zhukovskii had campaigned for the creation of a large scientific and technical institute specialising in aviation, and it was obvious to the new rulers in the Kremlin that aviation would be important to both national defence and the national economy. Zhukovskii repeated his plea at the All-Russian Aviation Congress in 1918, and on 23 June 1918 the Congress passed a resolution endorsing this.

Despite his youth, Tupolev was already a respected member of the tiny group of aviation engineers available in Moscow, and he became a member of the management board of the NTO-VSNKh. After much further discussion it was decided on 30 October 1918 to form a joint aerodynamic and hydrodynamic section of the NTO, and a committee of three was appointed to work out how this should be done. The three were Zhukovskii, Tupolev and I A Rubinsky.

Following fifteen meetings at Zhukovskii's home, regulations, an organisation and estimated financial costings for this institute were agreed and sent to N P Gorbunov, the Director of NTO. On 1 December 1918 the proposals were agreed, and a few days later the sum of 20,000R (Russian spelling *Rubl*) was allocated, together with a commandeered house, 21 Voznesensky Street, which had belonged to a fur merchant, Mikhailov. Today this house is No. 17 Radio Street (renamed because of the location of Moscow's first radio station), and houses the Scientific-Memorial Museum for Zhukovskii. By December the title of the organisation had been established as the Central Aero-Hydrodynamic Institute. The phonetic English rendering of the Cyrillic initials is TsAGI, but it is simpler to call it CAHI.

CAHI began operating on 14 December 1918 with a staff of thirty-three. Apart from the Director (Zhukovskii, of course) and Tupolev, the senior members included B N Yur'yev, N V Krasovsky, V P Vetchinkin, A A Arkhangel'skii, A I Putilov, N S Nekrasov, K A Ushakov, A M Chyeremukhin and G M Muzinyants. Almost all had been Zhukovskii's pupils. From this small group, working in an ordinary house in central Moscow, with little except a bench and a few items of laboratory equipment – some liberated from the MVTU and other schools, but at the outset not even having a small wind tunnel – CAHI was to grow until by the 1930s it was beyond question the largest and best-equipped aeronautical laboratory in the world.

At first much of the work was directly concerned with assisting the embryonic Red Air Fleet (VVS), for example by designing mountings for guns and bombs for the motley collection of aircraft that were left from the Great War, or had been imported since or (to the tune of several hundred) captured from the White Russian or interventionist forces. But on 1 January 1919 an Aviation Section was formed, and this did have a wind tunnel and began testing wing profiles and undertaking many other aerodynamic tasks, besides doing stressing calculations and strength testing of struts, wires and, later, of parts made of a new light alloy.

A diversion in 1920 was the formation of KOMTA (Commission for Heavy Aviation), on which Tupolev served under the chairmanship of Zhukovskii. There was an obvious need for an aeroplane large and powerful enough to do a useful transport job, replacing the obsolete and 'tired' *Ilya Muromets* bombers. In the end KOMTA produced a quite modest triplane, powered by two 240hp Fiat engines, and it was not only patently inadequate but also a most unsuccessful design. Tupolev was at pains to assure the author that he did not participate in the design, only in the initial planning.

In fact, as described in the first entry in this book, Tupolev had already acquired a general idea of how aeroplanes should be designed which was remarkable for its perceptiveness and essential rightness. Whereas in Britain the RAF and Imperial Airways insisted on biplanes, with wooden structures, Tupolev thought in terms of cantilever monoplanes with a structure of metal, and if possible of aluminium alloy. He did his best to read foreign magazines reporting on developments in other countries, and was already aware of the cantilever monoplanes produced by Junkers, which not only had light-alloy structures but were even skinned with corrugated sheets of the alloy, so that the skin could bear a major part of the load and also withstand rough handling.

Hugo Junkers had begun to talk to the Russian Government as early as 1917 to see if there was a market for his aircraft, and after the collapse of Germany these discussions became more serious and urgent. Tupolev played a role in persuading the new Soviet authorities that Junkers' wish to set up a design and manufacturing works in Moscow should be considered carefully. It took a very long time to get an agreement hammered out, but as described later it did come to pass, and it exercised a profound influence on

The first products: aero-sleighs in Voznesensky (later Radio) Street. (ANTK Tupolev)

Tupolev, which lasted for a decade from the early 1920s until the early 1930s. Without the technology already explored by the German designers it is probable that Tupolev's early designs would have been more difficult to achieve and would have taken longer.

Meanwhile, Tupolev's growing diversity of research and design effort was suddenly focussed on 28 August 1919 by the receipt of a suggestion by the Defence Council that CAHI should start building aero-sleighs. These were already a familiar kind of vehicle, mainly quite small (even single-seaters) propelled at frightening speeds across ice and compact snow by an air propeller. Zhukovskii appointed Tupolev vice-chairman of the committee to organise this work, and by 1922 six successful sleighs were produced. In 1921 Tupolev led the design and construction of the GANT-1 high-speed hydroplane, and he never lost contact with the technology of high-speed vehicles for travel over water, ice and snow.

In 1919–21 CAHI and its various sub-sections still presented the outward impression of an amateur group of enthusiasts. The designers, who often worked not on drawing boards but on ordinary tables, were in what had been the bedrooms on the upper floor of 17 Radio Street. All construction was done on the ground floor, and also in No. 16 across the road, and in a former inn, the *Rayek*, next to No. 17 on the corner of Nyemetskii (later Bauman) Street. This imposed severe limits on the size of what could be made, and final erection had to be done outside.

On 17 March 1921 Zhukovskii died, and S A Chaplygin succeeded him as Director of CAHI. Tupolev was appointed deputy director in charge of the AGO (*Aviatsii i Gidrodinamiki Otdel*, Aviation and Hydrodynamics Department), which was responsible for the design and construction of aeroplanes, seaplanes, aero-sleighs and fast torpedo-boats. By this time the Soviet Union was painfully and against terrible difficulties trying to get its act together, and (for example) rebuild its aircraft industry and eliminate the total dependence on imported machines.

The concession to Junkers, which was finally signed on 29 January 1923, provided for the establishment of a design and production factory at Fili, in the Moscow suburbs, where the German company would not only design and produce aircraft but also train a Russian work-force. Junkers also agreed to organise the production of the German aluminium alloy Duralumin, and its fabrication into standard rolled sections and corrugated sheets. Junkers further agreed to set up plant for making their large water-cooled inline and vee engines. Unfortunately, it was obvious from the start that Junkers and the Soviet Union had slightly different objectives. Junkers merely wanted to evade the harsh terms of the Treaty of Versailles which prohibited any German production of powered aircraft. The Russians wanted to become self-sufficient.

This dichotomy of views had been the main reason why it took from 1917 until 1923 to agree the terms on which Junkers set up shop in Moscow.

It was all too obvious to Ivan Ivanovich Sidorin, the chairman of the commission for negotiating with Junkers, and in 1920 he had recommended that something should be done to organise domestic production of a copy of Duralumin, which is a generic term for alloys of aluminium and copper (around 4 per cent) with smaller amounts of magnesium and manganese. It is only slightly denser than aluminium, but very much stronger.

Engineers V A Butalov and Yu G Muzalevskii led a small team which studied the problem, and in August 1922 the first batch of small ingots was cast at the factory of GosPromTsvet-Meta (State production of non-ferrous metals). As this plant was in the village of Kol'chugino, in Vladimir *oblast* (district), the alloy was named Kol'chug-aluminiyem. Unlike Dural, it contained a little nickel, and different proportions of copper and manganese, but it proved to be outstandingly successful. By October the alloy was being produced in pilot quantities in rolled sections and sheets. By 1923 Kol'chug alloy was gradually becoming available in production quantities, including Junkers-style corrugated sheet.

Obviously, the next thing to do was to design aircraft to use the alloy, and the leading crusader in this work was A N Tupolev. He was picked to head a CAHI commission to study the design of all-metal aircraft, the other members being Sidorin (vice-chairman), G A Ozerov and I I Pogosskii. Supporting the committee were designers A I Putilov, V M Petlyakov, B M Kondorskii and N S Nekrasov. While this was going on, Tupolev used the first available Kol'chug materials to build six improved aero-sleighs, which he designated ANT-1 to ANT-6, and the GANT-2 (G-2), a second fast hydroplane. This was of great value in unearthing and solving various unexpected fabrication problems, and also in training the shop-floor workers.

By early 1922 Tupolev could not wait any longer but got down to the design of his first aeroplane. There were at this time in the emergent Soviet Union more than thirty men who had designed successful aircraft, but apart from Grigorovich and Polikarpov none had achieved anything really significant, and none was ever to produce a design built in quantity. In sharpest contrast, a six-teen year-old boy in his last year at school, one Aleksandr S Yakovlev, was shortly to claw his way up from nothing to become a deputy Minister of Aviation Industry and the most politically powerful of all designers. Tupolev, on the other hand, was 'on the inside track' from the outset, with both professional and practical qualifications and a heritage of Marx-ist activity. Not least of his contribu-tions was to pick such men as Putilov, Kondorskii, Arkhangel'skii, Petlyakov, Myasishchev and Sukhoi to lead his design teams, and who in most cases were later to form their own OKB (*Opytno Konstruktorskoye Byuro*, ex-perimental construction bureau) and reach the supreme rank of General Constructor.

It had become accepted practice for every Soviet aircraft to have a letter/number designation based on its function. Thus, the first mass-produced machine, based on the D.H.9A, was called the R-1, from *Razvyedchik*, reconnaissance). It says something about Tupolev's stature that, instead of calling his first aero-plane something like LL-1 (flying laboratory 1) or K-1 (Kolchug 1) he designated it with his own initials as the ANT-1, which he had already done with the first all-metal aero-sleigh, and this heretical designation was accepted. Except in the field of light aircraft, no other designer had the

Taking a break from building the ANT-1. From the left: A I Putilov, I I Pogosskii, A N Tupolev, N S Nekrasov and Ye I Pogosskii. (ANTK Tupolev)

nerve to use his own initials for another twenty years.

The story of the ANT-1 and its many successors forms the body of this book. During the 1920s both the Soviet Union and Tupolev's work-place changed out of all recognition. The growing AGO department was renamed AGOS, the S standing for *Stroitel'stvo*, construction. Led by the dynamic Tupolev, AGOS was never going to be content merely to under-take research and testing, and the design of small research aircraft for use by CAHI. Indeed it was positively

encouraged to create powerful aircraft for military and naval purposes and for transport. This could hardly be done at 17 Radio St, where in order to get the parts of the little ANT-1 outside they had to take the door off its hinges. Subsequent aircraft were much larger, and their success inevitably led to CAHI relocating the AGOS group in a completely new purpose-designed building, with proper facilities for design, testing and construction. This was built on the site of 16 Radio Street in 1925. This was too late for the creation of two much more powerful

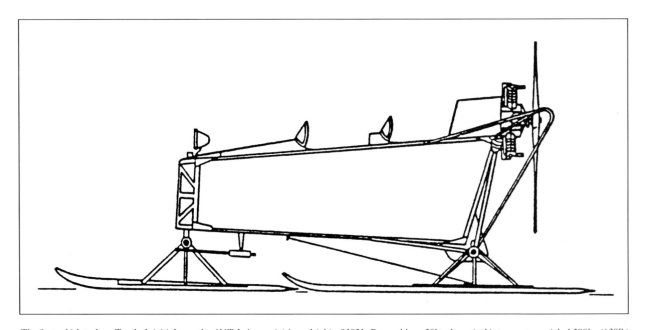

The first vehicle to bear Tupolev's initials was the ANT-1 Aerosani (Aero-sleigh) of 1921. Powered by a 38hp Anzani, this two-seater weighed 290kg (639lb) empty and reached 41km/h (25mph). Some of the numerous production successors could reach 100km/h (62mph).

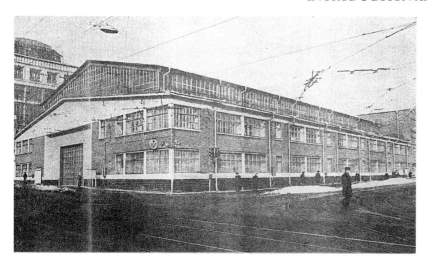

The new AGOS building seen across Radio Street, with Bauman Street at right. (ANTK Tupolev)

prototypes, the ANT-3 reconnaissance biplane and the ANT-4 monoplane heavy bomber. To get the main parts of the ANT-4 out into Radio Street the wall had to be demolished, but this was no problem as the old building was being pulled down completely.

The ANT-3 and -4 made international flights which put the Soviet Union, and Tupolev, on the aeronautical map. The ANT-4 in particular – which as an all-metal cantilever monoplane was precisely how Tupolev considered aircraft of the future should be designed – bore comparison with the best contemporary foreign aircraft, and provided a basis for much further development. These aircraft put Tupolev firmly in the saddle, and made him one of the most important men in the USSR. Later this very fact was to put him in peril, though he was more fortunate than some aircraft designers in that he was merely

imprisoned. But in 1925 there was no inkling of Stalin's future terror.

Tupolev organised AGOS efficiently, and was fortunate in that funds voted for experimental prototype aircraft and high-speed marine craft including torpedo-boats, could not be diverted to any other CAHI purpose. Thus, the flow of prototypes not only kept appearing but the aircraft became ever better, faster and more powerful, with the accent heavily on large bombers, transports and flying-boats, plus the occasional fighter.

From 1925 Tupolev organised his designers into groups (later called brigades), each specialising in a particular task. B M Kondorskii was charged with arriving at the general shape (including the first three-view drawing) and constructing the mock-up; A A Arkhangel'skii and A I Putilov designed the fuselages; V M Petlyakov designed the wings; N S Nekrasov was

responsible for tail units; the brothers I I and Ye I Pogosskii handled engine installations; N I Petrov designed landing gears; I P Tolstykh was assigned armament and equipment; and V N Belyayev was responsible for strength calculations. In addition to their basic work, Ye I Pogosskii and N I Petrov had in the past served as test pilots, though for 'factory testing' of future prototypes AGOS employed professionals.

Though this book is concerned with aircraft, the design of fast marine craft, called *Katera* (cutters), was very important to Tupolev until 1937. In 1929 AGOS produced the prototype G-5 torpedo-boat, with a 675hp engine designed by A A Mikulin, who was to provide power for most of Tupolev's larger aircraft for many years. With a speed which in some versions reached 59.8kts (100.8km/h), these 15 tonne craft served well into the Second World War in the Baltic and Chyernomorsk fleets.

In the second half of the 1920s Tupolev and his principal structural aides, Petlyakov and Arkhangel'skii, calculated the stressing and material thicknesses needed for cantilever monoplanes similar to the ANT-4 in configuration but much larger. For the immediate future the logical step was an aircraft with four engines in the 500hp class, as used in the ANT-4. This came out to a span of about 40m (130ft) and weight of some 16–18 tonnes (35,000–40,000lb). Beyond this no obvious arithmetical barrier could be seen, and it appeared possible to consider an aeroplane weighing at least 150 tonnes, with a span of 200m (over 650ft). No other design team since

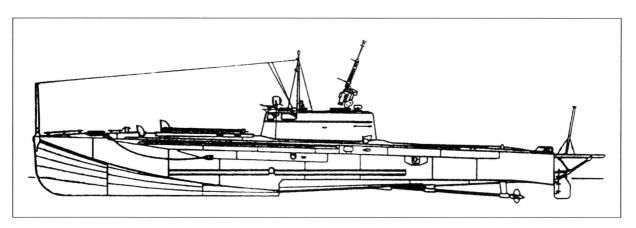

Until after the Second World War, Glissery *(hydroplanes) and* Katera *(fast boats) were important Tupolev products. This drawing, from* Neizvestnyi Tupolev *by M B Saukke (KTsNTI 'Original' 1993), shows the most important, the G-5. Based on the GANT-5 of 1925, this 14.5t craft had two 1,250hp Mikulin GM-34 engines and reached 52–55kts. In 1931–43 no fewer than 330 were built, most carrying two 533mm torpedoes.*

then has seriously considered aeroplanes of this size.

The different objectives made the relationship between Junkers and the Kremlin increasingly strained, and in any case by 1926 the deal had served its purpose. A particular cause of friction was that Junkers failed to train Soviet workers in the technology of metal aircraft. Thus, on 1 March 1927 – which by chance was the author's date of birth – the Fili factory was hung with red banners and slogans and named as the Soviet Union's GAZ No. 22.

As early as 1926 design began on the four-engined aircraft in the 40m category. The prototype, the ANT-6, flew at the end of 1930. It established the configuration that was to become universal in all countries, with the engines spaced across the leading edge, all driving tractor propellers. Remarkable as it may seem today, when the ANT-6 was designed such an arrangement was exceedingly rare. So successful was this design that it sustained a production run of 818 aircraft. Each was bigger and more capable than any comparable aircraft in other countries, and like all great aircraft the ANT-6 proved capable of continual modification to keep abreast of rapidly developing technologies. Sadly, in the Soviet Union's hour of need in 1941–45 these great aircraft had become outdated, and for various reasons there was no powerful force of heavy aircraft to replace them.

Just before the first ANT-6 took the air, on 3 December 1930, the Director of CAHI, S A Chaplygin, confirmed Tupolev as Chief Constructor and Director of AGOS. Tupolev at once organised within AGOS the ZOK (*Zavod Opitnyik Konstruktsyii*, factory for experimental construction). At this time the only other large and powerful aircraft design organisation was the Central Construction Bureau, at GAZ No. 39 named for Soviet pioneer V P Menzhinskii, where the NKVD (the secret police) were organising a special 'internal prison' for aircraft designers who had lost Stalin's favour. There were various attempts to combine AGOS and CCB, under their technical leaders Tupolev and S V Ilyushin, but what finally happened, taking effect in May 1932, was the transformation of both, AGOS becoming KOSOS.

On 13 January 1933 GUAP (aviation ministry) director P I Baranov

agreed this revised structure which, among other things, grouped CAHI aircraft design and experimental construction into ZOK, occupying GAZ No. 156, and KOSOS (*Konstruktorskii otdel opitnogo samolyet ostroyeniya*). In effect, Tupolev's work continued as before, though he formed his staff into nine brigades (omitting No. 4): No. 1, heavy aircraft, V M Petlyakov; 2, marine aircraft, I I Pogosskii, replaced after his death in the ANT-27 by A P Golubkov; 3, fighters and record aircraft, P O Sukhoi; 5, fast military aircraft and passenger derivatives, A A Arkhangel'skii; 6, experimental aircraft, V M Myasishchev; 7, propellers, V L Aleksandrov; 8, engine installations, Ye I Pogosskii; 9, landing gear, M N Petrov; and 10, torpedo-boats, N S Nekrasov.

Thanks to his outstanding team of senior designers, the list of successful 'ANT' products grew rapidly, notwithstanding the fact that almost without exception they were in the top bracket of speed and power. One type deserves particular mention: the ANT-20, biggest aeroplane in the world. The other world-beater was the ANT-40, which as the SB fast tactical bomber formed 94 per cent of Soviet front-line attack strength when Hitler struck east on 22 June 1941.

One of the many anecdotes told about Andrei Nikolayevich is that, during a meeting at the Kremlin in 1935, he was unwise enough to say to Stalin something like 'There is really little wrong with the SB, and all these defects are trivialities'. At which Stalin

is alleged to have replied, in most forceful terms, 'There are no trivialities in aviation. Everything is serious, and any of your trivialities could, if uncorrected, lead to the loss of an aircraft and its crew'.

Tupolev was the most respected designer in the entire Soviet aircraft industry. This was a time when, for several reasons, Stalin was about to embark on a policy of terror, which affected virtually everyone who was of any importance. For example, he had over 40,000 senior army officers put through show trials and executed, including fifty of the fifty-seven corps commanders, 154 of the 186 divisional commanders, all sixteen army commissars and 401 of the 456 senior colonels. Among various other groups, the terror extended to aircraft designers and factory managers (those of proven ability), though only a few, such as Kalinin and Chizhevskii, were executed. Most – including Polikarpov, Bartini, Myasishchev, Petlyakov, Nyeman and Putilov – were merely put behind bars, possibly in a belief they could be made to work harder, though they were described to the public as 'enemies of the people'. At the stroke of a pen, they lost all authority, as well as their liberty, and essentially ceased to exist.

It was Tupolev's turn on 21 October 1937. His show trial was as quick as the others, the charge being that he had 'given the secret design of the Bf 110 to the Germans'. For almost a year he was held in Lubyanka and then Butyrkii prison, while projects from

The top team in 1957, from the left: D S Markov, A M Cheremukhim, A N Tupolev, A A Arkhangel'skii and S M Yeger. (ANTK Tupolev)

The KOSOS building, which is still the Tupolev head office, on the corner of Radio Street and what is today Naberezhnaya Akademika Tupoleva. (ANTK Tupolev)

the ANT-46 onwards either fell by the wayside or (ANT-51) were taken over by Sukhoi. At first he and fellow designers were either put to hard labour or kept in idleness, but eventually Tupolev was allowed a drawing board. He busied himself with studies and calculations, but there was no way he could do productive work until in 1938 it was decided to organise the designers into Central Construction Bureau 29, often rendered in English as TsKB-29.

One day in late August 1938 Tupolev was taken out of Cell 58 at Butyrkii and escorted to the office of Lavrenti P Beria, head of the NKVD. He was curtly told to 'take work'. He was then taken to a fenced-off enclosure at Bolshyevo, near Moscow, where he found three huts, one for the detainees and guards, a second a kitchen/dining hut and a third a drawing office. He was met by A A Alimov, senior of the detainees, and led to a place of honour near the stove. He at once began to lead the small team of seventeen. On 10 November, his 50th birthday, they had a special treat: apples from the prison orchard.

Their first assignment was to design a four-engined dive-bomber; this, the '57', was soon cancelled. Tupolev himself was convinced the greatest need was for a modern frontal dive-bomber, to replace the SB. He eventually got permission to begin work on 'Aeroplane 103', though his work was

frequently interrupted by his being summoned to report to Beria on what had been accomplished. In April 1939 he told G Kutepov, the NKVD head of CCB-29, that the tiny team in the hut could get no further; they needed a much larger team, with aerodynamic and engineering specialists, and much bigger facilities. Kutepov asked for a list of names. Tupolev did not know who was still at liberty, and in any case was fearful that those listed would be arrested.

Ironically, Tupolev's *Spetzkontingent* (special team) next found themselves, though still in close detention, back in their proper offices at KOSOS CAHI. The entire building had been converted into a prison, with barred windows, locked doors and movement possible only past turnstiles guarded by sentries. The big Oak Hall, planned by Tupolev for lectures and meetings with customers, was fitted with twenty-four bunks, one of them for Tupolev.

The detainees were nameless, without identity; each living space was known not by the number or name of the occupant but by the names of selected trustees, who were responsible for internal order. As 'enemies of the people' the designers could not sign drawings, but each designer was given a numbered rubber stamp for this purpose. Any movement from one room to another required meticulous investigation and issue of passes.

They were escorted everywhere, but Tupolev's stature was such that his escort always walked a respectful distance *behind* him. On Sundays everyone was taken to the factory showers, and at 7 p.m. they were allowed to stroll on the caged-in flat roof. The list of crimes against each detainee varied, but one day Sergei P Korolyev said 'It makes no difference; they will snuff all of us, without obituaries'. (Later Korolyev was to be showered with honours as architect of the Soviet ICBM and space programmes.)

The prototype '103' flew on 29 January 1941. On 22 June Germany invaded, and it was soon decided to evacuate. Tupolev was released from detention on 21 July 1941, followed by some of his *Spetzkontingent* a few days later. By this time they were on their way to build a new factory, GAZ No. 166, at Omsk. Here, working round the clock, Aeroplane 103 was transformed into the production Tu-2.

Today the Tu-2 is recognised as one of the best tactical bombers of the Second World War, yet the story of its design is ignored even in 'corrected and improved' editions of standard Soviet aviation histories, which (for example) merely say it was 'a response to the demands from the front', ignoring the fact the prototype was flying six months before the war began. Like countless other examples of Stalin's repression, the truth is now on record, and an article by Leonid Kerber and Maksimilian Saukke in *Kryil'ya Rodiny* for June 1988 calls Tupolev and his *Spetzkontingent* 'people of the greatest value and genuine patriotism who, in spite of slander and degradation heaped upon them, provided our country with the best weapon for the destruction of the enemy'. Incidentally, Tu-2 histories written in the West in the 1980s still have details of numbers built and dates which are pure fiction.

Tupolev told the author he bore no grudge against the Soviet State for his cruel treatment, saying 'I was merely required to breathe purified air for a few years'. During this time he could no longer use ANT designations, and in fact throughout the 1940s his projects were identified by a number only. Gradually, those that went into production became prefixed by Tu.

From 1943 Tupolev gradually managed to restore the former KOSOS building into a suitable habitation for a free team of designers. While the Tu-2 led to a profusion of derived aircraft, Stalin increasingly urged the creation of a modern strategic bomber, whilst at the same time campaigning for Lend–Lease Boeing B-29s. Tupolev worked on Type 64 to meet the need, but in 1944 he was told to produce a copy of the American bomber itself, three examples of which had fallen into Russian hands. The resulting story has no parallel in aviation. First to fly was an unarmed transport, the Tu-70, followed by series batches of successive versions of the Tu-4 strategic bomber.

These led in turn to heavier and more capable bombers, one of which, the Tu-85, could reach the USA and so almost went into production. What stopped it was that, while it had engines of twice the power of the Tu-4, a propeller-turbine was developed which for similar bulk and weight could develop more than six times the power, 15,000hp! The result was one of the most remarkable aircraft in history, the Tu-95. By combining great power with effective propellers turning in very coarse pitch, in a swept-back airframe, it rewrote the textbooks and combined jet speed with the propulsive efficiency of propellers.

Assisted by many of his long-time aides, A N Tupolev used this super-powerful propeller-turbine and large turbojets created by the Soviet engine designers to extrapolate the technology of the B-29 into large modern jets. The first and greatest success was the Tu-16 bomber, whose configuration then proved suitable for conversion into a passenger transport, the Tu-104. The latter in turn led to the 124, 134 and 154, and at the time this book was written in early 1995 the 134 and 154 were together producing 65 per cent of the passenger-km in the Russian Federation. In turn, they have led to the Tu-204, which is likely to be built in significant numbers, some all-Russian and other versions with increasing Western content. Tupolev was also the most successful pioneer of the supersonic bomber, the versatile and long-lived Tu-22 leading to the Tu-22M which is today one of the most useful theatre attack aircraft in the world.

Andrei Nikolayevich died in harness at 84 on 23 December 1972. At that

Most former Soviet design bureaux have showcases full of models in the entrance foyer. (N.A. Eastaway/RART)

time the deputy chief of the bureau, entitled Assistant General Director, was Andrei I Kandalov, whose entire career had been spent essentially in this role. Until 1995 Kandalov was still in this post, and the author owes him a debt of gratitude for his co-operation. In 1972 it appeared likely to outsiders that the son of A N Tupolev, Dr Aleksei A Tupolev, would take over from his father. In fact, this did not happen until 1988, and four years later Dr Tupolev was replaced by the present General Director, Valentin Klimov.

When the Soviet Union was terminated in 1990 the giant factory at Kazan, originally called GAZ No. 124, had delivered most of its order for the Tu-160, the heaviest and most powerful combat aircraft ever built. The OKB was reorganised as ANTK im A N Tupoleva (Aviation Scientific–Technical Complex named for A N Tupolev). Today it has little funded military work, and – having doubled in size in 1975–85 – has now had to shrink dramatically and strive in the harsh world of the marketplace to find viable programmes.

To offer a total service to customers it has acquired the giant Samara (Kuibyshev) factory (one-time GAZ No. 18), the equally large production plant at Ulianovsk and the Taganrog aviation production corporation on the Black Sea. Cryoplanes using liquid hydrogen are being studied

with Deutsche Airbus (now Daimler-Benz Aerospace) of Germany. Current Tupolev projects are of necessity almost entirely for civil aircraft, ranging from light aircraft for passengers and agricultural use to business jets, heavy freighters, 600-seaters, a superior SST and an aerospace-plane.

Acknowledgements

I would like to thank two meticulous Russians, Yefim Gordon and the former Tupolev OKB archivist, Vladimir Rigmant, for making available a wealth of original material which was by no means easy to unearth. As always, Nigel Eastaway, Trustee of the Russian Aviation Research Trust, a registered charity, has done his utmost to ensure I lacked nothing relevant to Tupolev in the Trust's possession. The debt is enormous. I am grateful for the powerful support of Valentin Klimov, who bears the heavy responsibility of being General Director of ANTK Tupolev, and of his Deputy General Director, Yuri Kashtanov. I also acknowledge the careful editing on behalf of the publisher by my lifelong friend John Stroud and the publisher's in-house editor Julian Mannering. Of course, I welcome additional information and the correction of any remaining errors.

The Aircraft

ANT-1

The first aircraft designed by A N Tupolev and his team at AGOS was predictably a simple sporting aircraft, designed to do no more than 'get daylight under the wheels'. It was designed partly for the team to gain experience, partly because there was a chance it might prove useful and be built in series, and partly in order to use some of the first samples of Kol'chug aluminium alloy.

Work began in April 1922. At this time no Kol'chug alloy existed, so the basic design was all-wooden. There were many good reasons for wood being a popular material, among them the limitless supply from Siberian forests, the low cost, the existence of established technology and skilled woodworkers, and the ease of construction and repair. The UVVS (Red Air Force technical directorate) issued a manual in 1922 entitled *Technical conditions for use of wood in aircraft construction*.

Against these factors were the fact that every piece of wood is unique, with its own particular properties, so that it is impossible to establish (or confirm by testing) standardised strength for any part. Wood has a directional grain, so its strength varies greatly depending on the direction of the applied load. It also deteriorates rapidly in harsh climates, and so did the adhesives available in 1922. Tupolev also believed that, once the technology was established, metal aircraft could be made lighter. His objective was ultimately to make aircraft entirely out of metal, with almost all the structure of Kol'chug alloy.

Nevertheless, he told the author that the limited use of this alloy in the ANT-1 was not because there was no more in existence. He sanctioned only a limited use as a measure of prudence, because despite its apparently excellent properties it was still a completely new material, and it might have hidden shortcomings.

Tupolev and his assistants studied single-seat and two-seat designs, and monoplanes, biplanes and triplanes, making various models and testing them in their recently acquired wind tunnel, and making numerous calculations of strength, weight and flight performance. One feels that perhaps Tupolev knew in his heart all along that the ANT-1 would be a clean cantilever monoplane, because he wrote 'If we were after all to choose a biplane it would have meant that we had learned nothing at all from our work with Zhukovsky!'.

In any case, it is pointless to get too far into a design without knowing what engine is to be used, and for much of 1922 this crucial factor remained a question mark. The team – especially the Pogosskii brothers – studied the use of 14hp and 18hp Harley-Davidson engines and the 698cc Blackburne Tomtit of nominal 20hp, but not even the political 'clout'

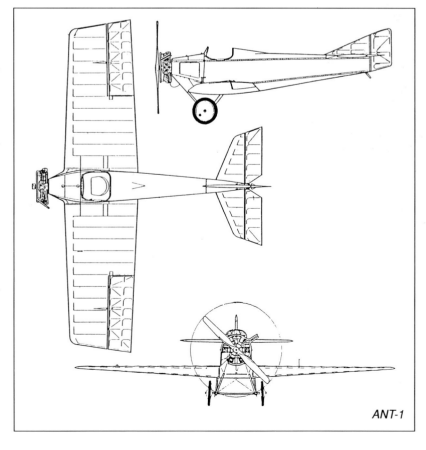

ANT-1

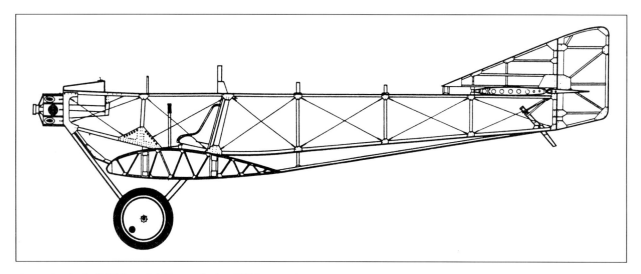

A recent drawing by K Udalov and A Morozov showing ANT-1 construction.

of CAHI could secure one for use in the ANT-1. This is strange, because examples of each of these engines were used by several contemporary Russian constructors, notably including V P Nevdachin who used all three!

Eventually, in early 1923 the team settled for an ancient six-cylinder Anzani. Dating from 1910, it had probably originally been installed in one of the primitive machines of Gakkel, Kudashyev or Sikorsky. Sadly, the engine was worn out and in a terrible state, but it was better than nothing, which was the alternative.

The wing appeared to have a profile similar to Clark Y, with a flat underside. Tupolev had no hesitation in making it a true cantilever, and it was made as a single unit, with two tip-to-tip wooden box spars. Towards the end of 1922 the first pieces of Kol'chug alloy arrived, and some of these, in the form of beaded channel sections, were used for the wing ribs, the completed wing then being fabric covered. The simple fuselage was based on four spruce longerons, the lower pair being joined to the wing spars by four steel bolts. Kol'chug alloy was used in the form of several small plates and angles to join the various wooden fuselage parts together. Eight panels of Kol'-chug sheet, the only such parts then in existence, were used to cover the fuse-lage from cockpit to engine (but without being required to bear stress), the rest of the fuselage being fabric-skinned. The tailplane was wooden, bolted to the upper longerons, but all the other tail surfaces were boldly made from Kol'chug sections and strip, the whole being fabric covered.

Four steel tubes formed the landing gear, one pair going diagonally up to the lower engine mounts on the front of the lower longerons and the others being connected to the longeron/rear-spar joint. Rubber bungees gave limited vertical movement to the axle. The engine had two exhaust collector rings, each serving three cylinders and discharging on the port side, and drove a carved wooden propeller. The cockpit had a small windscreen and head fairing, and contained an ignition switch, rpm counter and oil pressure gauge.

Construction was considerably de-layed by the availability of additional pieces of Kol'chug alloy which enabled the usage of this material to be in-creased, which in turn meant extensive restressing of the structure. Eventually

in October 1923 all the parts were completed in the ground-floor of 17 Radio St and they were carried outside and assembled. On 21 October the completed ANT-1 was trundled in triumph across the Dvorets bridge and along Krasnokazarmennaya St to Cadet Square, which was considered big enough for the first take-off. Petrol was put in and Ye I Pogosskii climbed in and took off successfully, landing at the Central Aerodrome (which had been named for Comrade Trotsky, who a year later on Lenin's death was defeated in a bid for power by Stalin, disgraced and later murdered on Stalin's orders, his name being erased from Soviet life).

Subsequently the ANT-1 was flown several times by Pogosskii and by N I Petrov, but its engine was utterly worn

A well-known photograph of the ANT-1. From the left: P P Sherstnev, N V Svistunov, A I Barulemkov, A S Komalenkov, A N Tupolev, S M Chugunkin, V N Il'in, Ye I Pogosskii and S I Lazarev. (ANTK Tupolev)

out. Tupolev asked BS Stechkin to refurbish it, to be told that it was quite beyond repair. Contemporary reports said the ANT-1 flew well, and there is no reason to doubt this. When the OKB moved into large new buildings in 1932 the ANT-1 was displayed in the entrance hall. Sadly, during Tupolev's imprisonment in the 1930s it disappeared.

Span 7.2m (23ft 7½in); length 5.4m (17ft 8½in); height 1.7m (5ft 7in); wing area 10sq m (108sq ft).

Weight empty 229kg (505lb); fuel/oil 42kg + 9kg (93lb + 20lb); loaded weight 355kg (783lb).

Maximum speed 125km/h (78mph); ceiling, calculated 4,000m, (13,123ft) but did not exceed 600m (1,968ft); duration 4hr (equivalent to c400km, 250 miles); landing speed 70km/h (43.5mph).

ANT-2

Tupolev's second design was the first to be built under contract. Apart from the unsuccessful KOMTA, it was also the first new type of aircraft to be ordered by the Government of the USSR. The contracting authority was the UVVS-RKKA (directorate of the air fleet of the workers' and peasants' Red Army), and it was to be a simple machine able to carry two passengers or, in a military rôle, to have a crew of two and two machine-guns. CAHI was sent a directive to build such an aircraft in early 1923, long before the ANT-1 was completed.

At this time the first all-metal aerosleigh (also called ANT-1) was already being tested, having been completed in February 1923. Its success encouraged Tupolev to go all out and design the ANT-2 as an all-metal aircraft, in the Junkers tradition. He and his team now had an adequate basis of test results on hundreds of test-pieces, and had established ultimate strengths for rolled sections and sheets in three grades from 36 to 40 kg/sq mm. After careful comparison with pieces taken from the captured Junkers aircraft Tupolev and his team felt confident to go ahead with an unbraced monoplane design using similar sections and jointing techniques.

The lead designers under Tupolev were I I Sidorin, the Pogosskii brothers, G A Ozerov, A I Putilov,

V M Petlyakov, B M Kondorskii, N S Nekrasov, N I Petrov and A I Zimin. As well as designing the aircraft the AGOS team made preparations for series production. Tupolev eventually decided that this would have to be done in the small workshops at Kol'-chugino, where the light alloy was produced. Nobody there had any knowledge of how the metal should be used; their work stopped at the production of ingots. Accordingly Tupolev and Ye I Pogosskii arranged for teams of Kol'chug workers to visit AGOS at 17 Radio St and learn how to convert ingots into standard strip, sheet, tube and sections, and then how to fabricate these into an airframe.

It was logical for Tupolev to combine his wish to build a second aircraft almost entirely from Kol'chug alloy with the requirement to produce a civil/military machine meeting the demands of the UVVS. Clearly it had to be bigger and stronger than the ANT-1, and of about 100hp, but it could have a low or high or even a parasol wing, and the passengers could be in open cockpits or a cabin. It so happened that in 1922 Junkers had flown a small transport designated K 16, with a cantilever high wing, an open pilot cockpit in the nose and an internal two-seat passenger cabin. Tupolev considered this ideal for his ANT-2. All he had to do was start from scratch with the same layout.

Detail design began in September 1923. Structurally and in most aerodynamic respects K 16 practice was followed, but in the basic matter of wing profile Junkers had by this time adopted aerofoils whose underside was either flat or convex, while Tupolev used a CAHI profile of considerable

depth (16 per cent) with a concave underside giving substantial camber. In plan the wing was pure Junkers, with straight taper modified by ailerons projecting far behind the basic trailing edge. Tupolev was to continue this practice for a decade, long after Junkers had switched to the slender, separated 'double wing' aileron/flap which Tupolev never adopted.

Structurally the wing was based on two straight parallel spars each made in three sections, the middle one spanning the fuselage. Each spar comprised upper and lower booms of seamless Kol'chug tube joined by front and rear webs of sheet pressed into a shallow channel profile. The thirteen ribs in each half-wing were formed from straight channel sections forming a Warren truss. The outermost ribs carried underside handles used as picketing loops and to allow ground crew to guide the aircraft when taxiing. The leading edge was flat sheet, but all the remainder of the wing and ailerons was covered in sheet with 8mm (0.315in) corrugations spaced 40mm (1.575in) apart. Four large bolts with Junkers-type hemispherical locating couplings joined the half-wings to the bridging spars across the fuselage.

The latter had an extraordinary, almost triangular, cross-section with sides which sloped in to form a deep sharp keel along the ventral centre-line. Tupolev adopted this profile because 'triangles are light and rigid'. It enabled him to avoid the need for heavy bulkheads, though of course channel and angle-section frames were needed at intervals to join the longerons. The grotesque depth of the fuselage also provided direct

The ANT-2 (original tail) with pilot and passengers. (Jean Alexander/ RART)

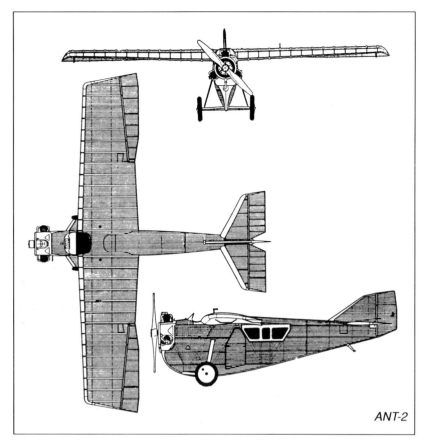

ANT-2

attachments for pivoting the horizontal mountings for the wheels. Each of these resembled a wing, tapered sharply on the leading edge, made with corrugated sheet wrapped round two tubular spars (the rear one carrying the wheels) and two full-depth channel webs. The aircraft's weight was carried by near-vertical struts with bungee shock-absorption at their upper ends inside the fuselage. Track was only 1.75m (69in).

As with the ANT-1 there were two main longerons each side, but because of the odd cross-section there were two more, progressively shorter, lower down, with a single curved keel member along the bottom which at its rear end carried a pivoted steel tailskid sprung by bungee. The top longeron carried the upper engine mounts, landing gear struts, wing and tailplane. All remaining longerons were broken by a door on the left side. The tailplane, which was mounted on pivots for trimming, and elevators were made like the wing, with a smooth leading edge, the span being 2.7m (106in) for the tailplane and 2.9m (114in) across the elevators. The small fin and rudder merely had a channel member along the leading edge pro-

viding the front attachment for the 8mm by 40mm skin.

The chosen engine was the Bristol Lucifer, a British product designed by Roy Fedden by using three of the nine cylinders of the Jupiter (an engine later used by the USSR as the M-22) but with slightly shorter stroke. It was rated at 125hp, but the Tupolev archives give the power as 100hp. It was notorious for the vicious torque-reaction from its firing strokes, which had a habit of breaking the engine mountings, but this was not a problem with the ANT-2. It was attached to a steel firewall which, by undoing two of the four bolts, could be hinged open to assist maintenance of the engine or cockpit. It could be started 'by magnetos or handwheel' and drove a CAHI carved-wood two-blade propeller of 2.2m (86.6in) diameter. Exhaust was collected in a ring from where a pipe at 6 o'clock led out through the starboard side of the fuselage. Fuel was housed in two 36kg (79lb) tanks in the wing roots.

The pilot gained access to the open cockpit by standing on the landing-gear sponson and clambering up two recessed footrests on the left side. He had a Celluloid windscreen, and

an altimeter, ASI, rpm counter, oil pressure/temperature gauges and fuel level indicator. His control column drove the ailerons by tubes and bell-cranks, all other surfaces being moved by cables. Tailplane trim was by a handwheel, cables and screwjack.

Behind the cockpit a corrugated bulkhead isolated the cabin, which had the door on the left and three windows on each side. In the passenger version (the only type built) there were three wicker passenger seats, a backless one immediately behind the front bulkhead facing two side-by-side at the rear resting on a corrugated rear bulkhead. As the requirement was to carry two passengers, three was considered an overload condition.

Detail parts were made in a warehouse shed. The aircraft was then erected outside the shed under an awning, and finally taken to the Central Aerodrome on 24 May 1924. N I Petrov made the first flight on 26 May, two sandbags simulating two passengers to get the CG (centre of gravity) in the right place. Official acceptance tests, watched by UVVS-RKKA and CAHI observers, took place on 28 May. From 11 June two passengers, and occasionally three, were carried. The only serious deficiency was inadequate directional stability, so the fin and rudder were enlarged (the three-view drawing shows the modified aircraft).

Further testing by the NOA (Scientific and Research Aerodrome) pilots Andreyev, Filippov, Rastegeyev, Zakharov and Savelyev 'revealed the need for further modifications'. However, as five Lucifer engines were available, a further four ANT-2 aircraft were built. At least one of these flew on skis, the rest of the landing gear being unchanged. In 1930 one was re-engined with a Wright Whirlwind radial rated at 200hp. One of the five aircraft, rebuilt after a fire, has been restored and is indoors at the Monino museum. The military variant was never built. This would have had the cabin replaced by a cockpit behind the wing, reached by two recessed steps on the part side, with a Scarff ring mount for one or two PV machine-guns.

The ANT-2 was important in demonstrating that Tupolev and his designers could with confidence design further aircraft made almost entirely from Kol'chug alloy. It also showed that such aircraft could be competitive in weight and performance, and that the unbraced monoplane was a safe configuration.

Span 10.45m (34ft 3⅛in); length 7.6m (24ft 11¼in); height 2.12m (6ft 11½in); wing area 17.89sq m (193sq ft).

Weight empty 523kg (1,153lb); fuel/oil 72kg + 8kg (159lb + 17.6lb); pilot and passengers 240kg (529lb); loaded weight 837.5kg (1,846lb).

Maximum speed 170km/h (106mph); ceiling 3,300m (10,826ft); climb (full load) 8½min to 1,000m, 21½min to 2,000m and 48min to 3,000m; range 750km (466 miles); landing speed 78km/h (48.5mph).

ANT-3, R-3

This was Tupolev's first aircraft built in series, and his first military design. It was also the first aircraft designed in the Soviet Union to be used in action, with guns and bombs, though its purpose was actually reconnaissance. The VVS designation was R-3 from *Razvedchik*, reconnaissance.

The order for the prototype was placed by P I Baranov, C-in-C of the VVS-RKKA, in early 1924. The R-1, based on the D.H.9A, was obsolescent, and A A Krylov's R-2 (flown in 1925) was likely to be even poorer, having the same technology and a wartime Maybach engine of considerably less power. Baranov had expected that a modern reconnaissance aircraft would be produced by Junkers at their Fili, Moscow, factory. However, the H 21 and A 20 (also designated R-2) again had outdated wartime engines, and in any case were made of Duralumin imported from Germany.

Baranov could see that the future lay with all-metal aircraft made of home-produced Kol'chug alloy, and he believed the AGOS team could produce a good replacement for the R-1, especially if money was found for a powerful modern engine. In any case, he expected that such an order would spur expansion of Kol'chug alloy production and of its widespread supply to aircraft factories skilled in its use.

In the design of the ANT-3 Tupolev followed exactly the same principles as in the ANT-2, the main difference being that, to reduce overall dimensions, he chose a sesquiplane configuration (biplane with a small lower wing). This led to the wing profiles being thinner, the thickness/chord ratio along the untapered part of the upper wing being 11.6 per cent. This time the concave undersurface was less pronounced, the CAHI profile being almost the flat-bottom Clark Y. Both wings had a straight trailing edge at 90 deg to the longitudinal axis, and a parallel leading edge over three-quarters of the semi-span, then tapering gently to the tip, which like the ANT-2 was straight but positively raked.

Each wing had two parallel spars, made with upper and lower tubular booms joined by two sheet webs. Each wing was made in one-piece left and right panels, with a smooth leading edge with an internal sub-spar of strip with flanged edges, and 8mm by 40mm corrugated sheet covering. Each upper

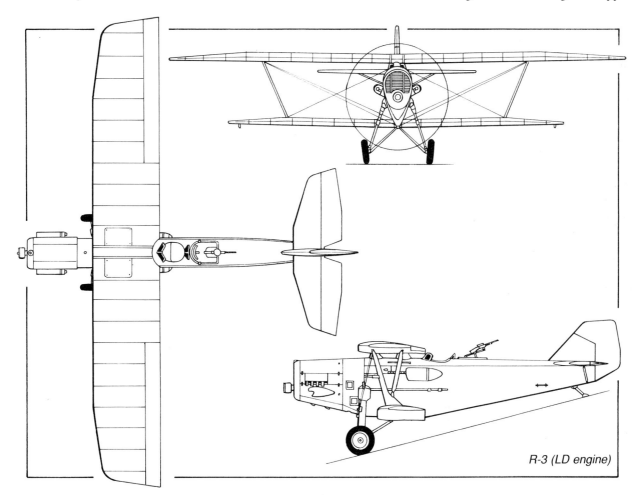

R-3 (LD engine)

The Lion-engined second ANT-3, at Berlin. (Lufthansa)

wing had eleven ribs and each lower wing eight, formed from a peripheral member braced by diagonal struts forming a Warren truss. These struts had flat ends to facilitate jointing, between which they were pressed to a flanged ⊃-section.

The lower wings were bolted by the spars to the bottom of the fuselage, while the upper wing was mounted only just above the top of the fuselage, carried on two shallow streamlined pylons picking up the spars on the centreline. The single N-type interplane struts sloped out sharply in front view. Bracing wires of streamlined profile joined the struts to the top and bottom of the fuselage. Under the wingtips were grab handles for ground staff and for use as picketing loops.

Far from thinking the strange fuselage cross-section of the ANT-2 a mistake, Tupolev repeated it in this military aircraft. As before, there were two full-length longerons on each side and a single keel member. Apart from a strong frame carrying the steel firewall, the front upper-wing pylon and the main landing gear, and steel or Kol'chug tubes supporting a transverse Kol'chug sheet at mid-depth (the rear part of which formed the floor of the pilot's cockpit), the only transverse members were channel-section frames supporting the 8mm by 40mm skin. The upper longerons carried the braced tailplane. With this aircraft Tupolev used plain constant-chord ailerons, worked by rods and cranks. Similar connections drove the elevators, but the rudder was moved by cables. The elevators and rudder were notable for their large horn balances. Tupolev's severe tail geometry was to become almost a trademark.

Fuselage areas subjected to propeller torque, including the cowling back to the firewall and the top decking back to behind the rear cockpit, were skinned with non-corrugated sheet. The deep keel provided direct pivots for the bracing struts of the main landing gears, the weight being taken by near-vertical struts (steel tubes faired by Kol'chug sheet) pivoted to the main firewall bulkhead and sprung by ten loops of bungee inside a faired box. The steel tailskid was pivoted and sprung by bungee.

The only engine available for the ANT-3 prototype was a US-built Liberty water-cooled V-12 of nominal 400hp. A total of 343kg (756lb) of fuel, approximately 476 litres (105gal), was housed in tanks in the left and right upper wings and above and below the fuselage floor, all centred on the CG. The aluminium oil tank ahead of the firewall held 34kg (75lb). The floor terminated at the rear of the pilot's cockpit, which had a comprehensive panel and a PV-1 machine-gun scabbed on the left side of the fuselage. The observer's floor was at a lower level so that he could stand conveniently and train twin Lewis guns on a Tur-4 (Scarff) mounting. He could also be provided with a forward-facing folding seat, basic flight instruments and dual controls. Behind him was a Potte-1 camera mounted vertically. Under the lower wings could be hung eight bombs of 32kg (70.5lb) each.

The prototype was built as a dual-control civil aircraft, with two forward-facing pilot cockpits with the same floor level, each with a windscreen. It was completed at Moscow Central Aerodrome in July 1925. It was first flown by V N Filippov on an unknown date not in August as often reported but in late July. It was then flown to the NOA where it was tested by M M Gromov (already a rising star among Soviet aviators) and pilot-observer V S Vakhmistrov. Vakhmistrov was later to design the optical gunsight for the R-3's fixed gun, and later still to be famed for *Zvyeno* combinations in which fighters or dive-bombers were to be carried by Tupolev heavy bombers and launched in flight.

The aircraft impressed this experienced test crew. The only adverse features were the rather poor view ahead because of the position of the upper wing, and a strong tendency to tail-heaviness because the CG was too far aft.

By December 1925 several ANT-3 pre-production aircraft had been built, the most notable modification being to replace the interplane struts by a novel new type resembling a letter K. The main strut was assembled from a straight leading-edge tube joining the two front spars and a trailing edge curved to join the upper front spar to the lower rear spar, these tubes being joined by transverse members and the whole then skinned with non-corrugated sheet. About one-third of the way up the trailing edge was a bracket carrying a pivot for a sub-strut, made of a tube with wrapped skin giving a streamlined profile, which was joined to the upper rear spar.

From the birth of the Soviet Union there had been a fervent national wish to demonstrate superior capability, for example by making long-distance flights and breaking records. Tupolev aircraft were to play their full part in this endeavour, and the ANT-3 was the first nationally-designed machine with adequate performance. For this purpose a more modern engine was needed, and Gromov suggested the Napier Lion, because of its economy and reliability. Despite its high price, this was agreed, and one of the pre-production machines, assembled at GAZ No. 5 in Leningrad, was fitted with the imported engine, rated at 450hp. The engine was cooled by water circulated through two of the patented French Lamblin radiators, mounted externally on each side of the cowling. It drove a propeller rotating in the opposite direction to the Liberty and increased in diameter from 3m to 3.1m (122in). This aircraft was tested on 20–21 August 1926 by A I Tomashyevsky, who reported it 'fully met all demands in the matter of handling and flight performance. It was then fitted with auxiliary tanks which increased fuel mass to some 840kg (1,852lb).

This aircraft was given civil registration RR–SOV, and the slogan *Aviakhim–SSSR*, *Proletaryi* (Proletariat). Between 30 August and 2 September 1926 it was flown by Gromov and flight engineer Yev Rodzyevich on a tour of European capitals. The route was Königsberg, Berlin, Paris, Rome, Vienna, Warsaw, Moscow, the 7,150km (4,443 miles) being flown in 34hr 15min.

By this time orders had been placed for series aircraft for the VVS inventory. Several sources state that these also were built at GAZ No. 5, but in fact all were constructed at GAZ No. 22, Moscow Fili, beginning as soon as Junkers had moved out on 1 March 1927. The cost ruled out the Lion engine, and initial series aircraft had the M-5, the 400hp copy of the Liberty, retaining the external Lamblin radiators. The first aircraft to be completed was given civil registration RR-INT, plus the name *Nash Otvet* (our answer). Between 20 August and 1 September 1927 it was flown by S A Shestakov, with engineer D V Fufayev on a tour of the Far East. The route was Moscow, Sarapul, Omsk, Novosibirsk,

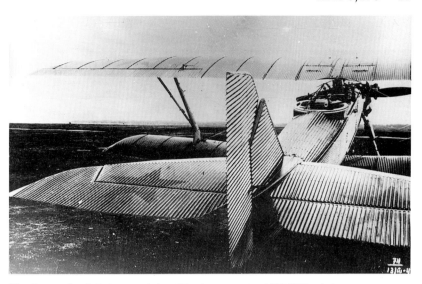

This close-up of an R-3 gives a good idea of Tupolev construction. (ANTK Tupolev)

Krasnoyarsk, Chita, Irkutsk, Blagoveshchensk, Spassk, Nan'yan, Okayama, Tokyo. The distance of about 22,000km (13,670 miles) was flown in 153hr.

In 1927 about 100 French Lorraine-Dietrich 12Ed engines were imported, and most of these were used in the final batches of ANT-3s. These engines were mounted further forward, the fuselage being extended, which at last got the CG in the right place. They had a flat frontal radiator and a large blister on each side fairing in the carburettor inlets. Total series deliveries from GAZ No. 22 comprised two in 1927, twenty in 1928 and seventy-nine in 1929, a total of 101. Designated R-3, they were cleared for operational service, and the 35th Independent Air Unit made numerous strafing and

bombing missions against Basmachi tribesmen (described by Moscow as bandits) who were resisting the incorporation of Tadzhikstan into the USSR. Most ended their careers flying with Siberian (especially Yakutsk) directorates of Aeroflot carrying mail and cargo.

In 1928 one R-3 was fitted with a 500/730hp BMW VIz, with a frontal radiator, but according to historian V B Shavrov 'the increase in speed and ceiling reduced the margin of safety and range'. One of Tupolev's designers, A I Putilov, killed off the proposal to produce an armoured attack version by calculating that to protect the crew and engine would need at least 400kg (882lb) of armour. In 1930 Tupolev produced a derived aircraft, the ANT-10, R-7.

The first series production R-3, with M-5 engine. (RART)

Span 13m (42ft 7¾in); length (Liberty, Lion) 9.4m (30ft 10in), (M-5) 9.5m (31ft 2in), (LD) 9.89m (32ft 5⅜in), (BMW) 9.7m (31ft 10in); height (over propeller) 3.72m (12ft 2½in); wing area 37sq m (398sq ft).

Weight empty (Liberty) 1,335kg (2,943lb), (Lion) 1,390kg (3,064lb), (M-5) 1,377kg (3,036lb), (LD) 1,340kg (2,954lb), (BMW) 1,547kg (3,410lb); fuel/oil (M-5) 387kg (853lb), (LD) 322kg + 30kg (710lb + 66lb); loaded weight (Liberty) 2,085kg (4,597lb), (Lion) 2,400kg (5,291lb), (M-5) 2128kg (4,691lb), (LD) 2,090kg (4,608lb), (BMW) 2,297kg (5,064lb).

Maximum speed (Liberty) 207km/h (129mph), (Lion) 226km/h (140mph), (M-5) 194km/h (120.5mph), (LD) 204km/h (127mph), (BMW) 229km/h (142mph); service ceiling (Liberty) 4,400m (14,435ft), (Lion) 5,110m (16,765ft), (M-5) 5,000m (16,400ft), (LD) 4,920m (16,142ft); (BMW) 6,200m (20,340ft); range (LD) 900km (559 miles); take-off run (LD) 160m (525ft); landing speed (LD) 85km/h (53mph), landing run 140m (460ft).

ANT-4, TB-1

This was possibly the most significant aircraft Tupolev ever designed. His first three offspring made little impact on the world, and had counterparts in many other countries. In contrast, his fourth design was a twin-engined low-wing monoplane of outstanding merit. It provided the basis for a propaganda aircraft which made a remarkable flight from Moscow eastwards to New York, for a heavy bomber used in large numbers, for ski- and float-seaplanes and for useful cargo aircraft. It was also an ideal stepping-stone to much bigger successors, all with similar distinctive severe outlines.

At the same time the ANT-4 was not, as widely stated in the Soviet Union (not least by the Tupolev OKB), 'the first all-metal monoplane bomber in the world'. Historian V B Shavrov even claimed 'All subsequent Flying Fortresses and Superfortresses were developments of the TB-1'. Indeed, one of its predecessors was the Junkers K 30C, which as the R 42 was to be sold in numbers to the Soviet Union. The ANT-4 was based like its forebears on Junkers technology, and in 1926 Junkers brought an unsuccessful action against CAHI, AGOS and Tupolev personally, claiming that this aircraft infringed his patented form of wing construction. He had expected the Soviet Union to be a lucrative market, not a competitor.

Though this was precisely the aircraft Tupolev wanted to build next, he had to have a customer. In fact one had appeared in 1923 when the *Ostekhburo* (Special Technical Bureau of Military Inventions) in Leningrad expressed a need for a heavy aircraft to test the air-launched weapons being developed by V I Bekauri's bureau. It was proposed to buy a heavy bomber from Handley Page, but the price quoted was high and the eighteen-month delivery too long. In any case, Tupolev argued (rightly) that AGOS could produce a more modern aircraft.

In October 1924 the NKAP (People's Commissariat for Defence) accepted Tupolev's offer. A month later it signed a contract with AGOS, stipulating that construction should take not more than nine months. Work began on 11 November, and precisely nine months later the parts were taken out through the demolished wall of 16 Radio Street and trucked to the Central Aerodrome. Bearing in mind the appalling working conditions, shortage of skilled Kol'chug workers and the very demanding complexity and quality of this prototype, the achievement was remarkable. Moreover, a model was tested in a CAHI tunnel and for the first time AGOS constructed a full-scale wooden mock-up, which was found so useful it became standard practice. Team leaders were Arkhangel'skii, Putilov, Petlyakov, Petrov, Nekrasov and Kondorskii, plus newcomers D A Baikov and I I Pugarovskii.

Tupolev called the wing profile A°, though it was virtually a thick (20 per cent) CAHI section based on Clark Y. Structurally the wing was a centre section and two outer panels. The centre section was horizontal, the leading edge untapered to just outboard of the engines and then tapered at almost the same angle as the trailing-edge taper from the root. Its span was 13.5m (44ft 3½in), and it had five spars and eighteen ribs. These all followed previous practice, the spars – which were all vertical, instead of being diagonals as was Junkers' practice – having upper and lower booms of elliptical tube section tapering from 72.6mm by 65.6mm (2.86in by 2.58in) at the root, joints along the booms each having four rows of five rivets. As before, the spars were built up into Warren girders by multiple vertical and diagonal channel sections. The ribs likewise comprised what Tupolev called A-section peripheral members with vertical and diagonal channel sections forming a truss, with intermediate L-section ribs. The skin panels were of a less-coarse corrugated form than before, the pitch being only 32mm (1.26in). Each was cut to fit between the ribs, riveted to the rib on each side with a top-hat section on top covering the projecting ridge of the A-ribs. Skin thickness was 0.3mm or 0.33mm except at the root where the upper surface was a walkway 0.8mm thick. Lack of confidence in skin stability led to addition of strips 1.5mm thick and 100mm or 150mm wide folded at 45 deg being added spanwise under the skin. Later these strips were found to be unnecessary, and they were omitted as was the fifth spar.

Starting at the ailerons, each outer panel had a span of 6.825m (22ft 4¾in).

The ANT-4 prototype, with its underslung dummy weapon. (G F Petrov)

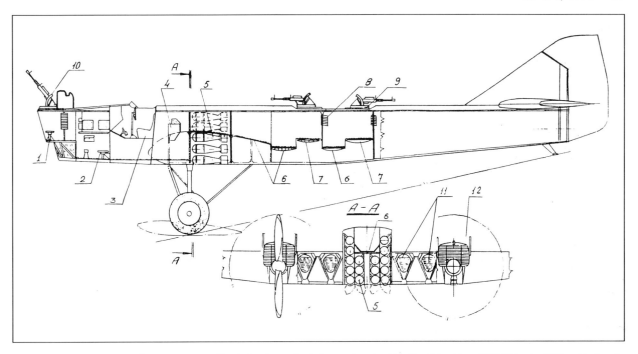

A standard TB-1: 1, nose gunner's seat; 2, navigator/bomb aimer's seat; 3, pilot's seat; 4, mechanic's cabin; 5, bomb bay (FAB-82 bombs, 181lb each); 6, walkway; 7, gunners' floors; 8, ammunition drums; 9, TUR-3 mounts for DA-2 guns; 10, nose ring for DA-2 gun; 11, fuel tanks; 12, M-17 engine.

Each had ten ribs, and structure similar to the centre section. The joint was made not with Junkers-type union nuts but with tapered bolts in reamed holes in plain sockets in each spar boom.

The fuselage again followed established practice, but instead of continuing to slope the sides down to a narrow keel the cross-section of the front and centre sections was almost a rectangle. The lower part of the sides only tapered modestly until behind the wing it quickly became the familiar near-triangle with slightly convex sides. Most of the length had A-section upper longerons and tubular lower longerons, twenty-one main frames of A-section, and tubular and channel diagonal internal bracing, with sheet (typically 2mm) gussets at the joints. The fuselage was made in three parts, F-1 the nose, F-2 from the pilots' cockpit and leading edge back to the fifth spar, and F-3 to the tail.

All control surfaces had a tubular structure with 20mm by 3mm corrugated skin. The ailerons had curved trailing edges extending behind the projected line of the wing, and a large horn balance projecting beyond the wingtip. Apart from the corrugations nothing broke the surface, and the same was true of the fin. The rudder, however, had two projecting ribs to which the hinges were attached, while

the tailplane and elevators had multiple ribs each pressed from sheet with flanged lightening holes, projecting beyond the surface profile with the skins riveted through on each side. At the root the tailplane had two built-up ribs resembling those of the wing on which it was pivoted above the fuselage top longerons, with a screwjack worked from a cockpit handwheel for trimming. The ailerons were pivoted to brackets under the wing and worked by push/pull rods. The tail controls, which again had large horn balances, were moved by cables.

Each main undercarriage comprised three simple struts of Ni-Cr steel of 130kg/sq mm strength machined to various profiles and pivoted to brackets on the fifth rib and two of the spars. The lower end of the main strut was braced to the V by multiple rubber cords and carried a half-axle for an unbraked wire wheel with a 1,250mm by 250mm Palmer tyre, though when first completed the prototype was mounted on Junkers skis.

Because of its excellent performance in the ANT-3, the prototype was fitted with two imported Napier Lion II engines, each rated at 450hp. These were mounted on steel-tube trusses bolted to the steel firewall mounted on the front spar, with the aluminium oil tank of flat-oval section under the top of the cowling. Each engine had short

exhaust pipes from each cylinder, drove a 3.25m (10ft 8in) diameter laminated wood propeller and was cooled by two Lamblin drum-type radiators hung under the wing. A total of 2,010 litres (442gal) of fuel was housed in ten interconnected metal tanks in the wing centre section, with overwing gravity fillers and a central drain cock under the fuselage.

The ANT-4 prototype was devoid of military or navigational equipment, and had merely a glazed nose and side-by-side seats in the pilot cockpit. Entry was via a ladder behind the starboard wing and the walkway across the wing. The aircraft was assembled at the Central Aerodrome (Khodinka), with main and tail skis of Junkers type. Under the centreline was added a large tube with faired ends simulating one of Bekauri's air-launched weapons. The first flight took place on 25 November 1925, but NOA pilot A I Tomashyevskii was unhappy with the handling and landed after seven minutes. After modification the aircraft flew again on 15 February 1926, Tomashyevskii landing after 35 minutes to express complete satisfaction.

Official acceptance tests took place on 26 March 1926 before a commission of NOA, *Ostekhburo* and CAHI delegates. Tomashyevskii and an *Ostekhburo* representative, N N Morozov, made various recommendations, and after modifications the ANT-4 was handed over for NOA State trials. Between 11 June and 2 July 1926 Tomashyevskii made twenty-five test flights, resulting in the overall evaluation 'First class aeroplane'. On 10 July Tomashyevskii took off with a simulated bomb load of 1,075kg and stayed airborne for 12hr, covering about 2,000km (1,242 miles). On another occasion he carried a load of 2,000kg on a flight of 4hr15min.

Span 29.6m (97ft 1⅛in); length 17.3m (56ft 9¼in); height 5.1m (16ft 8¾in); wing area 121.5sq m (1,308sq ft)
Weight empty 4,014kg (8,849lb); maximum weight 6,200kg (13,668lb).
Maximum speed 196km/h (122mph); service ceiling 4,550m (14,930ft); climb to 3,000m 23.2min; range 1,900km (1,181 miles); take-off run 260m (853ft) in 17s; landing speed 85km/h (53mph), landing run 350m (1,148ft).

The C-in-C of the VVS-RKKA, P I Baranov, sanctioned production, and indeed was eager to receive such a machine, which had capability of a totally different order from anything in VVS service. The Service designation was to be TB-1 (*Tyarzhel Bombardirovshchik*, heavy bomber Type 1). Discussions then centred on changes to the series aircraft, the type of engine to be fitted (not the costly Lion) and details of the systems, equipment and armament.

Preparations were immediately begun for full-scale production, and from the start it was clear that the only suitable factory was that at Moscow Fili. This had been an RBVZ (Russo-Baltic wagon factory) but in 1926 it was occupied by Junkers, administered by Aviatrust. By mutual consent, Junkers were not staying; they were finding competition (from Tupolev) rather than orders, and the firm was liquidated on 1 March 1927, the plant becoming GAZ (State Aviation Factory) No. 22. In July 1926 Aviatrust had organised a group of forty workers, including all those trained by Tupolev in Kol'chug fabrication, and began duplicating the AGOS drawings of the ANT-4, rechecking all calculations and where necessary making suggestions to facilitate production.

Rather unexpectedly, though GAZ No. 22 was joyously taken over on 1 March 1927, and the second ANT-4 was assembled there, almost no tangible preparations for series manufacture were made until the second half of 1928. This was partly because

of further experience with prototypes, and resulting improvements, partly because the first M-17 engines did not come off the assembly line at GAZ No. 26 until 1929, and partly because there was still very little Kol'chug alloy, so the ruling material in the TB-1 had to be German Duralumin.

Though contracts had been signed to import Lorraine–Dietrich engines, these were assigned to the R-3. It had been decided that the German BMW VI V-12 water-cooled engine would be built in the Soviet Union, with the designation M-17, and accordingly it was decided to fit the German engines in the *dubler* (second prototype). This was virtually complete in February 1928, but had to wait until July for delivery of the chrome-steel axles from Sweden. Factory tests began in July, and State trials then followed from 15 August 1928 to 26 March 1929. The flight-test team comprised M M Gromov, S A Danilin and engineer Kravtsov.

Compared with the prototype this aircraft had a wing of reduced span and area, and a fuselage with structural refinements including a nose extended by 0.71m (28in). The nose (Section F-1) was completely redesigned, with only the lower front portion glazed; above this was an unusual forward-sloping bow cockpit for a gunner. In the centre fuselage was a bomb bay, described later, and Section F-3 had open deck rings and a floor for two dorsal gunners. Under the wings were four landing lights, electric power being provided by a windmill generator on both inboard leading edges. The BMW VI(7.3z) engines, each rated at 730hp, had wide flat water-cooling radiators immediately ahead of the cylinder blocks. The top of the cowling comprised front and rear upward-hinged doors, the back of the rear door being left open above the wing. Three cooling louvres were provided in the lower cowl on each side, and additional small ram inlets were incorporated in the removable bottom section. The propellers, without dogs for a Hucks (ground powered) starter, were unchanged in size but in series aircraft were either oak or ash.

This aircraft confirmed the good qualities of the ANT-4, though control forces were rather heavy, and it was decided to have dual control and two pilots in the series aircraft. The VVS–RKKA assessment concluded

'After eliminating the recorded deficiencies the TB-1 has sufficient lifting power, rate of climb, speed and handling performance to have it accepted for service in VVS units'.

Span 28.7m (94ft 1⅞in); length 18.01m (59ft 1in); height 6m (19ft 8¼in); wing area 115.8sq m (1,247sq ft).
Weight empty 4,172kg (9,198lb); fuel/oil 1,450kg + 90kg (3,197lb + 198lb); maximum weight 6,560kg (14,462lb).
Maximum speed 198km/h (123mph); service ceiling 4,700m (15,420ft); range 1,380km (858 miles); take-off run 16sec, 200m (656ft); landing speed 85km/h (53mph), landing run 150m (492ft).

ANT-4bis This pre-production aircraft was built entirely at GAZ No. 22, and had c/n (construction number) 601. Its engines were delivered from Munich in July 1929, and State trials were made from 1 August to 19 October 1929. This was in almost all respects an operational bomber. The F-1 section contained a nose cockpit with a Scarff ring with a Lewis gun, with a rack for ten forty-seven round drum magazines, the gunner also having a circular seat. On the left were a map case, trailing radio antenna handwheel and Potte A vertical camera. In the floor was the small bombing window, plus a circular seat and table for the navigator and the master AN compass. On the right were the radio receiver, microphone case, handwheel for operating the bomb doors and the Goertz bombsight, which could be swung out of the way in cruising flight. An outside air temperature thermometer was added on the right of the nose, visible through the rear window on that side. In the roof was a hinged hatch. At the junction with F-2 was a corrugated-sheet bulkhead with a central door admitting to the open cockpit for two pilots, with duplicated panels, windscreens and flight and engine controls, and a large wheel in the central gangway to adjust tailplane incidence. Two venturis were mounted on the right side of the fuselage. Behind the pilots was the engineer, with engine instruments and controls on both side walls. Immediately behind was the bomb bay with SBR-8 racks on each side for various loads up to 730kg (1,609lb) held horizontally (increased in production aircraft). Double rectangular doors on each side of the centreline were opened and closed by the handwheel in the nose. Along the centre was a catwalk so that a stooping man could gain access to F-3, where were the two dorsal gunners, the forward gunner being low on the left and the aft gunner higher on the right, each again armed with a Lewis. In series aircraft the rear gunners also had a radio. International navigation lights were fitted to the leading edges of the wingtips and trailing edge of the rudder, while under the rear fuselage were red and green signalling lamps. Electrical power for all services was supplied by a large DC generator above the centre fuselage driven by a single counterbalanced windmill blade (this replaced the small wing-mounted generators). Just behind this was mounted the main radio mast to which antenna wires extended to masts (a pyramid of slim tubes) above the outer wings.

From this aircraft production gradually built up, GAZ No. 22 delivering two aircraft in 1929, sixty-six in 1930, 146 in 1931 and the final two in 1932, a total (excluding the two prototypes) of 216. From c/n 604 the Scarff rings with Lewis guns were replaced by, in the nose, a Tur-6, and in both dorsal positions a Tur-5, in each case with twin Degtyaryev DA guns; another change was to fit wooden inserts into the wing-root walkway with 64mm (2.5in)-thick steps, while on the starboard wingtip (sometimes both tips) was added a holder for two flares, usually used to back up the four landing lights. Another early modification was to alter the bomb bay to carry up to 1,312kg (2,892lb) made up of sixteen FAB-82 bombs on DER-9 cassette racks, or for larger numbers of smaller bombs, and (as an alternative load) to provide four racks under the wings for FAB-250 (551lb) bombs; another modification was to fit long exhaust pipes extending back across the wing. Early in production Soviet mainwheels became standard, with wire spokes and taking larger (1,350mm by 300mm) low-pressure tyres.

For winter, five designs of ski (by CAHI, GAZ-8 and –25 and by N R Lobanov) were evaluated. The CAHI ski made of Kol'chug was excellent, but expensive, and the choice fell on a round-topped wooden ski almost resembling a small (3.4m by 1.25m, 134in by 49in) seaplane float, with a much smaller ski at the tail, with sprung mounts and guy ropes. Near the start of of production the design

was made longer and narrower (3.7m by 0.9m, 146in by 35in), each ski weighing 115kg (253.5lb), and 200 sets were delivered and used routinely.

From c/n 631 the flight engineer's portholes were replaced by slightly larger trapezoidal windows, and a

handrail was added along the upper right side of the fuselage, because the steep slope of the wing was treacherous in winter even with the wooden steps. At the next aircraft (632) the engines changed to Soviet M-17s, rated at 680hp, in an unchanged

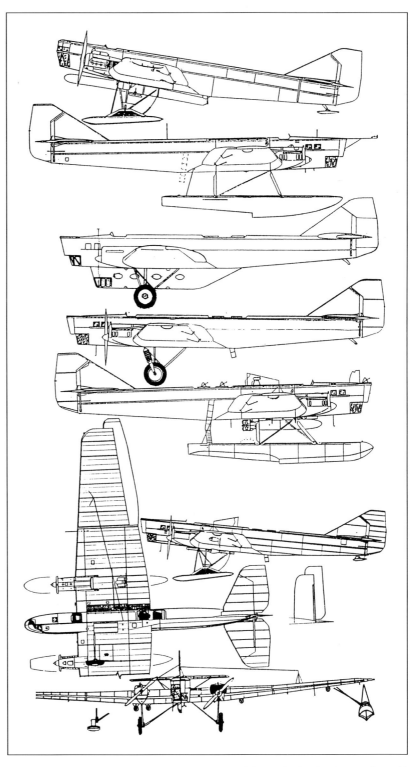

The three-view shows a series TB-1 on skis, with two types of float outlined in the plan view (inset, 4th series horizontal tail), and the front view showing (starboard side) the ANT-4 prototype (inset ski), (port side) series TB-1 (inset, float of TB-1P); the other side views are, from the top: ANT-4 prototype; Strana Sovyetov; TB-1 with desant (airborne troop) capsule; Zvyeno-1; TB-1P floatplane.

installation. At c/n 670 the right pilot's seat was changed, and the travel of his pedals increased, requiring the upper right part of the F-1/2 bulkhead to be omitted; a window was put in the bulkhead door; the navigator's thermometer was enlarged, making it easier to read when iced over; and the elevators were driven by a single bracket and push/pull rod on each side, instead of upper and lower brackets with cables. At c/n 690 four auxiliary tanks were added to increase combat radius to 800km (497 miles). From c/n 695 the lifting handles at the tail, Dural tubes attached by a row of brackets on each side beside the tail-skid, were lengthened so that at least two men could lift on each side.

Dimensions unchanged except length 18.06m (59ft 3in).

Weight empty (c/n 602) 4,427kg (9,760lb), (632 on) 4,520kg (9,965lb); maximum weight (602) 6,722kg (14,819lb), (632) 6,770kg (14,925lb).

Maximum speed (602) 202km/h (125.5mph), (632) 206km/h (128mph); service ceiling (602) 4,700m (15,420ft), (632) 4,920m (16,142ft); range (602) 1,380km (858 miles), (632) 1,000km (621 miles); take-off/landing distances little change.

The TB-1 proved to be a structurally strong and reliable aircraft, and almost every example built had a long and productive career. The famed *Strana Sovyetov*, seaplane versions and G-1 freighter are described later. Exhaustive NII–VVS trials with series bombers with M-17 engines took place in winter 1929–30, the pilots being Gromov, P M Stefanovskii and AB Yumashyev. Regular aircraft equipped all the first TB *Polks* (heavy bomber regiments), some of which saw active operations over Tadzhikistan and remained fully operational until 1936.

When the ice-breaker *Chelyuskin*, with Dr Otto Schmidt's expedition of 104 people on board, became trapped in ice in the Chukot Sea, north of the Bering Strait, aircraft proved the only means of rescue. On 5 March 1934 a TB-1 flown by A V Lyapidevskii was the first to reach the stranded ship, landing on rough ice beside it. He carried all the expedition's women and children back to Siberia. TB-1 bombers flew in supporting roles in 1938–39 in the 'undeclared wars' of Lake Khasan, Nomonhan and Khalkin Gol, a few continuing into the Winter War and finally as civil and military transports in the Great Patriotic War from 1941. A considerable number

were retained for mapping and survey duties, while others tested weapons and equipment.

The first and most spectacular tests were the *Zvyeno* (Link) experiments by LII engineer V S Vakhmistrov. He began by perfecting the launch of a glider from above the upper wing of an R-1, for use as an air gunnery target. From this he progressed to 'parasite' schemes, in which fighters and other small warplanes could be carried far into hostile territory and launched, if necessary then hooking on again for the journey back. The TB-1 was the ideal parent aircraft (later joined by Tupolev's bigger TB-3). With two fighters on board the total weight was about 12 tonnes, and the bomber had to have the main undercarriage legs augmented by adding on the outboard side shock struts from a TB-3. The first test was flown from Monino aerodrome on 3 December 1931. A TB-1 piloted by A L Zalevskii and A R Sharapov took off carrying an I-4 (*see* ANT-5) above each wing, the fighter pilots being V P Chkalov and A I Anisimov. Each fighter was loaded by a crew in front of the bomber hauling it by rope up a long wooden ramp to the bomber's trailing edge, until its axle could be secured by a hold-down link; then a pivoted rear tripod was swung up from the wing to a quick-release fixture under the fighter's cockpit. The combination, called Z-1, took off with the fighters' engines at full power. The plan was for Vakhmistrov himself to signal to the fighter pilots 'Release rear attachment'; he was then to release the axle hold-downs, both fighters flying off together, banking outwards to avoid collision. What happened was that at the last moment an official discovered that Vakhmistrov was not a trained pilot, and banished him to a rear gunner cockpit. Co-pilot Sharapov replaced him, but had not been briefed what to do, so he released one fighter's hold-down (confused records report that it was that of Anisimov and also that of Chkalov) without giving the 'release' signal. The result could have been catastrophic, but the fighter pilot, probably Anisimov, was quick enough to pull the rear release as his aircraft reared up, and he got away safely. The TB-1 flew on, Vakhmistrov briefed Sharapov and the other fighter was released with the correct procedure.

In September 1933 the Z-1a was

tested. This comprised the TB-1, flown by Stefanovskii, and two I-5 biplane fighters flown by V K Kokkinaki (later a famed test pilot for Ilyushin) and I F Grozd. This series of tests was trouble-free, and led to more ambitious *Zvyeno* trials using the TB-3, as described in the ANT-6 chapter.

In October 1933 two aircraft were used to test V I Dudakov's jettisonable solid-propellant assisted-take-off rockets: c/n 614 had three mounted above the rear part of each wing behind the nacelle pointing diagonally downward, while c/n 726 had two above and one below mounted at midchord pointing horizontally. The extra weight was 469kg (1,034lb), but pilot N P Blagin (later infamous, *see* ANT-20) got airborne in times as short as 6.5sec, compared with the normal time of 17sec.

P I Grokhovskii used a TB-1 for what were possibly the first tests in the world of cargo parachutes; individual loads of up to 1,000kg were dropped using several parachutes (much later, not with TB-1s, also with final braking rockets). In 1933–35 military engineer A K Zapranovannyi and pilot I Beloziorov made a successful series of flight-refuelling experiments, initially taking on fuel piped from an R-5, then passing fuel to another TB-1 and finally piping fuel to an I-15 and an I-16 simultaneously. In 1935–39 a major programme of remote piloting – one of the first in the world – was successfully completed, the objective being to use a TB-1 either as a gunnery target or, more importantly, as an explosives-packed cruise missile guided into its target. Initial testing used preset 'bang/bang' commands, in which each signal briefly puts a control surface hard-over, but later true modulated analog control was used, undoubtedly the first in the world. The remote pilot flew in a TB-3 which maintained control from a range of up to several kilometres.

Documents describe the mounting of a 76.2mm (3in) field gun, without carriage, for firing trials. It is doubtful if these trials took place, but a similar installation was certainly tested on a TB-3, as described in that chapter.

Strana Sovyetov (Land of the Soviets) was the first production aircraft, completed as an unarmed civil propaganda machine, unpainted except for its name, its Aviakhim sponsors and registration URSS-300. Long beforehand it had been planned that this aircraft should make a flight in stages eastwards from Moscow to New York, the overland portions being in the charge of Petlyakov, and the over-water parts, with floats fitted (see TB-1P) being managed by R L Bartini, an Italian Communist who played a major role in Soviet aircraft design.

This aircraft was in most respects a stripped-down bomber, with the choice of the first 1,350mm by 300mm wire wheels or extended Junkers floats. The only external difference was the fitting of an air-data sensor on a long braced bracket ahead of the nose. It left Moscow on 8 August 1929, the crew comprising S A Shyestakov (pilot) and F Ye Bolotov (copilot), B V Sterligov (navigator) and D V Fufayev (engineer). Unfortunately on the following day failure of the BMW engines resulted in a forced landing north of Chita in which the aircraft was badly damaged. The crew returned by rail, and by the time they reached Moscow a replacement aircraft – presumably the second production TB-1 – had been prepared with the same equipment and markings.

Taking off again on 23 August, the crew once more encountered incessant engine trouble, but coaxed the aircraft over the legs to Omsk, Novosibirsk, Krasnoyarsk, Chita and Khabarovsk, completing this portion of the flight on 5 September. Here the wheels were replaced by the floats, and on 12 September the aircraft took off on the most dangerous part of the mission, the 7,950km (4,940 miles) to Seattle.

The take-off from the River Amur was normal, though loaded to well beyond TB-1 weights it took over 30sec. The subsequent long flight to Petropavlovsk was almost entirely through low cloud which for long stretches forced the aircraft down to 15–20m (50–65ft) above the sea. The next leg, to a re-fuelling at Attu, required flying to an island just 10km (6 miles) in length, with no other land anywhere near, with no radio aids. A further challenge was refuelling from a boat off the island of Unalaska; to lighten the aircraft and ease take-off with maximum fuel for the long next stage to Seward, Shyestakov and Fufayev disembarked and went on by boat, rejoining the aircraft at Sitka.

South of Sitka a forced alighting had to be made in a small bay and an engine (weighing 545kg, 1,201lb) changed without any specialised equipment. Seattle was reached on 13 October, where at the US Navy station at Sand Point the floats were replaced by wheels. Leaving on 18 October the ANT-4 flew down the coast to San Francisco and then turned east, crossing the Rocky Mountains to Salt Lake City. The final legs took it via Chicago and Detroit to New York, which was reached on 1 November. The aircraft and crew then returned by sea. Everywhere, and especially from Seattle onwards, the *Strana Sovietov* was greeted by thousands of excited people, and this flight did far more than any other to put Soviet aviation on the world scene. The great-circle distance flown (appreciably shorter than the actual figure) was 21,242km (13,200 miles), covered in 137 hours' flying time, an average of 155km/h (96mph).

Dimensions (landplane) unchanged; (seaplane) length 18.9m (62ft).

Weight empty (landplane) 4,630kg (10,207lb), (seaplane) 5,130kg (11,310lb); fuel/oil (landplane) 2,350kg + 155kg (5,181lb + 342lb), (seaplane) 1,921kg + 90kg (4,233lb + 198lb); maximum weight (landplane) 7,928kg (17,478lb), (seaplane) 7,989kg (17,612lb).

Maximum speed (landplane) 207km/h (129mph), (seaplane) 190km/h (118mph); service ceiling (both) 3,600m (11,810ft); no other data.

TB-1T This designation identified a small number of landplanes equipped to launch a torpedo. Few details survive, and it is not certain than any 1T aircraft entered MA (naval aviation) service.

TB-1P Also in one document called TB-1A, the twin-float seaplane version was widely used by both the MA and by Aeroflot. It resulted from the original request by the *Ostekhburo* for the ability to drop torpedoes. Negotiations began on 15 January 1926, but though by this time both CAHI and AGOS had improved towing tanks for experiments with high-speed marine craft, neither had made any floats for seaplanes. A decision about floats was postponed until 1928, but in the meantime Junkers floats of the type fitted to the JuG-1 (K 30C/R-42) were tested, and when their forebodies had been extended 0.4m (15¾in) to match the greater weight of the ANT-4 they were acceptable, though their hydrodynamic behaviour was considered poor.

A TB-1P with a windmill generator serving the radio and two alighting lamps under the left wing. (RART)

In October 1929 the UVVS purchased a small selection of floats from Short Brothers, made of Dural with steel joints, with a 15mm overlap at plate joints sealed with red-lead tape and two rows of snap-(brazier-)head rivets. They performed well, and after prolonged CAHI tests a copy called Zh was selected for production, with the nose made broader on top and round-head rivets throughout. The Zh float was 10.66m (34ft 11⅝in) long and 1.25m (49.2in) wide. Each weighed 299kg (659lb) with fittings, and the complete aircraft installation weighed 816kg (1,799lb). Series production of these floats began in 1932, and 'over 100 pairs' were delivered.

In 1932 a total of sixty-six TB-1s were converted with these floats into seaplanes. Initially sixty were equipped as torpedo droppers, carrying a single 45-12-AN or similar torpedo under the centreline. Defensive armament and radio was unchanged, though the trailing antenna under the nose was fed out through a long vertical tube to avoid fouling the torpedo. The tailskid and lifting handles were removed, the ladder was fixed in place and extended down to the starboard float, and a 32kg (71lb) anchor was carried in the bomb bay.

Dimensions as TB-1 (including length) except height 6.6m (21ft 7⅞in).

Weight empty 5,016kg (11,058lb); fuel/oil 884kg (1,949lb); maximum weight 7,500kg (16,535lb).

Maximum speed 186km/h (115.5mph); service ceiling 3,620m (11,877ft); climb to 3,000m (9,843ft), 33.2min; range 1,220km (758 miles); take-off run 660m (2,165ft) in 26sec; alighting speed 90km/h (56mph); alighting distance 150m (492ft).

G-1 This designation, from *Gruzovoi*, cargo, applied to about ninety tired aircraft gutted of military equipment and transferred to Aeroflot for use as freighters approximately over the period 1938–46. The position is complicated by the fact that almost all other TB-1s and TB-1Ps were likewise simplified and used as cargo and paratroop transports during the Great Patriotic War. Some of these were equipped with large cargo boxes which could be pre-loaded and attached under the fuselage, and many of these containers were of Grokhovskii type equipped for paratroops, with four windows on each side, a front glazed area resembling that of the bomber and a floor hatch.

The standard G-1 was distinguished by flat Dural plates, sometimes hinged, covering the gunner cockpits and a shallow canopy over the cockpit for the pilots. This canopy had a vee windscreen, with four fixed roof windows, left and right glazed hatches hinged upward from the centreline, and a rear fairing of almost flat sheet the full width of the fuselage. The internal cargo compartment, previously the bomb bay and gunner area, had various small windows and was accessed through the top or a door on the right. A normal payload was 1,500kg (3,307lb).

Most of the civil aircraft were painted (at least on the upper surfaces) bright orange for visibility in Arctic areas, though one fleet brought sulphur from the mines at Ashkhabad. On 9 February 1937 G-1 registered N-120 took off from Moscow commanded by F B Farikh on a trans-Arctic flight over the general route Sverdlovsk, Irkutsk, Anadyr, Cape Wellen, Arkhangyelsk and back to Moscow. The total distance was 23,000km (14,292 miles), covered in forty-seven sectors in air temperatures between –40 degC and –70 degC. No ANT-4 suffered inflight structural failure, though many exceeded 4,800hr, possibly a record for the pre-1945 era. Restored ANT-4s survive at Monino and at Ulianovsk (Simbirsk).

Dimensions as TB-1 and TB-1P.

Typical weights empty (landplane) 4,500kg (9,921lb), (seaplane) 4,800kg (10,582lb); fuel/oil 1,500kg + 110kg (3,307lb + 243lb); maximum weight 7,500kg (16,535lb).

Maximum speed (landplane) 201km/h (125mph), (seaplane) 190km/h (118mph); service ceiling (landplane) 2,850m (9,350ft); range (landplane) 950km (590 miles), (seaplane) 700km (435 miles).

ANT-5, I-4

This was Tupolev's first fighter, and the first Soviet fighter to have either all-metal construction or an air-cooled radial engine. It resulted from a UVVS-RKKA decree of 1925 calling for the speedy elimination of dependence on foreign designs of combat aircraft. This led to an order placed on CAHI-AGOS on 21 September 1925 for a fighter to be constructed of Kol'chug alloy.

Tupolev assigned the task to the brigade headed by P O Sukhoi, as the latter's first major responsibility. In turn, Sukhoi put Putilov in immediate charge, tasking him also with the fuselage and undercarriage. Other responsibilities included: wings, Petlyakov; tail, Nekrasov; engine and propeller, I I Pogosskii; struts, Petrov; and equipment, Ye I Pogosskii; Sukhoi assigned himself the engine mounting.

After discussion with Tupolev and preliminary model tests the decision was taken to go beyond the ANT-3 configuration and choose a sesquiplane with an exceedingly small lower wing. Because of the demand for

This G-1 of Aviaarktika was converted from a TB-1 bomber. (John Stroud)

The ANT-5 prototype. (Courtesy Philip Jarrett)

tre section, a larger horizontal tail with square-tipped horn-balanced elevators, and a vertical tail of greater area with a horn-balanced rudder. Back in December 1927 it had been decided to build the ANT-5 in series.

Span (upper) 11.4m (37ft 4¾in), (lower) 5.7m (18ft 8⅜in); length (1st) 7.23m (23ft 8⅝in), (2nd) 7.28m (23ft 10⅝in); height 3.35m (11ft 0in); wing area 19.8 + 4 = 23.8sq m (256sq ft).

Weight empty (1st) 921kg (2,030lb), (2nd) 941kg (2,075lb); loaded weight (1st) 1,343kg (2,961lb), (2nd) 1,363kg (3,005lb).

Maximum speed (1st) 250km/h (155mph) at 3,500m (11,480ft), (2nd) 257km/h (160mph) at sea level; service ceiling (1) 8,200m (26,900ft), (2) 7,650m (25,100ft); time to 5,000m (16,400ft), (1) 11.4min, (2) 11min; range (2) 840km (522 miles); take-off run (1) 70m (230ft); landing run 140m (459ft).

agility the single-seat fighter was one of the few types of aircraft where Tupolev was prepared to consider something other than a monoplane, and the sesquiplane was considered a fair compromise. As finally agreed the areas were: upper wing, 19.8sq m, lower wing, 4sq m, so the prototype ANT-4 was almost a parasol monoplane.

An RAF flat-bottom aerofoil profile was selected, the upper wing having a thickness/chord ratio of 16 per cent at the centreline (where to improve pilot view both thickness and chord were sharply decreased) and strut joints, decreasing to 12 per cent at the tips. Structurally the upper wing was made in left and right halves, joined on the centreline. Each half had three spars and a typical truss structure with ten A-profile ribs and intermediate L-profile ribs skinned by chordwise panels of 32mm-pitch corrugated sheet attached by closed rivets through top-hat strips covering the A-ribs. The small lower wing had a single spar with four A-profile and three intermediate ribs on each side. The upper wing was mounted on two sloping struts to the front spar and a small streamlined pylon on the centreline attached to the rear spar and extended back to the windscreen. The upper wing was braced by V struts sloping sharply out to link the first A-rib of the lower wing to the fifth of the upper.

The rest of the airframe followed previous practice, except that the top decking of the fuselage, above the upper longerons, was skinned with smooth sheet. Flight controls comprised long, almost untapered ailerons on the upper wing only, moved by both cables and push/pull rods, a braced tailplane carrying horn-balanced elevators moved by rods, and a triangular fin carrying a parallel-chord unbalanced rudder moved by cables. The main wheels had 750mm by 150mm (29.5in by 5.9in) tyres and were carried on a transverse steel axle rigidly attached to a braced leg on each side which for the first time had the bungee shock-absorber incorporated into it. A steel-tube truss carried the Gnome-Rhône Jupiter GR9B rated at 420hp, driving a two-blade laminated wood propeller with a very large Dural spinner whose profile was continued back enclosing the engine to leave the upper part of each cylinder exposed. Dural tanks held 236kg (520lb) of petrol and 25kg (55lb) of oil. There was a prominent fairing behind the pilot's head, and in front of him were two 7.62mm (0.300in) Vickers guns each with a 1,000-round belt.

The prototype was built at GAZ No. 1, at the Central Aerodrome (Khodinka), and was completed in July 1927. Factory testing took place between 10 August and 25 September, NII-VVS testing following on immediately in the hands of Gromov, Anisimov, Yumashyev and I F Kozlov. The consensus was that the ANT-5 was the equal of any foreign fighter. A second prototype, ANT-4 dubler, was authorised, and this was flown in August 1928. This was powered by a 480hp Jupiter VI (CAHI history states Jupiter IV) in a helmeted cowling, had two fuel tanks in the upper wing totalling 303kg (668lb), a lower wing made in three parts with a small cen-

I-4 Predictably, the series fighter was heavier than the prototypes and had a lower performance. Contributory factors were rather high drag from the corrugated skin, unfaired strut ends and unfaired wheels with wire spokes. As early as 1928 Tupolev had realised that the air did not flow neatly along the corrugations but at various angles depending on the location, and he determined eventually to use smooth skin for the fastest aircraft. Series aircraft had the locally built M-22 (Bristol Jupiter) rated at 480hp, PV-1 guns (still very similar to the Vickers), a smaller pointed spinner and full VVS equipment. Aircraft were painted dull green. GAZ No. 1 delivered 170 in 1928–29, followed by GAZ No. 82 which produced two in 1929, 163 in 1930 and twelve in 1931, a total of 349 including prototypes. The I-4 had only a short front-line career (1928–33) because of the rapid pace of fighter development.

There were numerous modifications. A few I-4s operated on skis, and one aircraft was tested from 10 August 1931 with a pair of AGOS Dural floats, though interest in this quickly faded. In December 1931 one aircraft began a long programme of firing trials with L V Kurchevskii's 76mm (3in) recoilless cannon, of both DRP and ARK types. These were positioned under the upper wing 1m (39.4in) from the strut joint, thus well outside the arc of the propeller. Trials were also made with the cannon outboard of the struts. During one ground test a gun exploded, but the I-4 began years of trials which later involved

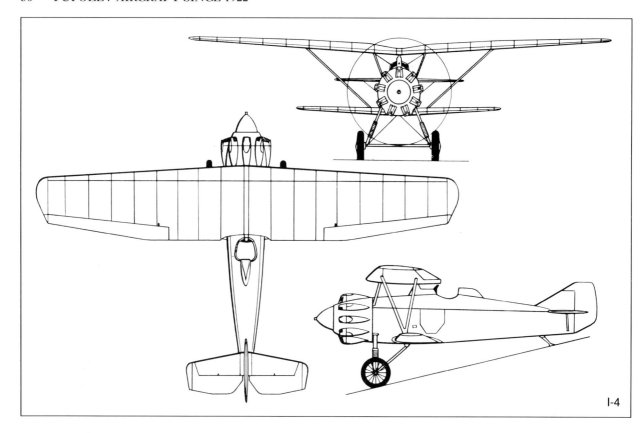

I-4

other platforms and led to Tupolev's ANT-23. Another I-4 was tested with a machine-gun in the upper wing (archives show drawings of a four-gun I-4), and also with four 50kg bombs. In 1932 an I-4 was the platform for the first firing trials of RS-82 spin-stabilised rockets, though this work was interrupted by transfer of the project from the GDL to the RNII. In 1935 an I-4 was chosen for AGOS tests in which solid-propellant rockets were used to boost the speed of fighters for short periods (not to boost take-off). The RNII rockets were 600mm long and 110mm diameter (23.6in by 4.33in) and weighed 5kg (11lb), of which 1.3kg (46oz) was propellant. Each gave a thrust of 450–500kg (up to 1,102lb) for 2.5sec. A cluster of three rockets was mounted on each side of the rear fuselage, fired in sequence. This programme ceased in 1936, but later resumed with an I-15.

Dimensions, unchanged.
Weight empty 978kg (2,156lb); loaded weight 1,430kg (3,153lb).
Maximum speed 220km/h (137mph) at sea level, 231km/h (143.5mph) at 5,000m (16,400ft); service ceiling 7,000m (22,966ft); climb 4.3min to 2,000m (6,560ft), 14.3min to 5,000m; range 840km (522 miles); take-off run 90m (295ft); landing speed 95km/h (59mph), landing run 210m (689ft).

I-4 Zvyeno Several I-4 fighters were modified. The first were two examples taken off production in 1930 for Vakhmistrov's *Zvyeno* experiments, as described in the preceding (ANT-4) entry. These aircraft, which in some documents are called I-4Z, had the lower wings removed outboard of the strut ribs, the tips being faired over. Shavrov, who comments that this version was 'more difficult to handle', correctly says that 0.5m of lower wing was left on each side (many accounts erroneously say 0.5m was removed on each side). Tunnel testing at AGOS had shown that this would make the fighter sit firmly on the bomber's wing until release. The only other modification was to add the quick release for the rear attachment, with a pull toggle in the cockpit. Many have copied Shavrov's statement that the I-4 Zvyeno had Jupiter IV or Jupiter VI engines, but there is every indication these were normal series aircraft with the M-22.

An I-4 with Kurchevskii recoilless cannon. (Malcolm Passingham/RART)

Dimensions, unchanged except lower wing area 1.8sq m (19.4sq ft).

Weight empty 940kg (2,072lb); loaded weight 1,362kg (3,003lb).

Maximum speed 253km/h (157mph) at sea level; service ceiling 7,650m (25,100ft); climb to 5,000m (16,400ft), 10.9min; range 686km (426 miles).

I-4 bis A single example was built in 1931 of this parasol monoplane version. The lower wing was removed completely, simplifying the fuselage but requiring a bridge at the frame to which the rear undercarriage struts were attached to react the wing-strut loads. The remaining wing was modified by fitting Handley Page slats over the outer 44.5 per cent semi-span (outboard of the struts). The slats were operated by the pilot. The engine installation was redesigned, the original enclosure and helmets being replaced by a Townend ring with blister fairings over the projecting valve gear and with two cooling air holes between each blister, exhaust being collected in a front ring leading to a pipe under the fuselage. This aircraft was faster but more positively stable and less agile. The need to operate the slats was a distraction, and after further testing automatic slats became common on Soviet fighters from 1940.

Dimensions, unchanged except total wing area 19.8sq m (213sq ft).

Weight empty 973kg (2,145lb); loaded weight 1,385kg (3,053lb).

Maximum speed 268km/h (166.5mph) at sea level; service ceiling 7,000m (22,966ft); climb to 5,000m (16,400ft) in 14min.

ANT-6, TB-3

First flown in 1930, the ANT-6 was an obvious next-generation heavy bomber beyond the ANT-4/TB-1. It used exactly the same structural materials and technology and, at least in the first series version, the same engines and propellers. It even used the same wheels, though it needed them in tandem pairs. Like a Beethoven symphony, though when it was finished it all seemed fairly obvious, it was actually the result of long and painstaking effort. As a by-product, it did more than any other aircraft to give the VVS element of the RKKA real muscle and striking power over a

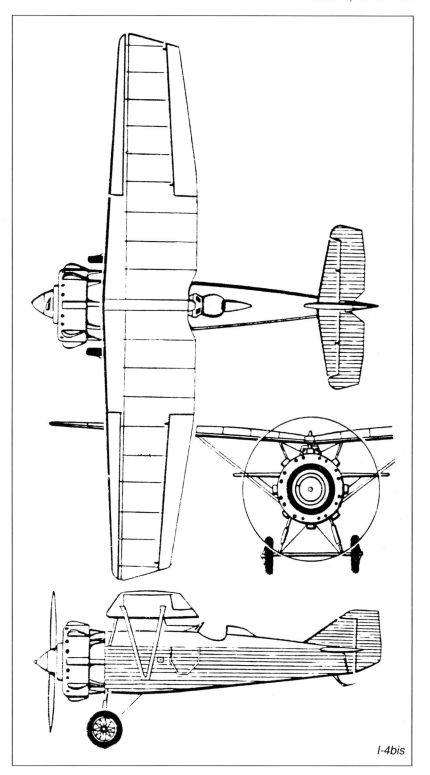

I-4bis

long range, which for much of the 1930s surpassed that of all other countries.

It was the first aircraft designed in the Soviet Union to be demonstrably ahead of the rest of the world. It established the configuration that ten years later had become almost universal for heavy bombers. It did more than any other aircraft to build

up the Soviet aircraft industry and its suppliers, the concept of paratroops and airborne forces, the dropping of heavy supplies and towing of gliders, the carriage of heavy freight to the most distant parts of the Soviet Union, and the opening up of the hostile Arctic.

Like all great aircraft it was steadily improved through many versions,

The ANT-6 prototype before its first flight. (ANTK Tupolev)

which were built in remarkable numbers (having regard to their size and power). Like all Tupolev aircraft the ANT-6 in all its versions proved to be tough and reliable – even if at first its engines were not – and able to go on flying for 5,000 or more hours in the harshest environments in the world, at a time when few aircraft in other countries could log one-tenth as much. Yet, so ignorant were many Western observers that, when they encountered these machines in the Second World War, they scornfully proclaimed that they demonstrated Soviet backwardness in aircraft design, forgetting that they were designed more than ten years previously.

Just for the record, the TB-3's counterparts in other countries were such machines as the Handley Page Heyford, LeO 20 and 25, Keystone B-4 and B-6 and Caproni Ca 89. All were fabric-covered biplanes in the 1,000hp class. In contrast, the ANT-6 stemmed from a request in 1925 that AGOS should build an all-metal bomber of 2,000hp, and with Tupolev at the helm it was certainly going to be a cantilever monoplane. It was also bound to be an aircraft of great 'character', though it could hardly be described as beautiful.

Having in 1923 asked CAHI, and through them AGOS, to produce what became the TB-1, the first mass-produced all-metal monoplane in the 1,000hp class, *Ostekhburo* came back two years later and in September 1925 requested landplane and seaplane versions of a four-engined bomber, to be made of Kol'chug alloy and with an installed horsepower of 2,000hp. As before CAHI handed the request to AGOS, where after building and

testing a simple tunnel model Tupolev launched the project in May 1926 as the ANT-6. It was to be managed under Tupolev's overall direction by Petlyakov, even though he was already burdened with the ANT-4.

In fact the two aircraft went through the basic design process very much in parallel. Both took longer than expected, the ANT-4 for reasons previously explained and the ANT-6 mainly because the customer kept altering the requirements. In any case, in the 1920s the Soviet Union did not possess factories in which to build TB-3s in quantity, nor engines to power them, nor even Kol'chug alloy from which to make them. In addition, the VVS wished to obtain experience operating the TB-1 before finalising the design of the TB-3.

Thus, while the ANT-6 prototype was taking shape, AGOS worked on several other aircraft and actually flew the prototypes of the ANT-7, ANT-9 and ANT-10 before the first ANT-6. It had also used major parts of the new bomber in the ANT-14 transport, with five engines, but the bomber had to have a gunner's cockpit in the nose and in 1929 it was decided to fly the prototype ANT-6 with four imported Curtiss V-1570 Conqueror water-cooled V-12 engines of 600hp each. By this time pressure for a seaplane version had faded, and in any case was being partly met by the ANT-8 flying boat. On 3 December 1929 the first detailed ANT-6 model was tested in a new CAHI tunnel, and a full-scale mock-up was agreed by the customer on 21 March 1930. The prototype then took shape rapidly at GAZ No. 22 and was completed on 31 October. Tests on the ground began on 20

November, and Gromov and engineer Rusakov made the first flight on 22 December 1930.

The wing, one of the largest ever built up to that time, repeated the Tupolev A° profile of the ANT-4, but the ailerons, extending from the outer engines to the tips, did not project to the rear beyond the straight even taper, though the angle of taper changed half-way along the aileron. Thickness/chord ratio was 20 per cent at the root, and as the root chord was 8m (26ft 3in) the maximum depth was no less than 1.6m (5ft 3in); at the tip the chord was 2.95m (9ft 8⅛in) and thickness/chord 10 per cent. Structurally the wing was made as a centre section of 7m (22ft 11⅛in) span integral with the F-2 fuselage section, ending at the inner engines, and four outer panels each of 4m (13ft 1½in) span, the inner pair ending at the outer engines. For rail transport all sections could be unbolted, as well as the leading and trailing edges, to restrict chord everywhere to under 4m.

Removing the leading and trailing sections left the main structural box, with four spars from tip to tip. Each comprised upper and lower tubular booms of elliptical section, tapering towards the tip from a ruling dimension over the five inner sections of 100mm by 90mm (3.94in by 3.54in). The booms were joined by vertical and diagonal box and channel sections, with reinforcing gussets in 3mm sheet at all junctions, the joints between the booms being steel connectors held by tapered transverse bolts. Truss ribs at 1.1m (44in) spacing repeated this structure, the wing profile between the spars being maintained by light spanwise trihedral girders formed from riveted channel and strip. Coarser skin corrugations (mainly 50mm by 13mm, 1.97in by 0.51in) matched the greater unsupported size of the skin panels, thickness being 0.3mm or 0.33mm except above the centre section where it was 0.5mm, and 0.8mm over walkways to the engines between spars 1 and 2. Chordwise top-hat strips covered the projecting parts of the seventeen A-type ribs on each side. On each side of each engine the leading edge could be hinged open and a folding ladder extended.

The fuselage was a direct extrapolation of that of the ANT-4, but with a normal rectangular cross-section showing only a hint of taper towards a triangular profile near the tail. Two prominent top-hat longerons projected on each side from nose to tail, the deck above the upper longeron having rounded edges, softening the otherwise severe lines. Six of the thirteen A-type frames were formed into corrugated-sheet bulkheads with 1.5m (59in) doors between sections. In the nose was an upper gunner cockpit and lower compartment for the navigator and bomb aimer with ten large 'chin' windows and two rectangular windows on each side. Next came the side-by-side open pilot cockpit, with a central gangway, followed by a compartment for the engineer with separate engine control and instrument consoles on each side of the gangway. Next came the bomb bay, with two pairs of flat rectangular doors in the belly, between the bulkheads to which the undercarriage struts were attached. Behind the wing were the two open cockpits for the radio operator/gunners. Like the nose cockpit, these were fitted with Scarff rings, but the prototype had no armament or military equipment.

The flight-control surfaces were skinned mainly with 25mm by 8mm (1in by 0.315in) corrugated sheet. The large ailerons had rectangular horn balances projecting beyond the wingtips, and were operated by cable circuits with two reduction gears and cable compensators to maintain correct tension and reduce pilot effort. The almost rectangular tailplane was pivoted above the upper fuselage longerons and braced by multiple wires. At the tailplane rear spar further upper and lower bracing wires served to adjust incidence through up to 14deg, moved by a cockpit handwheel.

The tapered elevators had long but narrow horn balances beyond the tailplane, and were driven by rods and bell-cranks. The tapered fin had a square top, above which was the large horn balance of the rudder, with six main ribs, driven by cables incorporating a multi-bungee tension device which, in the event of failure of an engine, could be released to free the pilots of the need to keep heavy pressure on one pedal.

The engines were mounted on welded trusses of M1 steel tube carried ahead of a steel firewall on the front spar. Each engine drove a Curtiss two-blade forged aluminium propeller, and its water radiator was of the flat block type mounted at 45deg under the front of the crankcase, the cowl under the wing being open at the rear. Exhaust was piped above the wing. Fuel was housed in four 1,950 litre D1 (Dural) wing tanks (total 7,800 litres 1,716gal), each subdivided into three and with all joints made with pan-head rivets and sealed with Shellac-coated paper gaskets. A total of fourteen liquid-level glass tubes calibrated in litres were placed so that they could be read in flight by crawling along the wing. These tanks outlasted the aircraft.

The main undercarriage followed ANT-4 practice. The KhGSA chrome-steel axles were joined by riveted collars to the end of a main strut running diagonally up to a pivot on the bottom fuselage longeron at the second wing spar. The weight of the aircraft was taken by a vertical strut from the axle to the second spar behind the inboard engine, the top of the strut incorporating faired boxes housing twelve rubber bungee shock absorbers. Triangular bracing was provided by a third strut between the axle and the bottom fuselage longeron at the third wing spar. The Palmer Tyre Ltd supplied the wheels, fitted with 2,000mm by 450mm (78.7in by 17.7in) tyres, but for the first flight special large wood skis were fitted, with multi-cord guy ropes at front and rear. The tailskid was pivoted to the tailplane frame and sprung by bungee cords.

A ladder was used to reach the trailing edge of the starboard wing root, and handrails were added along the fuselage. A venturi was mounted on the left outside the cockpit, and the pitot head was mounted on a braced vertical strut under the F-1/F-2 fuselage joint. Navigation lights were

fitted, as well as four landing lights in domes on the leading edge, but no radio. Electric power at 12V DC was supplied by a standard DOS-1 two-blade windmill generator above the fuselage on the centreline.

On the first flight the take-off was normal until vibration caused the starboard throttles to close. The aircraft slewed to the right, just missing a hangar; confused records then either state that Gromov removed one hand from the heavy loads on the control wheel to bang the levers open again, or (in the alternative version) frantically signalled to Rusakov, whose presence on board was simply to manage the engines. The rest of the flight was satisfactory.

Span 40.5m (132ft 10½in); length 24.2m (79ft 4¼in); height 5.52m (18ft 1¾in); wing area 231sq m (2,487sq ft).
Weight empty 9,375kg (20,668lb); fuel/oil 2,610kg (5,754lb) total; maximum weight 15,682kg (34,572lb).
Maximum speed 226km/h (140mph); service ceiling 5,100m (16,730ft); climb to 1,000m in 5.1min and to 3,000m in 19.9min; take-off run 18sec, 200m (656ft); landing speed 106km/h (66mph), landing run 150m (492ft).

By February 1931, though the Central Aerodrome was still snow-covered, the prototype was flying on its wheels. On 20 February the NII VVS announced that the ANT-6 would be produced in series as the TB-3, with M-17 engines. The aircraft was returned to AGOS for modification as nearly as possible to series standard. Among the changes were: the different engine installation, which surprisingly did not duplicate that of the TB-1 but retained the oblique ventral radiators, though in deeper and shorter ducts; a 120 litre (26.4gal) oil tank immediately behind each engine; 3.5m (11ft 5¾in) CAHI laminated-wood propellers with spinners; provision of compressed-air engine starting, using a ground supply or on-board bottles at 15 MPa, 2,176lb/sq in; redesign of the ailerons with slotted beak balances replacing the horns (reducing wingspan); increase in span of the tailplane by 1m (39.4in) and both span and chord of the elevators; redesign of the vertical tail, in effect transferring upper area from the rudder to the fin; cut away of the base of the rudder to leave room for a different tailskid pivoted to the rear of the stern frame; and installation of radio, navigation equipment and full armament.

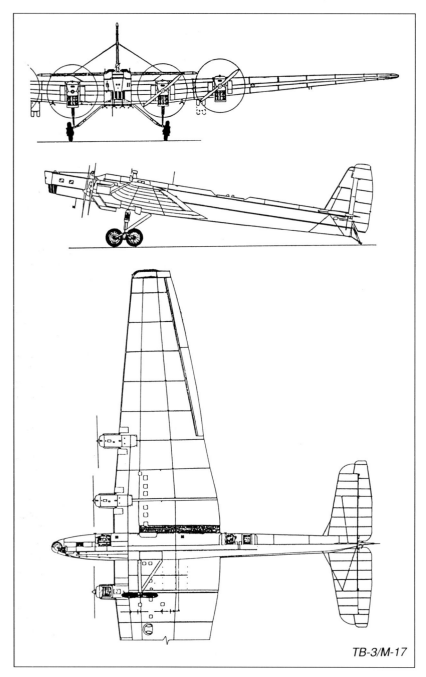

TB-3/M-17

the nose was a Tur-6, derived from the Scarff ring, mounting two DA machine-guns linked parallel, with twenty-four magazine drums of sixty-three rounds each. In the rear fuselage were two Tur-5 rolling-type rings at diagonally opposite corners (left front and right rear) of a large cockpit for two gunners. Each again mounted two parallel DA guns with twenty-four magazines. In addition the TB-3 introduced an idea Tupolev was to suggest for several even larger bombers: 'dustbin' gun turrets extended through the underside of the wing. These B-6 turrets had already been tested on the R-6 (ANT-7), extended from the fuselage. In the TB-3 the novel wing installation required the unfortunate gunner to crawl through the wing structure to reach his turret near the outer engine behind the No. 2 spar. Because of their location these turrets were manned by assistant engineers charged with supervising the adjacent outer engine and periodically reading the gauges on the fuel tanks. The B-6 was basically a Dural drum, open at the back, in which the gunner sat with his lower legs in two projecting streamlined tubes underneath. The turret could be manually cranked down until it was wholly below the wing, and then cranked to left or right while the gunner aimed his single DA up or down. He had side windows, and fourteen magazines. In an emergency the B-6 turrets could be severed from the aircraft. S I Savelyev headed the AGOS brigade responsible for TB-3 armament.

Two bulky radios were installed in the nose, under the control of the navigator with auxiliary controls in the rear-gunners' cockpit. The 11-SK covered HF bands from 60–120m, and the 13-PS the MF wavelengths 500–1,000m. The T wire antenna, so-called because of its shape, was strung fore and aft between wire-braced masts on the left side of the fuselage and from the front mast to wire-braced masts above the wing beside the outer engine nacelles. The MF set used a 150m (492ft) trailing wire. Telephone type intercom was not at that time provided, but some aircraft were fitted with a pneumatic system in which messages were enclosed in a cylinder 11mm (0.43in) in diameter and 55mm (2.17in) long, inserted into an aluminium tube and blown to its destination by a bellows,

Though M-17 engines were available, pressure on these going into series TB-1s and R-5s led to the ANT-6 being re-engined with BMW VI engines. The internal bomb bay resembled that of the TB-1, but with a full-depth central gangway and Der-9 racks on each side for twenty-eight (often twenty-six) bombs of up to 100kg (220.5lb) each. As in the TB-1 each Der-9 chute carried the bombs horizontally above twin rectangular doors. In addition two pairs of Der-13 under the fuselage allowed four 250kg (551lb) bombs to be carried externally. If the internal bay was not used, Der-15 or Der-16 external carriers could be added, giving various loads up to 5,000kg (11,023lb), for example four × 1,000kg + four × 250kg or eight × 250kg + two × 500kg + two × 250kg. No other bomber was to equal this for many years. The bomb-aimer sat on the right seat in the nose at his Boikov (licensed Goertz FL-206 or 110) sight, which in series aircraft looked through an optically flat window at the front of a small gondola in the underside of the nose. He could set up a switch panel through which the Sbr-9 electro-mechanical system could release any bombs in any sequence or salvo.

Defensive armament in the fuselage was the same as that of the TB-1. In

A B-6 under-wing turret with DA gun (note external bomb carriers). (M Passingham/RART)

each station having an electric buzzer to signal that a message had been either sent or received; tube length was 11.8m (38ft 9in).

The four 100W landing lights remained, imposing by far the biggest electrical load. The DOS-1 windmill generator had a nominal output of 350W at 12V DC, which was marginally adequate. All interior compartments had electric lighting with rheostat control.

Other loads included pitot heat, colour signalling lamps, navigation lights and the Sbr-9 system. The normal camera fit was to comprise two Potte-1B, one vertical in the F-1 floor and the other in the left rear gun cockpit. In the modified prototype two long Dural tubes were mounted projecting horizontally ahead of the nose, the lower one a standby pitot and the upper boom to give an index of angle of attack and lateral drift.

Still completely unpainted, the aircraft resumed flight testing in May 1931, the pilots being Gromov and CAHI's Volkovoinov, followed by NII-VVS pilots Stefanovskii, A B Yumashyev and engineer I P Petrov. Like the licensed M-17 the BMW engines suffered from serious faults, notably torsional fatigue failure of the crankshaft. In the late summer of 1931 the structural failure of similar wheels on the ANT-14 caused urgent redesign of the undercarriage with tandem pairs of Soviet 1,350mm by 300mm wire-spoked (later disc) wheels as used on the TB-1. Imported

steel was used for the welded frame which incorporated front and rear pivoted triangles carrying the axles, with multiple horizontal bungees pulling the axles towards each other. Data for prototype in 1931:

Span 39.5m (129ft 7⅛in); length (excluding nose booms) 24.2m (79ft 4¼in); height 5.6m (18ft 4½in); wing area 230sq m (2,476sq ft).

Weight empty 10,080kg (22,222lb); fuel/oil unchanged; loaded weight 16,387kg (36,127lb).

Maximum speed 213km/h (132mph) at sea level; service ceiling 4,660m (15,290ft); climb 6.2min to 1,000m, 22.8min to 3,000m; range 1,400km (870 miles); take-off run 220m (722ft); landing speed, 90km/h (56mph), run 170m (558ft).

TB-3/M-17F

As project leader Petlyakov was responsible for organizing production. From early 1930 he had spent half his time at GAZ No. 22, at Fili, which took over TB-3 production as TB-1 work tailed off, and produced over 90 per cent of the total of all versions. GAZ No. 39, named for V P Menzhinskii, and GAZ No. 18 at Voronezh made the remainder, as described later, and GAZ No. 38, at Moscow Khodinka, supplied rear fuselages and tail units.

Nothing like the TB-3 production programme had been seen before. Previous mass-produced aircraft had been much smaller, and in any case all-metal aircraft were still unusual even as prototypes. Despite this, the output of Dural (as Kol'chug alloy gradually became called) more than kept pace, and the imported KhMA steel engine mounts, undercarriage

struts and wing joints were replaced by Soviet KhNZA, only the chrome-steel axles being imported from Sweden for most of the production run. Far more than any other aircraft in history, the TB-3, especially in its initial M-17F version, forced the rapid growth of a supply industry for everything from rivets and washers to complete instruments and radios. The first aircraft from GAZ No. 22, No. 2201, was rolled out on 4 January 1932, and nine had flown from Fili by April. Aircraft 3901, from the Menzhinskii plant, was completed in November 1932.

Unfortunately, switching to local steel made a small contribution to the process of adding weight, which had become a problem before the first production aircraft had been completed. The causes were numerous: addition of combat equipment, steps, brackets, partitions and seats not on the prototype; plus-tolerances on all raw material, and use of heavy electric cable even to carry trivial currents; and rough hand-welds, generally poor workmanship and overthick and sloppily applied olive-green exterior paint.

Petlyakov instituted an urgent weight-reduction programme, and – though in early series empty weights varied 'by tens or even hundreds of kg' – the rise was checked and then reversed, so that while No. 3901 had an empty weight (without military equipment) of 11,207kg (24,707lb), the average by 1933 was 10,967kg (24,177lb).

Pilots had little difficulty converting to the TB-3, and when Stalin arrived at the Central Aerodrome in June 1931 to see the new bomber two TB-1 pilots who had never flown the TB-3 had to demonstrate it for him. Control forces were high by any standard, especially from the elevator, and towards the rear of the aircraft vibration and even resonance could be severe. Despite this, the stringent weight reduction caused no structural failures, merely various cracks which were arrested by drilled holes.

Development was intense on every part of the aircraft. Wherever possible the structure was simplified, the most obvious change here being to reduce the number of projecting chordwise strips over the A-ribs from seventeen to nine on each wing. The first few series aircraft were powered by imported BMW VI engines, but these

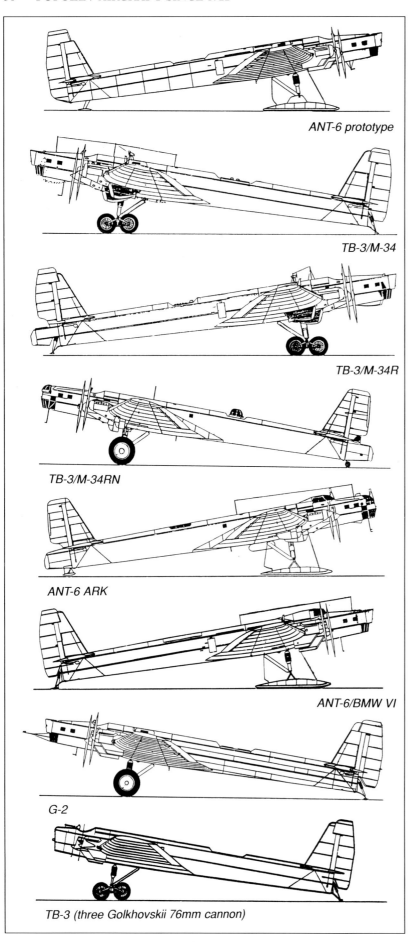

ANT-6 prototype

TB-3/M-34

TB-3/M-34R

TB-3/M-34RN

ANT-6 ARK

ANT-6/BMW VI

G-2

TB-3 (three Golkhovskii 76mm cannon)

were soon replaced by the M-17F, rated at 715hp for take-off, driving two-blade wood propellers (sometimes used in pairs at 90 deg to make a four-blade unit) with Hucks starter dogs on the projecting steel shaft. The top of the cowl was faired into the wing, with no open gap. The compressed-air starting was retained, the vital bottles being topped up in flight by a small compressor with cut-out. At aircraft 2225 the supply line from each fuel tank was extended 15mm (0.59in) up into the tank to try to eliminate drawing off water or solids. From 2254 this temporary fix was replaced by a redesigned fuel system in which any tank could supply any engine, with a proper filter unit in each line. Another change in this system (at 2271 and 3907, which shows how great was Fili's output) was to replace French diaphragm pumps by Soviet PM-18 gear pumps. Data for 1933 production by GAZ No. 39:

Dimensions unchanged except length 24.4m (80ft 0⅝in).

Weight empty 11,207kg (24,707lb); loaded weight 17,400kg (38,360lb).

Maximum speed 197km/h (122mph) at sea level, 177km/h (110mph) at 3,000m (9,842ft); service ceiling 3,800m (12,470ft); climb 9.2min to 1,000m, 41min to 3,000m; range 1,350km (839 miles); take-off run 300m (984ft); landing speed 110km/h (68mph), landing run 330m (1,083ft).

TB-3/M-34R By 1933 GAZ No. 22 was producing two TB-3s every three days, and altogether nearly 400 TB-3/M-17F bombers were manufactured. Minor improvements continued to be introduced to almost every system, those affecting ignition and carburation resulting in a significant improvement in range. Prolonged research had led to the one-type (and rather unpredictable) Kol'chug alloy of 1922 being replaced by a growing range of precisely controlled alloys of many kinds. In 1934 the introduction of D6 and D16 Dural enabled the wing structure to be restressed for higher weights, and also allowed the tips to be extended to improve field length and ceiling. Tested on an M-17F aircraft, this came on production in 1935 with the M-34 engines. Aileron gearing was improved, and the tail was refined to reduce elevator loads, and the rudder was given a neat servo tab along the trailing edge, which enabled the asymmetric-flight bias to be eliminated.

Despite all efforts, flight performance if anything tended to deteriorate, as is normal with military aircraft built in series. The VVS was naturally more than satisfied with what it rightly believed to be the best heavy bomber in the world, but it repeatedly asked for greater range. There were two obvious palliatives: increase power or reduce drag. Back in 1932 Petlyakov had headed a commission to study possible improvements.

In early 1933 one aircraft was modified for *Zadrayennyi* (battened down) trials. Everything possible was done to reduce drag, all armament and bomb carriers being removed, as well as antennas and other drag-inducers, nothing being left to project into the airstream except the pitot. The mainwheels were faired with aluminium spats. All this gave a gain in speed at different heights varying from 4 to 4.5 per cent, and thus a similar gain in range. This was clearly not worth while, though a later aircraft was tested with simulated smooth skin. More important was a change in engine, and in February 1933 the first TB-3 was flown with Mikulin M-34 engines.

These were again water-cooled V-12 engines, with unchanged swept volume (capacity), but with improved four-valve cylinders, a Hispano-Suiza rear wheelcase, Allison supercharger and Rolls-Royce Buzzard 'double herringbone' reduction gear. Take-off speed was increased from 1,630 to 1,800rpm, rising in the M-34R (R for reduction gear) of 1934 to 1,850. Take-off power was increased to 800hp, rising to 825 for the 34R. The test TB-3 was one of the first aircraft to be powered by the M-34, and the installation differed from that of the M-17 chiefly in the cooling. The water radiator was of an improved type, mounted vertically well back under the wing at the deepest point of an aerodynamically profiled duct. These radiator ducts were popularly called 'beards'. Immediately above each matrix was a separate oil cooler to handle the greater heat transfer (which rose further in the M-34R due to the big reduction gear). The cowling projected further ahead of the wing, and its underside sloped down from the spinner to the radiator. The exhaust pipes were extended back behind the leading edge.

In summer 1933 a second M-34 aircraft was tested with everything possible done to reduce weight. The wing 'dustbin' turrets were removed, reducing crew from eight to six, and many other items removed or modified. Empty weight was reduced from the original 11,725kg (25,849lb) to 10,956kg (24,153lb), while takeoff weight was increased by 300kg to 19,500kg (42,989lb). As a result, while leaving bomb load at one tonne, fuel mass could be increased from 5,000kg to 6,300kg (13,889lb), increasing range from 2,200km (1,367 miles) to 3,120km (1,939 miles). A major drawback was that to reduce propeller mass from 38kg to 33kg (72.7lb) their diameter was reduced to 3.18m (125in). At full power the tip speed due to rotation alone (ignoring aircraft forward speed) was 320m/s, higher than the speed of sound under most conditions (except sea level on a hot day) and the noise caused severe crew discomfort, especially to the pilots who were in the plane of the propellers.

NII testing of both M-34 aircraft was completed on 16 October 1933. It confirmed that, though the M-34 was basically more reliable than the M-17F, almost the only performance advantage was higher top speed at low altitude. Climb, ceiling, range and landing run were all worsened, and the heavier engine made it harder to achieve the primary aim of increasing the ratio of fuel mass to empty weight. No more aircraft were built or converted with the M-34, attention instead being concentrated on the geared M-34R. Though this was naturally heavier (670kg, 1,477lb, instead of 608kg, 1,340lb), it was hoped that driving a larger (and heavier) propeller more slowly would give improved flight performance, and remove the noise problem.

The first TB-3/M-34R took off in December 1933, and it was obvious that this was a superior aircraft. The propellers were still of laminated wood with two blades, but diameter increased to 4.4m (14ft 5¼in), weight rising to 50kg (110lb). Despite the greater diameter, the much lower rotational speed reduced tip speed to 250m/s (820ft/s), eliminating noise problems. Take-off run fell to 150–200m (492–656ft), climb to 3,000m (9,842ft) was reduced from 29min to 17.6min and speed at that height rose from 196km/h (122mph) to 211km/h (131mph).

This aircraft incorporated several further modifications which, under-taken in great haste, were to lead to more refined permanent changes. One was the addition of a tail turret. Previously the fuselage had terminated at the stern post carrying the tailskid and main fin spar. As a temporary installation the rear fuselage was modified to give a width at this station of 1m (39.4in) at the top, tapering to half this value at the bottom. The rudder was redesigned, the entire lower part being removed and replaced by a downward extension of the spar with projecting arms for the drive cables, and the top extended upwards to maintain area. The added upper section increased rudder area from 5.906sq m (63.6sq ft) to 6.31sq m (67.9sq ft), and a small area was added at the top of the fin. The new bottom of the rudder was itself cut away at the rear in a curve. A simple corrugated-dural tail end was then added, projecting behind the rudder, containing a floor and the new Tur-8 gun ring with twin guns. The added section incorporated the tail navigation light and a recessed step, but, apart from intercom, the gunner was isolated from the fuselage. This gunner brought total crew to eleven. Another change was that the ten defensive guns were of PV-1 type, though this was merely a heavier version of the DA.

The other major changes affected the undercarriage. The struts were improved in material, and the main vertical leg redesigned with shock absorption by an oleo-pneumatic tele-scopic cylinder replacing the rubber bungees. A closed-circuit hydraulic system was devised, energised by a hand-pump in the pilots' cockpit, which applied 3.04 MPa (441lb/sq in) pressure to expanding drum brakes on the rear wheels. A distributor connected to the rudder pedals transferred pressure progressively to either wheel to assist taxi-ing.

Both factories transferred to the TB-3/M-34R in February/March 1934. By this time the rear end had been properly redesigned, the F-3 section being smoothly continued to include a further modified tail end, slightly shorter and lighter. The braked wheels allowed the tailskid to be replaced by a castoring tailwheel (tyre 450mm by 200mm, 17.7in by 7.9in), and this not only made taxi-ing easier but also enabled most tail gunners, with difficulty, to clamber into the fuselage and along an added

Aircraft No. 22452 with No. 4 cowling open. (M Passingham/RART)

catwalk. The vertical tail was slightly modified to give a smooth outline with only a very small horn balance, the slot fairings and long servo tab having reduced rudder forces. To reduce drag the wind-driven generator was re-located on the left of the top decking and mounted on hinges so that it could be retracted inboard when not in use. In a few aircraft the mainwheels were enclosed in long spats, but when these became clogged with mud or snow they were removed.

In 1934 aircraft 22452 was taken off production before being painted, and flown to the Central Aerodrome for further CAHI drag-reduction tests. On the vertical tail was painted '1 IYe'. Though it retained the bomb-aimer's gondola, all turrets were removed and faired over, the rear wheels given fabric fairings and the wing leading edge skinned with taut fabric back to the front spar. Testing began on 1 January 1935, followed by further tests in which the fabric skin was extended over the upper surface to the rear spar, then the underside, then the whole wing, then adding the underside of the fuselage and finally the rest of the fuselage and tail. At the end of these tests, on 11 February, speed was increased by 5.5 per cent (thus, range similarly) and ceiling also by 5.5 per cent, while stalling speed was reduced by 10km/h (6.2mph) and rudder control improved. This aircraft made one flight lasting 18½hr. Later in 1935 it was fitted with four-blade propellers on the inner engines, resulting in better acceleration on take-off.

Span 41.85m (137ft 3⅜in); length 25.18m (82ft 7⅜in); height 5.6m (18ft 4½in); wing area 234.5sq m (2,524sq ft).

Weight empty 13,230kg (29,167lb); fuel/oil 2,020kg + 300kg (4,453lb + 661lb); maximum loaded weight 18,600kg (41,005lb).

Maximum speed 229km/h (142mph) at sea level, 211km/h (131mph) at 3,000m (9,842ft); service ceiling 4,500m (14,764ft); climb 4.1min to 1,000m (3,280ft), 17.6min to 3,000m; range with 2,500kg (5,511lb) bomb load 1,000km (621 miles); take-off run 300m (984ft); landing speed 127km/h (79mph), landing run 270m (886ft).

ANT-6/M-34RD

In 1933–34 nine aircraft were taken off production and modified as 'showing the flag' machines for goodwill overseas visits. These were basically simplified bombers, but bore the civil designation above. The M-34RD engine hardly differed from the RN, described later, though minor changes were made to the installation and all propellers were the large two-blade type. All armament was removed and the turrets and 'dustbin' holes faired over. A window was put in each side of what had been the compartment for the dorsal gunners. The tail end of the fuselage tapered to a narrow end fairing, but the cutaway rudder and bomb-aiming gondola remained. Additional radios were carried, and there were two windmill generators, one above and the other on the right side of the fuselage. Like series aircraft in 1934 the wheels were of disc type, not wire-spoked, the rear wheels being braked. No evidence has been found for the belief that these aircraft had additional fuel tanks. The aircraft were painted

white, with civil registration written in Arabic (Western) characters: URSS followed by a number beginning with the factory number 22. Only the first, 2236, bore red stars, and these were later removed. The upper surfaces of the wings behind the engines were blackened by exhaust carbon and oil.

The first mission was by 2236, captained by G P Baidukov, which flew to Warsaw in very bad weather in 8hr 40min on 28 July 1934, returning on 1 August. The success of this flight led to a tour by 2237, 2238 and 2239 to Paris and a single-aircraft flight to Rome, in each case with a high-level delegation. The four aircraft left Moscow together at 7 am on 5 August 1934 and flew in formation to Kiev, arriving at 11.08. The Rome aircraft continued via Lublin to Krakow, and on 7 August left Krakow at 14.23 and landed at Rome at 21.45 (all times GMT). The first return leg, on 15 August, took it to Vienna; this city was left at 7.43 on the 16th, on a non-stop flight to Moscow ending at 17.15, the average being 209km/h (130mph). The other three aircraft left Kiev on 6 August and flew via turning points at Lvov, Przemyzl and Krakow to Vienna, continuing next day to Paris. After various flights and functions the trio left Le Bourget on 13 August and returned via Lyons and Prague. Weather continued terrible, and on the 14th Zalevskii made a precautionary landing at Strasbourg, but overall the aircraft made a profound impression, not least because mechanical reliability and unaided navigation appeared perfect.

TB-3/M-34RN

Meanwhile, Tupolev and Petlyakov never ceased in their quest for improvements. By summer 1935 four-fifths of all TB-3s had been built, but there was plenty of scope for improvement in the final series blocks, and these improvements were retrofitted to many earlier aircraft. Most of the new features were tested on two aircraft built in February 1933, Nos. 22201 and 22202. The most significant was the installation of supercharged M-34RN engines, which dramatically increased power. Take-off (sea level) rating was slightly reduced, to 820hp, but above a height of 800m (2,625ft) the supercharged engine quickly became dominant. At its rated altitude of 3,500m (11,485ft) the power was still 750hp, compared with 490hp for the

M-34R. This transformed flight performance, even though on these two test aircraft the propellers (four-blade inners only) were unchanged.

A further change concerned armament. In 1932 a completely new machine-gun, the ShKAS (Shpital'nyi/Komaritskii), had begun to come into service. It used the same 7.62mm (0.300in) ammunition as before, but instead of having drums (which the gunner had to change when empty) it was fed by a long belt. More significantly, it fired at the remarkable cyclic rate of 1,800rds/min, so a single gun provided more firepower than the paired guns used previously. Accordingly, the defensive armament was completely revised.

The F-1 nose section was redesigned with an enclosed, fully glazed Tur-8 rotating on top of an extended lower section which moved the chin windows forward, the gondola being dispensed with. In place of the two mid-upper open gunner cockpits, a second enclosed Tur-8 was installed, on the centreline. The tail turret, already a Tur-8, was given a 'lobsterback' windshield with four pivoted sections of Plexiglas. These turrets all had a single ShKAS, with from 600 to 1,100 rounds. A fourth ShKAS was added on a pivoted mounting in a hatch under the rear fuselage, covering the blind spot under the rear. This enabled the heavy high-drag 'dustbins' to be at last omitted, reducing the crew to eight.

Normal bomb load was 2,000kg (4,410lb), in a KD-2 internal bay, using Der-19 or Der-21 carriers and either the previous Sbr-9 or electrical release ESbr-2. The bombsight could be an OPB-1, SPB-2 or KV-5. A very obvious change, made on both test aircraft and introduced during production in late 1935, was to return to single mainwheels with tyres 2,000mm by 450mm (78¾in by 17.7in). The hydraulically braked wheels were similar to the original Palmer type, but strengthened. Other changes included improved radios, addition of a D/F (direction-finding) loop antenna in a streamlined fairing under the nose, and improved pilot instruments including a Sperry horizon, new turn/slip, improved compass, US-400 ASI (airspeed indicator) reading to 400km/h, VD-10 altimeter reading to 10,000m, VAR-10 VSI (vertical speed indicator) reading to

10m/s and an AOA (angle of attack) indicator.

Series production of the TB-3/M-34RN began immediately the NII test programme was completed in October 1935. The first few aircraft retained the old tandem-wheel landing gear, but incorporated all the aforementioned changes plus four-blade wooden 4.1m (13ft 5½in) diameter propellers on all engines. Despite the weight of 118kg (260lb) per propeller, and 725kg (1,598lb) for each engine, the other changes enabled empty weight to be held down (see data), and overall performance was judged adequate for a heavy bomber of 1935. There was, however, a piloting difficulty which was sharply accentuated with the change to the single mainwheels.

On take-off the original M-17 aircraft had tended gently to swing to the left. With the geared M-34R, propeller rotation and the resultant swing had been in the opposite direction. The greater power of the RN engine accentuated this effect, and the change to single wheels made some aircraft almost uncontrollable, so CAHI undertook a comprehensive test programme. With a few aircraft the four throttles could be opened fully, the tail raised and the aircraft steered by the rudder. With most, rudder was ineffective below an airspeed of about 70km/h, and even using differential brakes the final take-off direction was typically 45 deg to the right of the start of the run. Formation take-offs became virtually impossible, and holding back the No. 1 (left outer) engine until 70km/h was reached lengthened the run. Many tests were made with different rudder control, different asymmetric powers and with wheels offset 0.75 deg to the left. The best answer was to tighten the Bowden cables from the cockpit to the engines, so that power could be controlled accurately, and then use modest asymmetric power.

About 140 TB-3/M-34RN were delivered, the engine designation being changed to AM-34RN in August 1936 in honour of A A Mikulin. Six aircraft from one of the 1936 blocks were supplied to China. They were to the latest standard but without guns. They were used by Chinese crews in the war against the invading Japanese.

Span 41.8m (137ft 1⅜in); length 25.1m (82ft 4¼in); height 5.6m (18ft 4½in); wing area 234.5sq m (2,524sq ft).

Weight empty 12,585kg (27,745lb); fuel/oil 2,460kg + 360kg (5,423 + 794lb); maximum loaded weight 18,877kg (41,616lb).

Maximum speed 245km/h (152mph) at sea level, 288km/h (179mph) at 4,000m (13,125ft); service ceiling 7,740m (25,395ft); climb 4.4min to 1,000m (3,280ft), 13.2min to 3,000m (9,842ft); operating radius with 2,000kg bomb load 960km (597 miles); take-off run 400m (1,312ft); landing speed 129km/h (80mph), landing run 280m (920ft).

TB-3/AM-34TK In 1936 a series aircraft, 22682, was fitted at the Central Aerodrome with special AM-34 engines equipped with TK-1 turbo-superchargers, the first to be qualified for flight in the Soviet Union. The turbosuperchargers were installed immediately ahead of the leading edge on the outboard side of each engine. This aircraft was also fitted with VRSh propellers with three forged Dural blades whose pitch could be adjusted on the ground. Performance at altitude was outstanding, but the TK-1 was available in prototype quantity only. A puzzle is that on 15 June 1936 a civil ANT-6, URSS-1216, took off with a top-level delegation headed by General Ya I Alksnis, head of all Soviet air forces, to fly via Jassy in Romania to Prague; it returned on 22 August. Painted white, this unarmed aircraft was not only to the latest build-standard but was fitted with TK-1 turbos and VRSh propellers !

TB-3/AM-34FRN Mikulin never ceased to develop the AM-34, and by 1936 had qualified the AM-34FRN (F, *Forsirovannyi*, intensive, forced). An unusual feature was that, while the left bank of cylinders retained the 160mm by 190mm bore/stroke, the right bank had stroke lengthened to 196.8mm, increasing capacity from 45.84 litres to 46.66 litres. More important was clearance to run at 2,000rpm, at which speed take-off power was no less than 1,200hp, for a tiny weight increase to 735kg (1,620lb). Rated power at height was a useful 1,050hp at 3,050m (10,000ft). While the FRN had twin carburettors of a new type, the FRNA had four, and the final production version, the FRNV, had six, each feeding a left/right pair of cylinders. The V stood for *vysotnyi*, high-altitude, because while take-off power was unchanged, power at high altitude increased considerably.

Most of the fifty aircraft built after autumn 1936 were powered by the FRN engine, driving unchanged four-blade wooden propellers. The other significant modifications, compared with the TB-3/M-34RN, were changes in the cowling and exhaust system, and removal of the radiator ducts downstream of the actual air/water matrix. Oxygen was provided from bottles for all crew-members. Additional fuel tanks were installed in the outer wings, as had been done with some earlier experimental versions, but there is a complete absence of weight information on this final sub-type.

The FRNV aircraft was used for FAI-homologated load/height records. On 11 September 1936 it took 5,000kg (250 20kg sandbags) to 8,102m (26,581ft). Five days later the same NII-VVS pilot, A B Yumashyev, reached 6,605m (21,670ft) with a payload of 10,000kg (22,049lb), more than double the previous record. On the 20th he took 12,000kg (26,455lb) to 2,700m (8,858ft), and on the 28th, accompanied as usual by Engineer A Kalashnikov, he broke his own 5,000kg record by a climb to 8,980m (29,462ft).

Aviaarktika ANT-6 No. SSSR N-169. (via Philip Jarrett)

Dimensions, unchanged. Weights, no data.
Maximum speed 300km/h (186mph) at 5,000m (16,400ft); service ceiling about 8,000m (26,250ft); no other reliable figures.

Total production of all ANT-6 versions was 820. GAZ No. 22 produced one in 1930, 155 in 1932, 270 in 1933, 126 in early 1934 (when work was stopped to speed up the start of SB production, *see* ANT-40), 74 in late 1935, 115 in 1936, twenty-two in 1937 and one in 1938. GAZ No. 31 produced five in 1932, thirty-seven in 1933 and eight in 1934. GAZ No. 18 produced five in 1934 and one in 1937. These figures include variants described hereafter.

ANT-6/AM-34R

Also known as ANT-6 *Aviaarktika*, four aircraft were specially prepared at GAZ No. 22 for exploration of the giant triangle bounded by meridians 32° 4'35"E and 168° 49'30"W (meeting at the North Pole) and north of the Soviet coastline, constituting the vast Arctic territory

claimed by the USSR. After several preliminary sorties by ship and air, an expedition was organized under Ivan Papanin with the objective of establishing a semi-permanent station on the drifting ice near the Pole. It was discovered that the task would require about 36 tonnes (79,365lb) of food, scientific equipment, fuel, radio and other items to be transported over 2,500km (1,553 miles); even with four aircraft, only the ANT-6 could do the job.

The four aircraft were taken off production when partially completed. The AM-34R engines were specially produced at GAZ No. 24, named for Mikhail Frunze, for operation at –50°C. They incorporated structural changes, carburettors heated by hot air from exhaust heat exchangers which also heated occupied parts of the fuselage, and coolant containing anti-icing additives. A special grade of lubricating oil was used, circulated in lagged piping from tanks which could be electrically heated. For starting, one engine would be encased in thick insulating wadding and the cooling liquid heated by a large blowlamp. Starting was then by the usual pneumatic method, the bottles being recharged on board, with an electric inertia starter as back-up. Ignition was supplemented by a powerful trembler coil, and to reduce radio interference all ignition leads were screened.

Fuel capacity was augmented by four additional main and two auxiliary tanks, all in the wings, to give an endurance of fourteen hours. A new design of metal propeller was used,

with a spinner and three blades adjustable in pitch on the ground.

All military equipment was removed, though the stripped bomb bay was retained so that cargo could be passed through the bomb doors. Gunner cockpits were sealed, and everything possible done to reduce drag. The F-1 section was specially demagnetised, steel parts being replaced, and fitted with upper and lower hatches and three openable windows and the front turret replaced by a fixed glazed section offering minimum drag. Navigation equipment included all the expected flight instruments, five different magnetic/gyromagnetic/radio compasses, a drift sight, a Sun-shadow compass (a refined sundial) and a remote compass driven by the navigator to tell the pilot the required course. The navigator also had the switch panel for the intercom, with seven terminals in the command aircraft and three in the other three. In addition, a pneumatic post system connected the pilots, navigator and radio operator.

The pilots' cockpit was completely enclosed, with a four-pane vee windscreen, two sliding windows on each side, and a roof of non-corrugated D16 incorporating an upward-hinged roof hatch above each pilot. Hot air from the engine muffs was piped from the wing along the outside of the fuselage and in at floor level. On the left side were triple venturis, supplementing the navigator's under the nose. One of the first Soviet autopilots was fitted, giving control in pitch and roll.

The engineer had several additional instruments and controls, and was also required to enter each wing to check on the engines and tank contents. The radio cabin contained a *Luch* shortwave transceiver for communication with other aircraft, and an *Onega* allwave radio for long-distance communication, using the trailing antenna. Windmill generators would have been difficult to de-ice, so both inboard engines drove generators, and a third was driven by a petrol-driven 3.5hp APU (auxiliary power unit).

Both wheel and ski gear was provided. The wheels were the hydraulically braked type with 2m tyres. The skis differed only in detail from the standard pattern used since the original ANT-6 of 1930. Whichever was not in use was secured under the fuselage in tandem on studs and lock-nuts. Another innovation was a braking parachute. This was a ring type, made of silk with a powerful elastic ring round the periphery to balance canopy diameter against airspeed. The folded parachute was stowed in the tail end, released only by safety interlocks opened by the engineer and pilot in sequence. The cable was anchored to the rear wing spar in the centre section and held the streamed canopy about 10m (33ft) behind the aircraft.

The completed aircraft were painted bright orange, picked out in blue, and carried civil registration N-169 to N-172. They were cleared to a take-off weight much heavier than any previous version. Equipment included parachutes for aerial delivery of cargo, and a magazine of small bombs for testing ice thickness before landing and smoke markers to indicate wind. The four aircraft left Moscow on 22 March 1937, with the required cargo and forty-two crew. On 21 May, after various adventures, the command aircraft, N-170, flown by M V Vodopyanov, landed on an ice floe about 18km from the Pole. After much Polar work, N-169 was left at Rudolf Island to support the station established there, while the other three aircraft returned to Moscow. Subsequently N-169 became an ANT-6/AM-34FRN, equipped with the high-altitude engines in a new installation, driving constant-speed propellers with large spinners faired into a new cowling with a long shallow radiator tunnel under the wing. On 5 March 1941 it began a 26,000km (16,156 mile) expedition establishing temporary scientific stations on ice floes between Wrangel Island and the Pole, returning on 11 May. These aircraft supported Arctic colonisation well into the Second World War.

Later a fifth aircraft was produced to a further upgraded specification and, given civil registration N-212, was operated by the Polar arm of the GVF (civil air fleet). It bore *Arktika* on the nose, and was used for navigation and communications research and for mapping. Date are for N-169/172 in original form:

Dimensions, unchanged.
Weight empty 12,500kg (27,557lb); fuel/oil (usable) about 3,200kg (7,055lb); loaded weight 24,050kg (53,020lb), but N-170 took off for the Pole at c27,000kg (59,524lb).
Maximum speed 240km/h (149mph) at sea level, 275km/h (171mph) at 3,000m; range with full payload 2,500km (1,553 miles); landing speed 120km/h (74.5mph).

G-2 Under this designation more than 170 TB-3/M-17Fs were converted into civil (in a few cases military) cargo aircraft, later joined by a smaller number of G-2/M-34RNs. Most of the conversion was done in the growing number of repair shops of Aeroflot, the all-Union civil operator formed in 1932. Apart from stripping out all purely military equipment, and skinning over the gunner cockpits, the task involved reinforcement of the structure in forty-six locations, usually by adding doublers or additional members or by an increase in gauge of a structural member or skin panel. This was because of the fact that the aircraft were already well-used and were now to be operated more intensively for typically 2,000 hours between major overhauls (one G-2 achieved 3,468).

The most obvious external change was that the nose, now looking like the prow of a ship, was faired back with a sharp upper chine to meet a canopy over the pilots' cockpit. The canopy was generally similar to that of the Arctic variant, and each pilot had a roof hatch. Cabin heat was provided, again as in the Arctic machines, but the radio and navigation equipment of most G-2s was more spartan. These machines, mostly unpainted, operated intensively in Central Asia, the Caucasus and Siberia, while more comprehensively equipped examples served *Glavsevmorput*, supporting ships in the Arctic. The introduction of G-2s from 1936 transformed Aeroflot cargo, from the 6,800/10,500 tonne level to 34,200 tonnes, rising thereafter to nudge 50,000 tonnes in 1940. Every conceivable cargo was carried, notably sulphur, copper ore, newspaper matrices and, after 22 June 1941 when special civil air groups were formed, ammunition, ballbearings, electronics and, for Leningrad, food.

Data for G-2/M-17F:
Span 40.5m (132ft 10½in); length 24.2m (79ft 4¼in); height 5.4m (17ft 8⅔in); wing area 230sq m (2,476sq ft).

One of the TB-3 bombers retrofitted with two MV-2 dorsal turrets. (M Passingham/RART)

Weight empty (typical) 11,179kg (24,645lb); fuel/oil 1,600kg + 160kg (3,527lb + 353lb); loaded weight 17,000kg (37,478lb).

Maximum speed 198km/h (123mph) at sea level; service ceiling 2,200m (7,220ft); climb 24min to 1,000m (3,280ft); take-off run 625m (2,050ft); landing speed 110km/h (68mph).

Data for G-2/M-34RN:

Span 41.8m (137ft 1⅜in); length 25.1m (82ft 4¼in); height 5.5m (18ft 0in); wing area 234.5sq m (2,524sq ft).

Weight empty 13,417kg (29,579lb); loaded weight 22,000kg (48,500lb).

Maximum speed 240km/h (149mph) at sea level, 280km/h (174mph) at 3,000m; service ceiling 4,960m (16,275ft); climb 8min to 1,000m, 16min to 2,000m; take-off run 525m (1,722ft); landing speed 120km/h (74.5mph), landing run 300m (984ft).

TB-3 tests Many TB-3s were used to test new equipment and procedures. As early as 1932 Stefanovskii used a series aircraft to test blind-flying instruments, radio compasses and, in 1935, a blind landing technique using marker beacons and finally a 20m (66ft) aluminium tube released at one end to hang pivoted below the aircraft. Another TB-3 was used for autopilot development.

P I Grokhovskii, besides designing aircraft, was a pioneer of air assault and transport systems. Following his work on TB-1s, he used TB-3s and G-2s to perfect paradropping of supplies, with or without special containers. He also pioneered carriage of vehicles, including armoured cars, and even designed the equipment for successful air drops of T-37 and T-38 light amphibious tanks into water. On 15 December 1934 ground testing began of a Model 1927 76mm (3in) field gun mounted under the nose, Grokhovskii arranging for the carriage to measure recoil loads. Air firing began two days later. In early 1935 this aircraft began further trials with two such guns in the outer wings and with the entire F-1 section replaced by a short box containing a long-barrel 76mm AA gun (the navigator station being moved to the mid-fuselage). The guns were loaded by hand and fired on receipt of lamp signals by the pilot who aimed the aircraft.

The TB-3 was the platform for the development of power-driven gun turrets, including the MV-2 and MV-5, and after 1941 some TB-3s were fitted with two MV-2 turrets on the closely spaced dorsal rings. CAHI managed complex testing of a centralised fire-control system governing the operation of up to three power-driven turrets, but this never reached the production stage. TB-3s were also used for countless tests of dropped stores, and in 1941–42 several were used as cruise missiles carrying a special charge weighing 6,200kg (13,668lb), the human pilots needed at take-off baling out and subsequent flight being controlled by radio from another TB-3.

In January 1937 a former M-17F aircraft designated TB-3D (*dieselnyi*) took off on the first flight of A D Charomskii's AN-1A diesel engine. Though this engine was rated at 900hp, the aircraft had inferior performance apart from range, which was increased to an estimated 4,280km (2,660 miles).

Tests with airborne troops and parachute troops grew in scope from 1932 onwards. The aircraft had their gun rings removed and were arranged for about thirty-five paratroops to drop either in a stick or a tight group from the gunner cockpits, the right door and the bomb bays, some even being carried externally above the wing roots holding the handrail. With reduced fuel fifty paratroops could be carried. The biggest exercises were near Kiev in 1935 when 1,200 paratroops, 150 machine-guns and eighteen 45mm guns were dropped, followed by 2,500 airlanded troops and heavy support weapons, and in Byelorussia in the same year when 1,800 paratroops were followed by 5,700 airlanded plus heavy weapons.

Most spectacular of the experiments were Vakhmistrov's *Zvyeno* trials (*see* ANT-4/TB-1). These took place from September 1933 to August 1935. Z-2 comprised a TB-3 carrying an I-5 fighter above each wing and a third above the fuselage. In Z-3 an I-Z fighter was hung under each wing; A V Korotkov, in one fighter, made serious mistakes and was eventually killed. Z-5 involved an I-Z taking off independently and hooking on, using some techniques borrowed from US Navy F9C/airship tests; on 23 March 1935 this achieved the first-ever hook-on to another aeroplane. Z-6 featured an I-16 under each wing. Finally the incredible *Aviamatka* was (after a hair-raising take-off) successfully tested on 20 November 1935. It involved an I-15 above each wingtip, an I-16 below

A TB-3/AM-34RN in 1941 converted to launch two SPB dive-bombers. (via Philip Jarrett)

each wingtip and an I-Z which hooked on under the centreline in flight. At Stefanovskii's signal, all five fighters departed safely!

These experiments led in the Second World War to numerous SPB (fast dive-bomber) operational missions flown by I-16s of the 32nd Regt dropped from under the wings of TB-3/AM-34RN bombers of the 63rd Brigade. Each I-16 carried two FAB-250 (551lb) bombs, aimed in a dive. Most of the missions were in the Constanta (Romania) region, but a particularly important series of attacks in August 1941 seriously damaged the great Danube bridge at Chernavoda.

There is no room here to outline the thousands of bombing and transport sorties flown by the TB-3/M-17F and /M-34RN in the Lake Khasan conflict of 1938, the Khalkin Gol war of 1939, the Winter War of 1939–40 or, above all, the Great Patriotic War. Not one TB-3 was preserved, but a crashed specimen was discovered near Arkhangelsk in 1992 and is to be restored in Moscow.

ANT-7, R-6

As early as 1924 the VVS–RKKA had considered the possibility of using long-range twin-engined fighters to escort its planned heavy bombers. Discussions over the subsequent three years, often involving CAHI and AGOS, refined the idea and increasingly became polarised around an all-metal monoplane with two 500hp engines. The obvious solution was for Tupolev's designers to create a smaller edition of the TB-1.

Over this long gestation it became apparent that such a machine, as well as serving as an escort to bombers, could be a useful bomber in its own right. It could also drop a torpedo, fly reconnaissance missions and be a 'cruiser' sent deep into hostile territory to gain command of the enemy's own airspace.

Actual design started in October 1926, but there was no urgency because even the first type of heavy bomber was still years away from operational service; in any case, there was no available factory capacity, engines or light alloy. AGOS at last received a

requirement on 8 January 1928, for 'an air fighting aircraft'. The general idea was exactly the same as the *avions multiplaces de combat* so popular in France in 1928–38, though the French designs were not only later but also more ungainly. In 1936 Britain's Handley Page thought it had something similar in the Hampden, which had a fixed gun and was touted as 'a fighting bomber'.

Tupolev never lost sight of the need for aerodynamic cleanliness, though to obtain all-round firepower he did sanction a bluff gun turret extended under the fuselage, as described later. He also decided to hang all bombs externally. Tupolev archives do not name the brigade leader responsible for overall direction under Tupolev, and it appears that Tupolev managed the design himself, assigning the various parts to the usual specialists. ANT-4 experience made most of the detail design straightforward, though at the outset there was argument over the choice of engine. AGOS suggested either the Bristol Jupiter air-cooled radial or the Hispano–Suiza water-cooled V-12 (both of which were to be produced in the Soviet Union under licence), but the final choice was to use the same engine as in the TB-1 and TB-3 bomber, the German BMW VI, licensed as the M-17. This decision was taken before the bombers had suffered from numerous failures with these engines.

The wing repeated the flat-bottom CAHI aerofoil, a Clark-Y analogue, used in the bombers. It was structurally a horizontal rectangular centre section and straight-tapered outer panels with dihedral resulting almost entirely from the taper in thickness. Thickness/chord ratio varied from 19.5 to 14 per cent, and because the centre section was untapered the A-type rib at its outer end was the same as on the ANT-4, but no other parts were common. As before, a deep channel section was riveted about 120mm (5in) back from the leading edge, ahead of which the skin was uncorrugated. The five principal spars comprised tubular booms at top and bottom joined by channel or angle sections. The ribs at 1m (39.4in) spacing comprised the usual A-section bounding member covered by a riveted top-hat section, braced by diagonals linking each bottom boom to the top boom of the next spar to the rear. At

the back were lighter spars carrying the ailerons, which extended over the whole span of each outer panel with a horn balance beyond the wingtip.

Throughout the aircraft the skin was mainly corrugated at a pitch of 32mm, thickness being 0.3–0.8mm. The fuselage again followed established principles, and was made in front, centre and rear bolted sections. The rectangular cross-section had a rounded top, and aft of the wing gradually became progressively more triangular in the Tupolev manner. As before, the main longitudinal members were four strong tubes, joined by vertical channels at the sides and transverse tubes. The skin was corrugated, except for the single-curvature sheet wrapped round the nose and the immediate vicinity of the joints between the three sections.

In the nose was a navigator's station with a Tur-5 turret, Goertz bomb-sight, vertical camera and drum magazines. The front section also included the cockpit for the pilot, on the left side, with a wide instrument panel extending across to the right and two large fire extinguishers piped to the engines. The centre section did not house a bomb bay, but instead incorporated a retractable gunner 'dustbin' designated PBSh-1. This was similar to the TB-3's B-6. It was carried on two long curved arms pivoted to the bottom of the rearmost mid-section bulkhead, and cranked up/down along curved guide rails by a cable passing over a pulley near the top of the fuselage and operated manually via pulley blocks. In the rear fuselage was the Tur-5 ring for the rear dorsal gunner, as well as the radio receiver, 13SK transmitter, batteries, red/green downward signal lights and, centrally on top, the windmill generator which, as on the TB-1, had a single counter-balanced blade.

The tail was again of familiar construction, using skin with 20mm by 5mm corrugations. The horizontal tail had a span fractionally greater than that of the ANT-4, though chord was much smaller. The tailplane was pivoted to the upper longerons, driven by a cockpit wheel through ±5 deg via the bracing struts on the underside. All control surfaces had projecting horn balances. The elevators were moved by push/pull rods on bell-cranks on the underside, while the rudder was moved by cables acting on

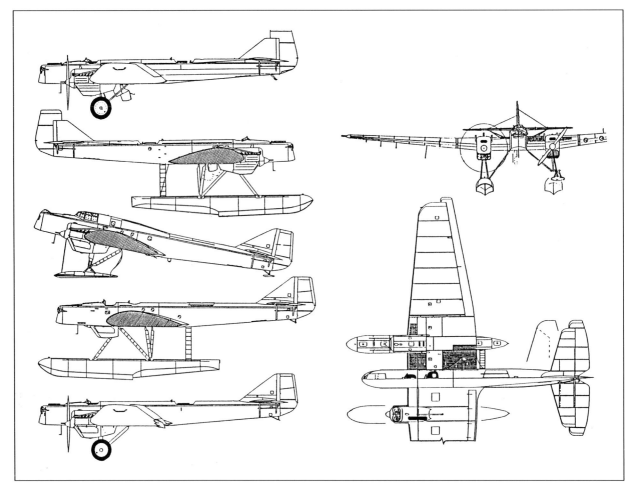

Variants of the ANT-7. From the top: 2nd prototype (dubler); *MR-6 (wing not shown); PS-7 (wing not shown); MP-6 (wing not shown); KR-6; front view (port half) MR-6 showing ski/float, (starboard half) MP-6; plan (above) MR-6, with inset of horizontal tail of later versions, (below) KR-6.*

both sides. Each aileron had a crank on the upper side driven by a rod faired into the top of the wing. Ball-bearings were used through the flying-control circuits.

The undercarriage was again a simple steel-tube pyramid. Each axle was an extension of the transverse strut pivoted to a double trunnion welded to the bottom longeron of the fuselage, and sprung by multiple bungee cords round the bottom of the bracing struts, vertical in front elevation, pinned to the first and fourth wing spars. The single wheels had 1,100mm by 250mm (43.3in by 9.8in) Palmer tyres, and were unbraked. The Dural built-up box-section tailskid had a steel shoe and was pivoted between the bottom longerons, sprung by bungee cords which permitted limited castoring.

The engine choice eventually fell on the BMW VIa, rated at 730hp. These water-cooled V-12 units were mounted on simple triangulated steel-tube trusses carried off the front spar.

The oil for each engine was contained in a Dural tank between the engine and the firewall built into the front of the front spar. About 1,000kg (2,200lb) of fuel was housed in four tanks of Dural reinforced by steel strips housed inboard and outboard of the engines between spars 2 and 3. A faired drain pipe extended below each wing just outboard of the root. The fuel piping was rigid dural, using the French AM-type fittings which became standard in the Soviet Union.

Exhaust gas from each bank was led into a pipe extended back across the wing. The water radiators were of the rectangular matrix type mounted on vertical rails and extended any desired amount out of their housing in the inboard wing, just ahead of the inner tank, by a pulley system worked by one of the crew. The two-blade fixed-pitch wooden propellers had a diameter of 3.2m (10ft 6in).

Though there was no bomb bay, provision was made for an external

bomb load. The usual arrangement on series aircraft was to attach six DER-7 tubular carriers under the fuselage and wing roots, each equipped for a single AF-32 (70.5lb) bomb. The release system was the standard electromechanical Sbr-8. Normal defensive armament comprised five DA machine-guns, two in each Tur-5 ring and one in the retractable BPSh-1. Each gun had seven (in the nose, six) spare ammunition drums. The Tur-5 installations each had fabric hoses down which empty cases were removed, and the rear ring could be traversed from one side of the fuselage to the other so that the gunner could fire straight downwards.

A single prototype was completed on 11 September 1929. No factory testing took place, and the first flight was the start of State NII testing on an unrecorded date in May 1930. During the intervening period a redesigned horizontal tail was fitted, with greater chord and reduced span. The first flight began with double engine failure, but Aleksei Konstantinovich Tumanskii managed to turn back and land. The mechanic, who had for several days been preparing the aircraft, had forgotten to fill the fuel tanks. Subsequent testing was satisfactory, apart from occasional serious tailplane buffet.

Span 23.2m (76ft 1⅛in); length 14.75m (48ft 4⅝in); height 6.92m (22ft 8⅜in); wing area 80sq m (861sq ft).
Weight empty 3,790kg (8,355lb); loaded weight 5,173kg (11,404lb).
Maximum speed 222km/h (138mph); cruising speed 192km/h (119mph); service ceiling 5,640m (18,500ft); climb to 1,000m (3,280ft) 3.4min, to 3,000m (10,000ft) 11.7min and to 5,000m (16,400ft) 30.3min; range 1,780km (1,106 miles); take-off run 120m (394ft); landing speed 100km/h (62mph), landing run 250m (820ft).

MP-6 No. SSSR Zh-1 about to take off. (G F Petrov/RART)

Later in 1930 the prototype resumed testing after various modifications. Called *Ispravlennyi* (corrected), it had a lengthened fuselage (though empty weight was reduced), increased fuel capacity, and a narrow-chord flap fixed at an incidence of 29 deg behind the trailing edge from the root to a point oversailing the inboard end of the aileron. Though this flap reduced maximum speed by 7–8km/h, it cured the tail buffet problem. Data show it also significantly raised the service ceiling.

Dimensions as before, except length 15.06m (49ft 4⅝in).
Weight empty 3,708kg (8,175lb); loaded weight 5,406kg (11,918lb).
Maximum speed at 3,000m (10,000ft) 227km/h (141mph); service ceiling 7,090m (23,260ft); landing speed 97km/h (60mph).

R-6 In 1931 a *dubler* (second prototype) was built, slightly longer and powered by production M-17F engines rated at 715hp. A major change was that the water radiators were no longer retractable into the wing but were fixed directly under each engine inside the cowling. The matrices were so designed that, installed with top and bottom horizontal, the front and back sloped back at 50 deg. With other minor changes the ANT-7 was accepted for production as the R-6, the designation reflecting the primary role of long-range reconnaissance. Other duties were long-range escort and fighting. Data for first (pre-production) aircraft:

Dimensions unchanged.
Weight empty 3,900kg (8,598lb); fuel/oil 1,440kg (3,175lb); loaded weight 6,130kg (13,514lb).
Maximum speed 240km/h (149mph) at sea level, 212km/h (132mph) at 5,000m (16,400ft); service ceiling 5,620m (18,440ft); climb 39.3min to 5,000m; range, 1680km (1,045 miles); landing speed 110km/h (68mph).

Production of the R-6 began at GAZ No. 22 in June 1931. This factory delivered fifteen in that year and thirty in 1932. Thus the total for the R-6 was forty-five, excluding prototypes. They had a fairly brief period in front-line VVS service, but were soon found useful as crew trainers.

A team led by Myasishchev used one aircraft as a test-bed for liquid cooling systems using radiators mounted inside the wing centre section. The inlet was in the leading edge, the air being ducted through the third spar and out through an aperture in the upper surface, curved vanes deflecting the hot air aft to give thrust. Speed was increased by 5km/h and ceiling by 300m, but the engines tended to overheat on the ground, or in hot weather. This aircraft tested various ducted radiator schemes, several of which were used in production aircraft of the Great Patriotic War.

It was recognised that the R-6 was obsolescent even while new examples were being delivered. Many were transferred to Aeroflot as PS-7s, described later. Others were used for armament development.

Dimensions unchanged.
Weight empty 3,856kg (8,501lb); fuel/oil

1,385kg (3,053lb); loaded weight 6,472kg (14,268lb).
Maximum speed 230km/h (143mph) at sea level, 216km/h (134mph) at 3,000m; service ceiling 5,620m (18,440ft); climb 16.7min to 3,000m, 24.0min to 5,000m; range 800km (500 miles) [this dramatic reduction is not explained]; take-off run 160m (525ft); landing run 250m (820ft).

MR-6, R-6a In 1932 GAZ No. 31 at Taganrog tested a series R-6 fitted with Zh-type floats identical to those of the TB-1P. This aircraft was also used for torpedo dropping, the BPSh-1 turret being removed. GAZ No. 31 then built the MR-6 in series, historian V B Shavrov also using the designation KR-6P. Similar aircraft with the ventral turret retained were designated R-6a. Even though they were the heaviest of the whole family these aircraft were powered by the M-17 engine, of 680hp. GAZ No. 31 delivered the pre-production aircraft in 1932, fifty in 1933 and in 1934, a total of seventy-one. Civil MP-6 conversions are described later.

Dimensions unchanged, except length 16.15m (52ft 11⅞in).
Weight empty 4,640kg (10,229lb); loaded weight 7,500kg (16,535lb).
Maximum speed 234km/h (145mph) at sea level, 215km/h (133.5mph) at 3,000m; service ceiling 3,850m (12,630ft); range 492km or 900km (306/559 miles) depending on tankage; take-off run 600m (1,970ft); alighting speed 130km/h (81mph), alighting run 350m (1,148ft).

Limuzin A single example was produced in RVZ workshops by modifying an R-6, though as it had BMW VIa engines it was probably the previously modified prototype. The aircraft was stripped of armament and all apertures sealed. The cockpit was enclosed, and the interior equipped lavishly for nine passengers. The R-6 *Limuzin* flew in July 1933 and, through no fault of the aircraft, crashed two months later killing P I Baranov, A Z Goltsman and others.

Dimensions, as 'corrected' prototype.
Weights, not recorded except fuel/oil 1,300kg+120kg (2,866lb+265lb).
Maximum speed 248km/h (154mph) at sea level; cruising speed 220km/h (137mph); endurance 8hr; landing speed 120km/h (74.5mph).

KR-6 The *Kreiser Razvyedchik*, cruiser reconnaissance, version carried no bombs, had the BPSh-1 turret removed, and instead had a total of ten fuel tanks extending over a considerable proportion of the span between spars 2 and 3. The chief external differences were that the tail unit was cleaned up, the large rudder horn being removed and all surfaces having a neater outline, the fuselage was shorter, fixed flaps were replaced by hinged Zapp-type flaps (under, not behind, the trailing edge), the M-17 (not 17F) engines had modified installations which in some aircraft had shorter radiator ducts, the main wheels had drum brakes and the tailskid was modified. One aircraft tested oleo-pneumatic shock struts.

GAZ No. 22 built all of this version. In 1934 it delivered 150, plus seventy-two of a version designated KR-6A which some documents describe as the floatplane version, though the seaplanes were the preserve of GAZ No. 31, on the Black Sea. In 1935 GAZ No. 22 delivered twenty KR-6 and twenty-eight KR-6As, making the total for this version 270, and bringing the overall ANT-7 series production to 406. Data for series KR-6:

Dimensions unchanged, except length 14.7m (48ft 2¼in).
Weight empty 3,870kg (8,532lb); fuel/oil 2,200kg+200kg (4,850lb+441lb); loaded weight 6,992kg (15,414lb).
Maximum speed 226km/h (140mph) at sea level, 210km/h (130.5mph) at 5,000m; service ceiling 5,120m (16,800ft); range 3,100km (1,926 miles).

PS-7, P-6 Both designations were used for conversions to the cargo and occasional passenger role for Aeroflot. At least 220 were delivered from 1935 onwards. Armament and military equipment was removed, apertures skinned over (usually by hinged hatches) and the cockpit enclosed and provided with exhaust-muff hot-air supplies. There is evidence that the P-6 was the rebuilt R-6 and the PS-7 the rebuilt KR-6. Payload of both versions was 700kg (1,543lb). Both served mainly in the Arctic, spending at least half the year on ski landing gear closely similar to that of the TB-1. In spring 1937 a P-6 flown by Pavel Golovin served in the weather reconnaissance role for the Papanin expedition (see ANT-6). It flew over the North Pole and, without landing, returned to Rudolf Island.

Dimensions unchanged.
Weight empty 3,880kg (8,554lb); loaded weight 6,250kg (13,779lb). No other data.

MP-6 This was the designation of civil transport conversions of the MR-6/KR-6P/R-6a twin-float seaplanes. These aircraft usually had the following features: armament removed and apertures skinned over or covered by loading hatches; refined tail, similar

An R-6 Limuzin, *the luxury passenger version of the R-6.* (RART)

to the KR-6; short radiator ducts; tandem open cockpits for the pilot (in front) and two passengers or crew (behind); and a tailskid and provision for changing to ski or wheel gear. Payload was again 700kg.

Dimensions, as MR-6.

Weight empty 4,457kg (9,826lb); loaded weight 6,750kg (14,881lb).

Maximum speed (M-17 engines) 211km/h (131mph) at sea level; cruising speed 184km/h (114mph); service ceiling 3,360m (11,025ft); climb 14min to 1,000m, 34min to 2,000m; range 700km (435 miles); take-off run 300m (984ft); landing speed 110km/h (68mph), landing run 200m (656ft).

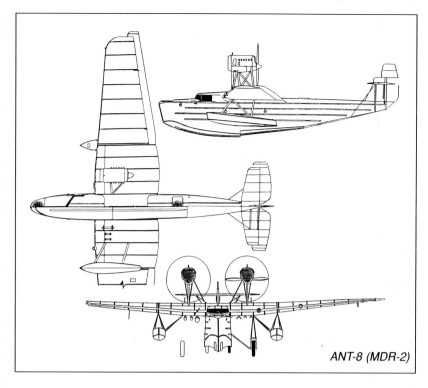

ANT-8 (MDR-2)

ANT-8, MDR-2

In 1925 the RKKA-VVS had made a preliminary request for a water-based version of the proposed twin-engined bomber. Tupolev was always interested in marine aircraft, and of course AGOS remained the national leaders in fast torpedo-boats, but for several years all available design manpower had to be applied to the landplanes, and their twin-float seaplane versions. Not until 1929 was Tupolev able to form a small design team under I I Pogosskii to study the design of a twin-engined flying-boat, the ANT-8. In February 1932 this team was to grow to brigade strength as KOSOS.

By 1929 the VVS (there was then no naval air force) had begun to feel that any aircraft stemming from the 1925

requirement would soon become obsolescent, but it funded a prototype of the ANT-8 flying-boat as the MDR-2 (marine long-range reconnaissance, 2nd type). Tupolev himself regarded the ANT-8 as more of a design exercise, and called it MER (marine experimental reconnaissance). AGOS had no experience of flying-boats, and indeed had used foreign (mainly Short Brothers) expertise in the design of seaplane floats. Accordingly the ANT-8 was a minimum-risk project, with an ultra-conservative approach to hull design.

Taking it for granted it would be a cantilever monoplane, the overall configuration was fairly obvious, and similar to established designs, notably by Rohrbach and Dornier, but using underwing stabilising floats instead of sponsons. To save time and cost, the wing was based on that of the ANT-7 (apart from the different engine mountings it was almost identical to the wing of the ANT-9). The wing comprised a horizontal centre section and tapered outer panels with dihedral. The ruling aerofoil profile was again CAHI A°, with a root thickness/chord ratio of 18 per cent, chord throughout the wing being slightly greater than that of the ANT-7. Though the centre section was only slightly greater in span than that on the ANT-7, there were five ribs on each side of the root instead of four, and the trailing edge was curved. The long ailerons also had curved trailing edges at the inboard end. Most of the skin panels had corrugations at 32mm pitch, with the usual smooth leading edge.

The tail was likewise based on the ANT-7, but with significant changes. The fin had reduced root chord but greater height, so that the rudder also had more height and less chord. The tailplane again had more span and less chord, and it was made in one piece mounted very high near the top of the fin, with its leading edge well in front

Pavel Golovin's R-6. (G F Petrov/ RART)

The ANT-8 on its beaching chassis. (G F Petrov/RART)

of that of the vertical surface. The tail-plane was fixed, not pivoted, and braced by two struts on the underside on each side. Like the tailplane, the elevators had an extra rib on each side, the horn balances had greater chord and reduced span, and another change compared with the ANT-7 was that the corners of the ailerons, tailplane and elevators were rounded. Tail skins had corrugations of the 20mm by 5mm type.

The hull was of course completely new. The cross-section was a deep rectangle, tapered at front and rear but at last having no tendency at the rear towards a triangular section. There were twenty-seven main and fifteen intermediate frames, twenty-five having 0.3mm or 0.4mm sheet doors to form watertight bulkheads. Longitudinals comprised two heavy tubular upper longerons, double stringers along the chines and a deep 3mm central keel. The planing bottom was based on Rohrbach geometry, being sharply concave on each side of the keel and curving downwards to the chines. A shallow skeg or auxiliary keel was riveted on each side of the centre-line from near the bow to the rear step. The steps were vertical and at 90 deg to the longitudinal axis. Both were reinforced with 2.5mm plate, the front step being at the CG and the rear step aft of the wing, behind which the underside was a very flat V to the tail. No water rudder was fitted.

Hull skin was mainly 32mm corrugated, and reinforced by four deep (40mm radius) longitudinal beads on each side with two more along the top. The structure was exception-

ally strong. The bare hull weight of 1,395kg (3,075lb) equated to 42kg/sq m of surface area, compared with 25kg/sq m for the ANT-6 and ANT-7.

The stabilising floats were situated well inboard, being attached under the outer ends of the centre section by steel-tube N-struts with two inboard lateral bracing struts. Each float had smooth skin and a convex stepless underside, and was set at a nose-up angle of 6 deg. Provision was made for attaching a beaching chassis. Each main unit had a single wheel with an 1,100mm by 250mm tyre carried on steel-tube struts to the second rib from the root, with bracing struts to two hull frames at the level of the bottom skin bead. A twin-wheel rear chassis could be attached under the rear step.

The cockpit for two pilots was completely enclosed, the windows on each side being able to slide open. In the bow was a Tur-6 with a single DA gun and navigation equipment, provision for a camera, Sbr-8 bomb release and an anchor. The midships section had two small portholes on each side and could be equipped for an engineer and/or radio operator, though radio was not actually installed. Aft of the wing was a Tur-5, able to traverse laterally as on the R-6, with a single DA. Under the centre section were two DER-13 racks for FAB-250 bombs and four DER-7 for FAB-100, total bomb (or depth charge or mine) load thus being 900kg (1,984lb).

The engines chosen were two 680hp BMW VI. Each was mounted high above the centre section on steel-tube struts, a V and two verticals attached

between ribs 2 and 3 and a V lateral brace to the wing root. Each engine was reversed, with a flat water radiator on the front and driving a two-blade 4.325m (170in) pusher propeller. Tanks in the centre section housed 1,100kg (2,425lb) of fuel. A 45kg oil tank was fitted between each radiator and engine. Two powerful lights were installed in the leading edge of the port wing, energised by a windmill generator on the inboard leading edge of the starboard wing.

A full-scale mock-up was approved on 26 July 1930, the single prototype was completed at GAZ No. 31 on 1 December, and the first flight was successfully made on 30 January 1931. S T Rybal'chuk did NII-VVS acceptance testing from 15 February to 17 March, as a result of which the floats and rudder were enlarged. Further State testing took place from 8 October to 14 November in the hands of M M Gromov, B L Bukhgolts and N G Kastanayev.

The consensus was that, while the ANT-8 was a tough and seaworthy aircraft, which could be safely operated in heavier sea states than any other marine aircraft known, its growth in structure weight (resulting in a rise in loaded weight from 6,665kg to 6,920kg) had made the design disposable load of 2.6 tonnes unattainable. It was also thought that the take-off was protracted, the alighting too fast and the time of 34sec for a 360 deg circle unduly long. Accordingly no recommendation for production was made.

With the benefit of hindsight, this was probably an error. Most of the supposedly more modern flying-boats, with the exception of the small Beriev MBR-2, were either failures or else much delayed in development, so that through most of the 1930s the AV-MF had no satisfactory long-range flying-boat in service.

Tupolev and Pogosskii studied a version of the ANT-8 with tractor engines, as well as the MRT-1 civil regional transport version.

Span 23.7m (77ft 9in); length 17.03m (55ft ½in); wing area 84sq m (904sq ft).
Weight empty 4,560kg (10,053lb); loaded weight 6,920kg (15,256lb).
Maximum speed at sea level 203km/h (126mph); cruising speed 166km/h (103mph); climb to 1,000m 7min, to 2,000m 16min, to 3,000m 35min; service ceiling 3,350m (11,000ft); range 1,062km (660 miles); take-off time 29sec; alighting run 15sec, alighting speed 115km/h (71.5mph).

ANT-9, PS-9

In the earliest years of the Soviet Union the only real transport aircraft were the wartime *Ilya Mouromets* bombers. The failure of the first attempt to build a Soviet transport, the KOMTA, is noted in the introduction. After this 1922 aircraft, various light single-engined transports were produced in the Soviet Union, but nothing that might be called an 'airliner' appeared apart from the German-designed Junkers G 24, a few of which were built at Fili. In March 1927 the Fili factory became GAZ No. 22, devoted initially to building Tupolev aircraft. On 8 October 1927 Tupolev was summoned to the Kremlin and told that there had long been a need for capable transport aircraft able to serve on the trunk routes of the airline Dobrolet and the Soviet/German operator Deruluft.

Of course, he replied that his twin-engined monoplanes then either flying or in advanced development would make the task relatively straight-forward. In December 1927 the GVF (civil air fleet) placed an order for a prototype with AGOS. The aircraft was to have a comfortable enclosed cabin for at least eight passengers with baggage, and preferably was to be powered by three air-cooled engines of 250–300hp each. A A Bessonov and

B S Stechkin were ordered to design such an engine, the M-26.

During the early years of the Soviet Union military aircraft enjoyed far greater priority than civil transports, but the Kremlin also rated propaganda highly, and thus needed a modern airliner just as it needed aircraft specially designed to break international records. Accordingly, the ANT-9 immediately became a priority programme, rushed through ahead of even the ANT-6, 7 and 8. Design took place in 1928, with the full-scale mock-up being approved in October of that year. Series manufacture was included in the first Five Year Plan, ahead of schedules to build the Tupolev bombers.

From the start Tupolev elected to place the wing above the cabin. The wing was almost identical to that of the ANT-7, though the span was increased by the wider fuselage. The trailing edges of the centre section and inboard end of the ailerons were curved, as on the ANT-8 (the number of centre-section ribs being as on the ANT-7). The tail was also based on that of the ANT-7, though the chords of the fin, rudder and elevators were all increased, and the height of the vertical tail reduced. A cockpit handwheel could drive the tailplane, via the single bracing struts, from +6 deg to –0.5 deg.

The fuselage was of course a fresh design, though following the established AGOS method of construction.

The cross section was a rectangle with a slightly rounded top and bottom, with four strong tubular longerons, multiple frames, diagonal tubular drag struts and skin corrugated at 32mm pitch. Aft of the cabin the frames were cross-braced. Along the cabin the frames were especially strong, and not only projected into the cabin but the transverse bottom sections formed a series of walls over which passengers had to step. Passengers also had to duck under the wing and then, to reach their seat, step over a longitudinal keel on each side of the central aisle, having the same height as the transverse frames. The seats, normally four on the left and five on the right, were inserted loose between the keels and the wall. On the outside of the skin on each side were riveted longitudinal beads, between which were secured the large celluloid windows on each side, one window being beside each chair. The frames formed the vertical boundaries between the windows. Each window had two sliding curtains. Higher, level with the bottom of the wing, were narrow net-type racks for light baggage, but suitcases were stowed in a small bay at the rear, with an external hatch. Next to this bay was provision for a simple toilet, and an integral ladder up a bulkhead to an emergency exit in the roof. The door was at the rear on the left, hinged at its forward edge.

At the front of the aisle two steps led up through a bulkhead door to the

The ANT-9 prototype before its first flight. (V B Klepacki)

Interior of the ANT-9 prototype. (G F Petrov/RART)

After only minor adjustments Gromov, Mikheyev and Spirin made a 4,000km (2,486 mile) flight on 6–12 June over the route Moscow – Odessa – Sevastopol – Kiev – Moscow. On 10 July, with engineer V P Rusakov and nine journalists, Gromov left on a flag-waving trip via Travemünde, Berlin, Rome, Marseilles, Paris, London, Paris, Berlin and Warsaw, returning on 8 August. A total of 9,037km (5,615 miles) was flown in 53 hours.

Span 23.71m (77ft 9⅛in); length 17m (55ft 9¼in); wing area 84sq m (904sq ft).
Weight empty 3,353kg (7,392lb); fuel/oil 700kg (1,543lb); loaded weight 5,043kg (11,118lb).
Maximum speed at sea level 209km/h (130mph); climb 6.8min to 1,000m, 15.5min to 2,000m, 30.7min to 3,000m; service ceiling 3,810m (12,500ft); range 1,000km (621 miles).

Production of the ANT-9 began in mid-1929, GAZ No. 22 delivering one in 1930, thirty-three in 1931 and twenty-seven in 1932, plus five produced in 1932 at GAZ No. 31 at Taganrog, for a total of this version of sixty-six. The engine fitted was the M-26, rated at 300hp but actually delivering about 240hp. They were uncowled, had exhaust collector rings at the rear leading to short pipes discharging above the wings and under the right side of the fuselage, and drove two-blade propellers without spinners. Capacity of the fuel tanks was considerably increased. Other changes included the addition of two landing lights which hinged down from the underside of the port wing, and a carrier for flares on the tip of that wing or, sometimes, both wings. A windmill generator was added above the fuselage to serve the lights and radio.

The supposedly more powerful engines were expected not only to save foreign currency but also to result in substantially higher performance. In fact the M-26 suffered from several defects which were so difficult to cure that production was halted.

Dimensions, unchanged.
Weight empty 3,950kg (8,708lb); fuel/oil 1,130kg (2,491lb), loaded weight 6,000kg (13,228lb).
Maximum speed at sea level 185km/h (115mph); climb 7min to 1,000m, 15.9min to 2,000m, 39.4min to 3,000m; service ceiling 3,400m (11,155ft); range 1,000km (621 miles).

After protracted trials it was decided to re-engine a number of these aircraft with imported engines. The choice fell

enclosed cockpit, with a sliding window on each side. In most ANT-9s the cockpit had dual flight controls, but the crew normally comprised a pilot and a *bort-mekhanik* (flight mechanic). The latter managed the fuel system, with a Dural tank on each side of each engine between spars 1 and 2, the total capacity being 972 litres (214gal). Each main undercarriage had a wire-spoked wheel (usually with a disc fairing on each side) with an 1,100mm by 250mm tyre. The axle was carried on a transverse strut pinned to the bottom fuselage longeron and braced by a drag strut pinned to the next frame ahead. Weight of the aircraft was taken by a vertical strut pinned under the front spar, with multiple rubber-block shock-absorbers in a telescopic box fairing just above the wheel. The wheel could be exchanged for an ANT-7 type ski. The tailskid was of the usual bungee-sprung pivoted type, with ability to castor.

The prototype was rushed to completion in time for it to be exhibited on May Day 1929. The parts were assembled in Red Square on 28 April, and on the great day it was finished, registered No. 309, bearing Dobrolet titles and the name *Krylya Sovyetov* (Soviet Wings). As the plannned engine was not ready, the prototype was powered by three 230hp Gnome-Rhône (Bristol design) Titan engines. Each five-cylinder radial drove a 2.59m (102in) two-blade wooden propeller with a spinner whose profile was continued to the rear in a metal cowl enclosing the crankcase but leaving the cylinders exposed. No exhaust pipes were fitted, but an oil cooler was positioned below and behind each engine.

The outer wings were then removed and the important prototype towed to the Central Aerodrome where Gromov made the first flight on or about 7 May. He was well satisfied with handling and performance, finding the controls light and well-harmonised. The ANT-9 would fly hands-off, and a height of 200m (656ft) could be sustained with a wing engine shut down, and 1,200m (3,940ft) without the nose engine. Official NII-GVF trials were completed before the end of May.

The ANT-9 prototype after being fitted with Wright Whirlwind J-6 engines and Hamilton propellers. (Jean Alexander/RART)

on the Wright Whirlwind J6, which really did deliver 300hp. The engines were fitted with exhaust collector rings on the front of the crankcase leading to pipes discharging high above the wing or to a long pipe extending far back under the fuselage. The propellers were of Hamilton Standard manufacture, unchanged in diameter but with three light-alloy blades whose pitch could be adjusted on the ground. All data show that wing span and area were slightly increased. Conversions began in 1933 and by 1934 most of the re-engined aircraft also had Townend-ring cowls.

Thus modified, the ANT-9 gave excellent service. Most had the rudder increased in height to give better handling after failure of a wing engine, in which condition it was claimed still to reach 170km/h (106mph). Performance of the re-engined aircraft was not very different from that of the prototype, but the aircraft was now very heavy on the controls.

In 1933 two re-engined ANT-9s were handed over to Deruluft. These aircraft had the skins over the wings and fuselage covered with fabric, which fractionally improved performance. The airline used them on stopping services from Berlin to Moscow and Leningrad, while Dobrolet (later Aeroflot) used them on numerous services from Moscow, including those to the Black Sea, Prague/Vienna and Transcaucasia.

From 1935 most tri-motor ANT-9s were replaced by PS-9s (see later). Many were passed on to the VVS (air force) as staff transports and paratroop research aircraft. Many survived to fly front-line supply missions after 1941.

AGOS schemed an ambulance version and a multirole military version with a Tur-5 gun ring aft of the wing, but neither was accepted for production.

Span 23.8m (78ft 1in); length 16.65m (54ft 7½in); wing area 84.9sq m (914sq ft).

Weight empty 3,680kg (8,113lb); fuel/oil; loaded 920kg + 80kg (2,028lb + 180lbs); weight 5,690kg (12,544lb).

Maximum speed at sea level 205km/h (127mph); speed at 3,000m (10,000ft) 170km/h (106mph); climb 8.5min to 1,000m, 16min to 2,000m, 28min to 3,000m, service ceiling 4,500m (14,765ft); range 700km (435 miles); take-off run 170m (558ft); landing speed 93km/h (58mph), landing run 150m (492ft).

PS-9 In late 1931 AGOS studied replacing the three imported engines

by just two home-produced engines of greater power, and one series aircraft was set aside to be fitted with two M-22 (Bristol Jupiter) radials. Eventually it was decided a simpler solution was to use the complete engine installation of the 680hp water-cooled M-17 as already developed for the ANT-7. By good fortune the major change in power, though it considerably increased the empty weight, left the centre of gravity in almost the same place.

The heavy V-12 engines were installed in almost exactly the same way as in most ANT-7 versions, with inclined radiators in short ducts underneath and driving 3.16m

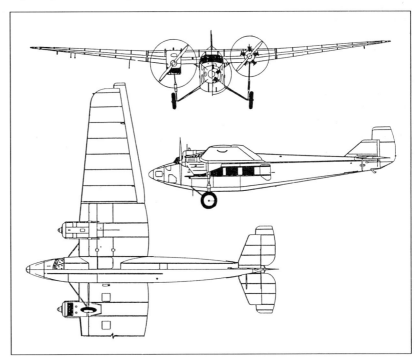

Front view (port) ANT-9/J-6, (starboard) PS-9; side and plan views PS-9.

A late series PS-9. (G F Petrov/RART)

(124in) two-blade metal propellers with spinners. The oil tanks were smaller than on most ANT-7s, and there were other minor changes. The nose was simply faired over by an aluminium end-cap and uncorrugated side panels. In most aircraft small hatches were fitted so that a limited amount of baggage or cargo could be put here, a sheet bulkhead sealing the nose from the cockpit. Almost the only other change was to redesign the cockpit glazing, with larger and more sharply raked front windows.

The twin-engined version was designated PS-9 by Aeroflot (which Dobrolet became in 1932), and quickly became the airline's chief international and trunk-route transport. A total of sixty were built, the engine changing in 1933 to the M-17F of 730hp, giving improved performance. Some archives thus give the total of all versions as 126 or 127 aircraft, but ANTK Tupolev gives the total as 130. These aircraft flew harder and longer than any other aircraft of their time, several exceeding 5,000hr and No. 114 reaching 6,170hr by 1942. Aircraft No. 125 was paid for by the satirical magazine *Krokodil*, and delivered with modifications by V B Shavrov, including an elongated nose painted with crocodile teeth and wheel spats painted with claws. It served from early 1935 on 'special civil missions' with the *Maksim Gorkii* propaganda squadron.

Span 23.8m as before; length 17.01m (55ft 9⅜in); wing area 84.9sq m as before.
Weight empty (M-17) 4,420kg (9,744lb), (M-17F) 4,530kg (9,987lb); fuel/oil; loaded 720kg + 70kg (1.587lb + 154lb) weight (all) 6,200kg (13,668lb).
Maximum speed at sea level (M-17) 215km/h (133.5mph), (M-17F) 228km/h (142mph); cruising speed (all) 180km/h (112mph); climb (M-17) 6min to 1,000m, 12.5min to 2,000m, 18.5min to 3,000m; service ceiling (M-17) 5,100m (16,730ft); range (all) 700km (435 miles); take-off run (M-17) 16sec, 165m (540ft); landing speed 110km/h (68mph), landing run 180m (590ft).

ANT-10, R-7

With this military biplane A N Tupolev again produced a prototype as an internally funded AGOS project, though eventually he obtained contract cover for it. Having achieved a good series run with the R-3, limited by the capacity of the industry, Tupolev felt he had a good chance of winning much larger orders for a later design able to replace the huge fleet of 2,447 Polikarpov R-1s based on the wartime D.H.9A.

The ANT-10 design was really a sesquiplane, as was the ANT-5 fighter, the upper wing almost being an enlarged version of the upper wing of that earlier aircraft. Structure was of course all-duralumin, with almost all exterior skin other than the engine cowling being corrugated, the ruling pitch being 20mm (0.79in). The lower wing was untapered until quite near the tips, and had dihedral. The big upper wing was horizontal, and had the only ailerons. It was mounted on

N-struts unusually high above the fuselage, and was braced to the lower wing by unusual struts resembling a mirror-image version of those of the ANT-3. A large strut of broad streamline section was pinned to the third spar of the upper wing and spread out at the bottom to link with spars 1 and 3. From this, near the bottom, a forward bracing strut sloped ahead to link with the front spar of the upper wing, pin-jointed at each end.

The tail was typical AGOS, with large horn balances and a tailplane pivoted to the upper fuselage longerons and driven through a small arc by a handwheel. The fuselage was remarkably slender, the cross-section being dictated by the 680hp BMW VI engine, which drove a fixed-pitch two-blade propeller. On the front was a radiator similar to that of the ANT-4/TB-1, the oil cooler being underneath the cowling to the rear and the exhaust stubs being led into long left/right pipes carried up above the upper wing. A total of 583 litres (128 gal) of fuel was housed in Dural tanks on each side of the centreline in the upper wing. The simple V-type undercarriage had faired rubber-block springing and 900mm by 200mm tyres. The similarly sprung tailskid could swivel through a useful arc.

A battery charged by a small windmill generator on the centreline above the upper wing would in series aircraft have provided power for navigation lights, radio and electromechanical bomb-release gear. To reduce drag and improve performance Tupolev put the bomb load of 300kg (661lb) in an internal bay. This resulted in the pilot and observer being seated surprisingly far aft. This was not felt to be a drawback because, together with the high position of the upper wing, it was considered to improve their all-round view. No armament was fitted to the prototype, but the intention was to fit two synchronised PV-1s firing ahead. Strangely, no mention is made of a defensive gun for the observer.

AGOS launched the ANT-10 project as a reconnaissance and light bomber aircraft in July 1928, and a contract was placed for a single prototype on 28 March 1929. Factory testing took place in January 1930 in the hands of Gromov, A B Yumashyev and V O Pisarenko. The aircraft looked attractive, though it had no markings; the cowling, top decking and wheel

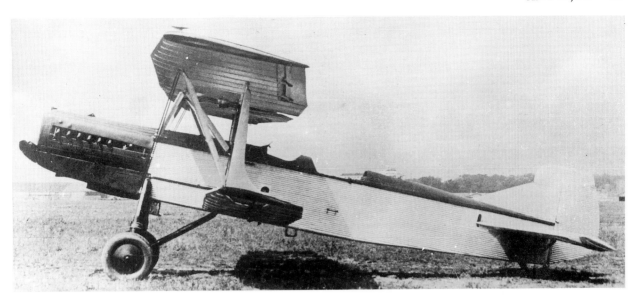

AHT-10   1930г

Two views of the ANT-10 prototype. (G F Petrov/RART)

discs were painted. On the whole the ANT-10 flew well, and it got as far as NII-VVS State testing in spring 1930.

Unfortunately for AGOS, it really had little chance of being accepted. The rival R-5 was cheaper, already in production, easier to repair and, except for sea-level speed, had higher all-round performance.

Span 15.2m (49ft 10½in); length 10.9m (35ft 9⅛in); wing area 49.04sq m (528sq ft).

Weight empty 1,725kg (3,803lb); fuel/oil 450kg (992lb); loaded weight 2,920kg or 2,575kg [both figures in archive] (6,437lb or 5,677lb).

Maximum speed at sea level 235km/h (146mph); speed at 3,000m 212.5km/h (132mph); cruising speed 184km/h (114mph); climb 3.1min to 1,000m, 6.6min to 2,000m, 10.9min to 3,000m, 16.5min to 4,000m and 26.5min to 5,000m; service ceiling 5,560m (18,240ft); range 1,100km (684 miles) [also given as 950km, 590 miles]; take-off run 150m (492ft); landing speed 90km/h (56mph), landing run 300m (984ft).

ANT-11

This 1929 project was for a multi-purpose flying-boat. The only record states that it was used as a basis for the ANT-22.

ANT-12, I-5

In 1927 Aviatrust, the original Soviet organisation formed in 1923 to manage aircraft manufacture and act as an interface with foreign licensees, prepared its first Five Year Plan. This was in conformity with the Soviet Union itself, which until the Great Patriotic War organised the entire country into such plans. The 1927 scheme, which was accepted on 22 June 1928, required AGOS to produce a single-seat fighter, the I-5, to be powered by either the Bristol Jupiter VI, for which Aviatrust had concluded a licence, or the locally developed experimental M-36. Construction was to be mixed.

The project was launched as the ANT-12, but A N Tupolev's designers were so preoccupied with other aircraft that work on the fighter languished. In any event, Tupolev was uncertain that it would be significantly better than the ANT-5. Little progress had been made by the delivery deadline of September 1929. A rival, N N Polikarpov's wooden I-6, also made slow progress, and it was this situation which probably gave Stalin the idea of rounding up designers – including

Polikarpov and eventually Tupolev – into 'internal prisons'.

In September 1929 the chief Central Design Bureau designers, Polikarpov and D P Grigorovich, were arrested, and three months later a 'special KB' was formed to take over the I-5. Polikarpov's VT-11 (internal prison No. 11) design, similar to the I-6, was accepted, and the ANT-12 was terminated.

ANT-13, I-8

In 1928 Tupolev had visited Krupps, Mannesmann and other German steel firms, and returned with specimens of three Ni-Cr stainless (called KhGSA in Russian) and two Cr-Mo high-strength steels (called KhMA). CAHI and the VIAM materials institute tested these, and supported Tupolev's proposal that these should be used in a suitable new aircraft. Tupolev worked out calculations for simple wing spars of stainless rolled strip and sheet, and a fuselage of welded Cr-Mo tubing.

AGOS were embarrassed and apprehensive after their failure with the ANT-12/I-5 programme, and engineer V M Rodionov suggested that they should build a small single-seat fighter using the new steels on their own initiative. Each member of staff volunteered to work 70 hours without pay. Accordingly Tupolev launched the project as the ANT-13, and the assignment for it was willingly issued by the UVVS as the I-8. The requirement was released on 30 December

1929. The design, led by P O Sukhoi, was approved on 30 January. It became known as the *Zhokei* (Jockey) after a small Vickers fighter of the day.

The ANT-13 was a small biplane, with basic fuselage and wing construction as described. Wing ribs, secondary fuselage formers, ailerons and tail were Dural, and apart from the engine cowl, front fuselage and corrugated-Dural tail, the covering was fabric. Unlike the ANT-5 the upper wing was only slightly larger than the lower, and bracing wires were omitted. To achieve the highest possible speed, the engine was an imported Curtiss Conqueror V-1570 of 700hp, with direct drive to a Hamilton propeller with two ground-adjustable Dural blades. The water radiator was in a duct under the mid-fuselage. Planned armament was two synchronised PV-1 machine guns, in a dorsal hump ahead of the windscreen.

Though one document states that Gromov flew the ANT-13 on 28 October 1930, Tupolev archives state that the aircraft was completed in November and first flown on 12 December. Performance was below expectation, and for 18 months the I-8 was subjected to various tinkering. The mainwheels were fitted with spats, the engine was changed to one fitted with a supercharger, the oil cooler was enlarged, the tailplane made adjustable, the tailskid made swivelling (with the rudder) and the propeller fitted with a spinner. Planned changes included an enclosed cockpit, an ethyl-glycol cooling system, wings of new profile and a reshaped tailplane. None of these modifications had

been effected when interest was lost in late 1932, mainly because no Soviet engine was suitable and rival fighters were clearly going to be ordered in quantity. Tupolev told the author the ANT-13 was adequate but not the winner everyone had hoped, and that it was always 'an outsider'. His personal view was that he should continue to meet the Soviet Union's need for large aircraft. Data as first built, where noted otherwise.

Span 9.03m (29ft 7½in); length 6.7m (21ft 11¾in); wing area 20.3sq m (219sq ft).
Weight empty 980kg (2,160lb); loaded weight 1,424kg (3,139lb).
Maximum speed 281km/h (175mph) at sea level, 250km/h at 5,000m; after rebuild with supercharged engine 303km/h (188mph) at sea level and at 5,000m, 313km/h, then a national record (194.5mph) at 3,600m; climb 13.1min to 5,000m (16,400ft), ultimately 8.2min to 5,000m; service ceiling 6,700m (22,000ft), ultimately 8,500m (27,900ft); range 490km (305 miles); landing speed 118km/h (73mph).

ANT-14

This large passenger transport was proposed by Tupolev in September 1930. At that time the prototype ANT-6 was about to fly, and Tupolev showed that for a relatively modest budget AGOS could produce a transport resembling a greatly enlarged ANT-9, using the bomber's wings and tail. It would seat thirty-two passengers two + two. To obtain maximum reliability Tupolev elected to use five engines, the choice falling on the Gnome-Rhône Jupiter 9Akx (Bristol Jupiter VI), each rated at 480hp.

The wings were of ANT-6 type but with a span 0.8m greater because of the wider fuselage. The latter followed established principles in having large tubular longerons at the corners of the 3.2m (10ft 6in) square cross section, with a mainly braced-strip structure aft of the wing. Forward of this point the frames had deep webs, and an unusual feature was that each frame had two vertical girders forming a 'cloister' down the centre, the seats being between these pillars and the wall. As in the ANT-9, passengers had to step over the deep frames, and over deep keels in line with the pillars. An ANT-9 type door was provided on each side at the rear of the cabin, small ladders being needed.

The ANT-13 prototype. The colour of the painted parts was red. (via Philip Jarrett)

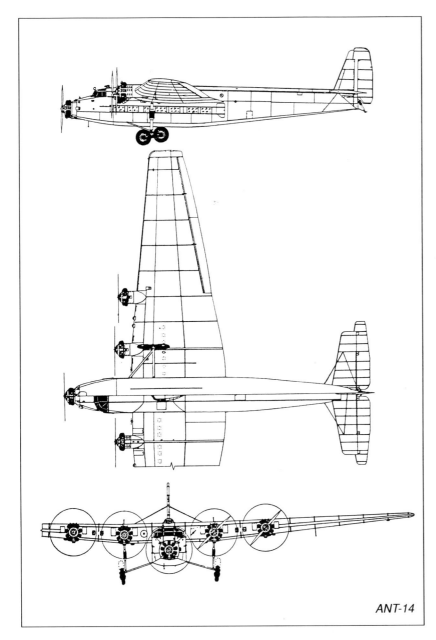

ANT-14

Along each side were seven large sliding windows, later fitted with curtains. Above the seats were false ceilings from which hung straphanging loops, and above which were voluminous spaces where light objects (not suitcases) could be stored. In the centre of the cabin the aisle was obstructed by a Dural ladder up which the engineer climbed to reach his enclosed fighter-type cockpit projecting above the fuselage. This was on the centreline between wing spars, with a glazed front portion and a metal rear fairing. The top of the canopy was hinged, enabling the engineer or ground personnel to step on to the wing. Here they could refuel the four tanks, with a total capacity of 2,765 litres (608gal) of fuel, or hinge down any of the eight leading-edge sections adjacent to the engines to perform maintenance.

Ahead of the main cabin, steps led to the crew compartment for two pilots and a wireless operator (though radio was not installed). Electric lighting was fitted throughout, together with navigation lights and four leading-edge landing lights, all powered by a windmill generator in the leading edge of the port wing root. The five engines had front exhaust collector rings discharging through short stubs at the bottom, and were fitted with tight ring cowls with bulged 'helmets' over the valve gears. Each drove a 4.5m (177in) two-blade wooden propeller.

The tail resembled that of the ANT-6 prototype, but with the later form of sprung tailskid mounted behind the

The ANT-14 in late 1931. (Jean Alexander/RART)

The ANT-14 with Pravda *titles in the mid-1930s.* (G F Petrov/RART)

stern frame, the rudder being cut away at the bottom, and the navigation light was mounted higher up the rudder trailing edge. The main undercarriage was a scaled-up version of that of the ANT-9, with the large (2,000mm by 450mm) Palmer Tyre wheels and tyres as imported for the ANT-6.

Construction of the ANT-14 was extremely rapid, and M M Gromov made the first flight on 14 August 1931. Few modifications were needed, and in fact it was soon decided to add an additional row of seats to make a total of thirty-six. The only untoward occurrence was structural failure of one of the wheels, as a result of which both this aircraft and the TB-3 bombers were fitted with a pivoted frame carrying two 1,350mm by 300mm wheels in tandem. Later in 1931 the ANT-14 was fitted with TB-3 type skis on the main legs, the tail-skid being unchanged. Earlier in 1931 some flying had been done with the nose engine uncowled, and though the cowl was later replaced it was decided in 1932 to remove all of them to facilitate maintenance. Other changes in 1932 included fitting metal propellers and long exhaust pipes leading back under the fuselage from the nose engine. The aircraft later bore registration SSSR-N1001.

Historian Shavrov's assertion that 'series production was impossible' merely meant that there was no requirement for so large an aircraft, and in fact contract cover for the prototype was not easy to arrange. An answer appeared in January 1933 when this, the largest Soviet aircraft at the time and the world's only active five-engined aircraft, was sponsored by the newspaper *Pravda* (Truth) and delivered to the *Maksim Gorkii* propaganda squadron as its flagship. Parts were painted red or black, with the name of the sponsor on the nose and registration URSS-N1001.

Subsequently the ANT-14 had a much more active career. In 1933–42 it made over 1,100 flights over Moscow, carrying more than 40,000 ticketed passengers. It visited Kharkov and Leningrad, in 1935 it took a parachuting team and supporters to Bucharest, and it was especially active during the 1937 election to the Supreme Soviet. At last, with its record unblemished, its wings were removed and the rest was given a place of honour in Gorkii Park, Moscow, surrounded by captured German equipment, and fitted out as a cinema showing documentary films.

AGOS produced drawings of a proposed heavy bomber/transport version, with four AM-34 engines and three gun turrets, one being in the nose, and a crew of eleven.

Span 40.4m (132ft 6½in); length 26.485m (86ft 10¼in); wing area 240sq m (2,583sq ft).

Weight empty (as built) 10,828kg (23,871lb); fuel/oil 1,990kg + 166kg (4,387lb + 366lb); loaded weight 17,530kg (38,646lb).

Maximum speed 236km/h (147mph) at sea level, 195km/h (121.7mph) at 3,000m; cruise 175km/h (109mph); climb to 1,000m in 4.9min and to 3,000m in 21.8min; service ceiling 4,220m (13,850ft); range 1,000km (620 miles); take-off run 250m (820ft); landing speed 105km/h (65mph), landing run 220m (721ft).

ANT-16, TB-4

In early 1930 Tupolev was called to the Kremlin to find that the VVS had made a case for gigantic bombers with a span of some 100m (328ft) and bomb load of 25 tonnes (55,116lb). He expressed the view that it would be too great a leap to attempt to build such an aircraft, and that it would be prudent to proceed in two steps by first flying one of intermediate size. Accordingly in April 1930 he was authorised to design such a 'half-way' aircraft, with roughly twice the weight and wing area of the TB-3. This was designated ANT-16 by AGOS, and received the VVS designation TB-4.

According to V B Shavrov, at this time the speed of heavy bombers was considered unimportant. Tupolev did not therefore attempt to provide double the TB-3's installed power, but merely selected six M-34 engines of 750/830hp each, and was content to accept a significantly poorer flight performance. He naturally retained the established configuration and methods of construction, though after much discussion with CAHI it was decided to add the two extra engines

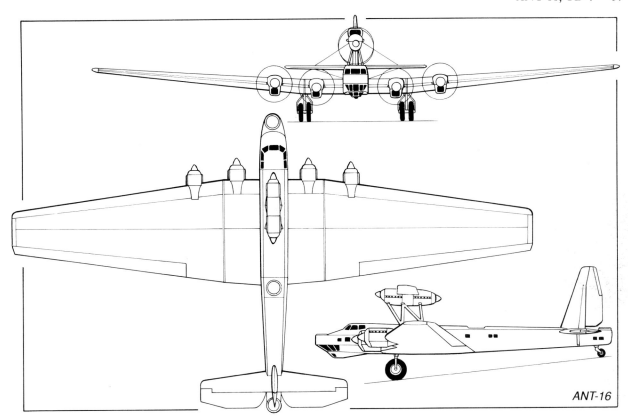

ANT-16

in a tandem nacelle carried high above the fuselage.

The wing was the largest and strongest designed for any aircraft in the world at that time. Its design was complicated by the need to place the spars between the cavernous bomb bays, which meant that virtually all wing loads had to be carried by three massive spars. The horizontal centre section, of 10.5m (34ft 5⅜in) span and maximum thickness/chord ratio of 17.5 per cent, was joined by multiple steel bolts to outer panels with dihedral, each with a span of 21.75m

(71ft 4¼in) and t/c ratio at the tip of about 14 per cent. The huge ailerons continued the line of straight taper at the trailing edge, and were without balancing horns.

The fuselage was only slightly bigger in cross-section than that of the TB-3, though the structure was significantly heavier because of the greater length and much higher tail loads. The side skins were stiffened by five half-round external beads as introduced on the ANT-8. The nose contained a Tur-6, intended to have two 7.62mm DA guns, beneath which

were windows all round as well as a glazed gondola for prone bomb-aiming. The dual pilot cockpit had windows all round but an open roof. Further back, at the lower level, the engineer could crawl through either leading edge to reach the back of each engine, as well as the oval-section riveted aluminium fuel tanks, with a capacity of 6,625 litres (1,457gal).

The four wing engines were all mounted close inboard in installations similar to those of the first series TB-3, except for the omission of long pipes to carry the exhaust above the

The ANT-16 prototype. (G F Petrov/ RART)

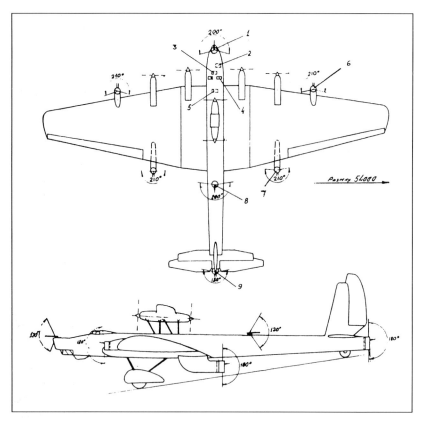

Planned armament of the TB-4 (never fitted): 1, 20mm Oerlikon; 2, bomb aimer; 3, radio station; 4, two pilots; 5, flight mechanic; 6, single or twin DA-2; 7, single or twin DA-2; 8, 20mm Oerlikon; 9, twin DA-2.

wing. The two extra engines were carried on steel/Dural struts and fully cowled, with a dorsal duct above the nacelle housing the water cooling radiators. Providing inflight access to this nacelle was considered, but abandoned because of the difficulty of opening the cowlings in the air. Throttle linkages went to the engineer's cabin. All propellers were wooden two-bladers.

The tail followed TB-3 practice on a bigger scale, though the rudder had a servo tab instead of a horn balance. The main undercarriage was entirely new, each unit having two 2,000mm by 450mm wheels side-by-side on an axle held in a massive frame of Cr-Mo tubing with faired rubber shock absorbers. Track between the mid-points was 10.645m (34ft 11in). Brakes were provided, because instead

of a skid a castoring tailwheel was fitted.

One of the major design tasks was to provide the two bomb bays. Each was in the form of an immensely strong box, approximately 5m (16ft 5in) long and 1.8m (5ft 11 in) square in section, with the bombs held horizontally in vertical racks above power-operated doors. The forward bay was designed by a team led by K P Sveshnikov, and the aft bay by a team under V M Myasishchev. They were the largest bomb bays attempted up to that time. Normal bomb load was to be 4 tonnes, but an overload total of 10 tonnes (22,049lb) could be accommodated, for example with twenty FAB-500 or forty FAB-250.

The ANT-16 prototype was never fitted with the planned armament, but merely with provision for Tur-5 and -6 gunner cockpits in the nose, mid-fuselage and tail. Drawings show that the scheme planned for the series TB-4 was more elaborate. In the nose there was to have been an open ring mounting a 20mm Oerlikon cannon, with an arc of fire of 200 deg. In the mid-upper position was to have been a Tur-5 with a single or twin DA, with a rear-firing arc of 200 deg. In the tail was to be a second 20mm Oerlikon, restricted by the elevators to the entire 180 deg rear hemisphere. On each outboard leading edge was to be a Tur-6 in a round-fronted box cantilevered well ahead of the wing, with twin DA guns covering an azimuth arc of 210 deg. Finally, two similar Tur-6 were to be carried on streamlined boxes well down below the wing

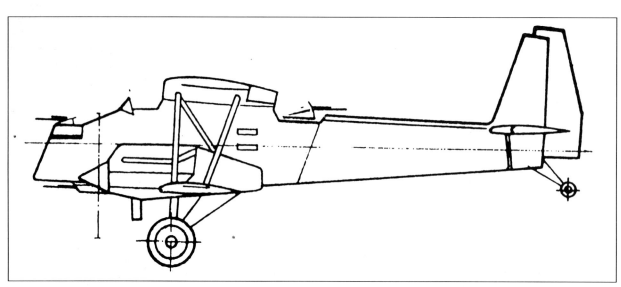

This drawing of the ANT-17 has been published by CAHI but is said to be unreliable.

(thus able to fire across under the fuselage) behind the trailing edge, in line with the outer engines, each with twin DA covering a rear arc of 210 deg. The vertical arc of these trailing-edge guns was to be 180 deg, but a tubular barrier was to protect the bomber's own rear fuselage and tail.

Altogether the TB-4 was a pedestrian design, and when Gromov at last flew it on 3 July 1933 he pronounced its performance poor, and the control forces unacceptably high. Tupolev's designers eventually achieved satisfactory control forces, but the overall performance remained most unimpressive. Gromov suggested that not only were the engines nothing like powerful enough but the wing propellers were blowing against a bluff wing more than 2m (6ft 7in) thick. The monster got as far as NII-VVS testing, by Ryazanov, Stefanovskii and Nyukhtikov, but this was abandoned on 29 September 1933. Tupolev's suggestion of fitting Mikulin's next engine, the 1,200/1,350hp M-35, was rejected. Though the aircraft remained secret, it appeared (and simulated a crash) in the 1935 film *Bolshiye Kril'ya* (big wings). Its main contribution was to facilitate design of the ANT-20.

Span 54m (177ft 2in); length 32m (105ft 0in); wing area 422sq m (4,542sq ft).
Weight empty 21,400kg (47,178lb); fuel/oil 4,950kg (10,913lb); loaded weight 33,280kg (73,369lb), (overload) 37,000kg (81,570lb).
Maximum speed 200km/h (124mph) at sea level, 188km/h (117mph) at 5,000m; cruise 159km/h (99mph); climb 12.4min to 1,000m (3,280ft); service ceiling 2,750m (9,025ft); range (at 37,000kg), 940km (584miles) with 8 tonnes of bombs, 2,000km (1,242 miles) with 2 tonnes; take-off run 36 sec/800m (2,624ft); landing speed 105km/h (65mph), landing run 400m (1,312ft)

ANT-17, TSh-B

In December 1929 AGOS-CAHI received a VVS contract for the prototype of a front-line ground-attack aircraft, designated TSh-B signifying 'heavy attacker – armoured'. Though based on the ANT-7, it was a sesquiplane, its structure being the usual light .alloy with corrugated skin. Though the accompanying drawing appears in the official CAHI history, experts have described it as 'pure

fantasy', so perhaps too much should not be deduced from it.

What is known is that the engines were to be two 900hp M-34FS and that the crew numbered four (pilot, navigator/bomb-aimer and two gunners). Armament is reported by CAHI to have comprised one 75mm DPK (one of Kurchevskii's recoilless heavy cannon), a total of eight 7.62mm DA machine guns (four firing obliquely down and ahead and two pairs for defence aimed by the gunners) and vertical cells housing up to 1,500kg (3,307lb) of bombs. The author believes this applied to the bigger ANT-18, and that the ANT-17's bomb load was 600kg (1,323lb). One account states that the DPK could be replaced by two conventional guns of 37mm or 45mm calibre (the drawing appears to show such a gun firing directly ahead from under the nose). Armour was to weigh 600kg (1,323lb) of which 380kg (838lb) was to be structural. The project was abandoned in early 1932.

Span about 20.5m (67ft 3in); length 13.5m (44ft 3½in); wing area 67sq m (721sq ft).
Weight empty 3,600kg (7,937lb); weight loaded 5,700kg (12,566lb).
Estimated maximum speed 255km/h (158mph).

ANT-18, ANT-19

Few records survive of this 1930 project for a heavy attack aircraft, which apparently was also designated TSh-1, an appellation used again for a biplane by D P Grigorovich. Again based on the ANT-7, it appears to have been a monoplane with estimated empty and loaded weights of 4,400kg and 7,000kg respectively. The author believes the CAHI history has confused this aircraft with the ANT-17, and that the heavy armament cited for the TSh-B actually applied to the bigger ANT-18.

ANT-19 No information.

ANT-20, MG

In October 1932 it was suggested that, to celebrate the fortieth anniversary of the start of the career of the famed

writer Maksim Gorkii, a colossal aeroplane should be created to serve as the flagship of the *Agiteskadrilya* (propaganda squadron) that was being formed in his honour. Journalist M Ye Kol'tsov, president of Yurgaz, launched a nationwide appeal for funds, and this eventually raised 6m Rubles (then about £2m, $8m). An all-Union committee was formed to manage the project, from which a technical council was selected to oversee the design. This council included representatives from about 100 institutes and factories, including of course CAHI, CIAM and VIAM. All actual design and construction was assigned to AGOS-CAHI, led by A N Tupolev.

Tupolev had already, in mid-1931, drawn a passenger derivative of the ANT-16 powered by four geared M-35R engines, designated ANT-20. To meet the new demand, which was from the start named for the writer it honoured (abbreviated to MG), Tupolev kept the ANT-20 designation but further extended the outer wings and made other minor changes (though as design progressed additional alterations were introduced). The most important modifications were that, lacking the desired engines, Tupolev had to add two more on the leading edge, outboard of the existing four. Then, finding that this was still only marginally enough power, he added a final pair, making eight in all, in a tandem push/pull nacelle exactly as on the ANT-16.

For the duration of this programme Tupolev avoided going abroad, though when he was absent from AGOS Petlyakov deputised. Arkhangel'skii was appointed director of the design team. General design was assigned to B M Kondorskii, stressing to a small group under V N Belyaev, propulsion to Ye I Pogosskii, equipment to A A Yengibaryan and N S Nekrasov, and B A Saukke was appointed chief engineer. Work proceeded extremely rapidly, other projects taking second place. Actual manufacture began on 4 June 1933, and on 3 April 1934 the huge sections of the ANT-20 were transported from GAZ No. 22 to the Fili airfield.

Aerodynamically the new giant was almost an extended-span ANT-16, but with huge wheel spats. Almost the whole exterior was covered in skin with corrugations of 50mm (2in) pitch and 16mm depth, thickness being the usual

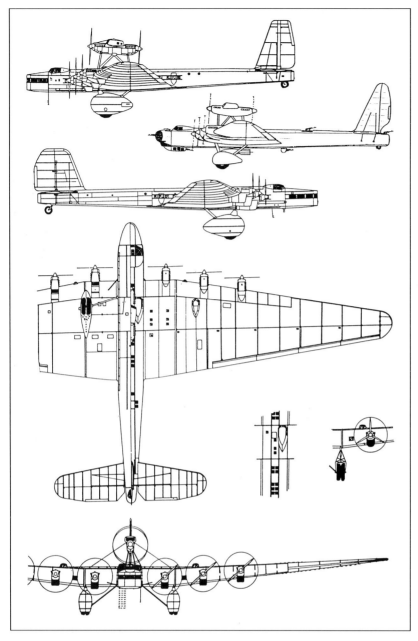

Upper side view, plan and front view, ANT-20 as built, before removal of spats; second side view, unbuilt ANT-20V; third side view, ANT-20bis; left inset, ANT-20 mid-fuselage with centre-line engines removed; right inset, ANT-20bis, showing different undercarriage, engine and propeller.

span, had an untapered leading edge but the same trailing-edge taper as the outer panels. Chord at the joint with the outer panels was 11m (36ft 1in).

Each outer wing carried two engines and had a span of 26.18m (85ft 10¾in), over which the chord tapered to 3.2m at the final rib, carrying the screwed-on tip. Including the double rib at the side of the fuselage there were ten A-ribs on each side of the aircraft, interspersed by lighter L-section ribs and the multiple spanwise stringers which inboard were built-up triangular Warren girders. Experience with the ANT-16 indicated that the very long ailerons should be made in four sections, to prevent the hinges from binding due to flexure of the wing. For rail transport the wing could be unbolted into ten sections.

The fuselage had a cross section fractionally wider than that of the ANT-16, namely 3.5m (138in), the height being 2.5m (98.4in), or the same as the depth of the wing at the root. It was made in sections F-1 to F-5, with steel eye/lug bolted joints at the four principal tubular longerons. Most of the structure was made from standard profiles, but highly-stressed members were thick tube or special tetrahedral profiles built up from riveted sections, sometimes with lightening holes. Again the skin was mainly 50mm by 16mm corrugated, with small smooth areas of stressed skin.

F-1 contained the navigator's station, with a panoramic view around the nose through three single and four twin windows of birdproof glass. In front was a single seat with two tables, and such equipment as a drift meter, sextant and basic flight instruments. Later the navigator was moved to a small room to the rear, and the nose furnished for VIPs with two seats in front and two pairs behind. F-2 contained an upper section housing the dual-control enclosed cockpit, with a plywood roof and sliding side windows. At the lower level was a cabin for four pairs of passengers, with the radio cabin immediately to the rear. Along each side was added a shallow box covering the cable runs for the flying controls, that on the right side being easily disconnected.

F-3, between the massive wing spars, housed the telephone (intercom) switchboard with sixteen lines, secretaries, a writer's cabin and toilets.

0.3–0.8mm. The enormous wing, by far the biggest ever designed up to that time, had CAHI-6 profile, thickness/chord ratio decreasing from 20 per cent at the end of the tapered centre section to 10 per cent near the tips which, unusually for Tupolev, were rounded. To achieve the best possible lift/drag ratio the aspect ratio was 8.2 (a high figure for 1932, though Tupolev exceeded 13 with the ANT-25).* This was the first time such an aspect ratio had been used on a heavy aircraft.

The basic spacing of the three spars and the ribs was 2m (6ft 7in), dividing

the inner wing into useful compartments which were used as small rooms, the available ceiling height also being never less than 2m. The spars were mainly trusses built up from diagonal or vertical angles, with booms of 160mm by 190mm elliptical section (sometimes circular 160mm diameter) formed by wrapping and riveting 6mm sheet. As usual, the A-ribs projected beyond the wing profile, the corrugated skin being applied in strips to fill the 2m gaps between them, with underlying spanwise members. The centre section, of 10.645m (34ft 11in)

F-4 housed a buffet dining area, store-room, cinema projection room, film library, racking for the medium-wave and long-wave radios and other items. F-5 was a cross-braced area not used other than to house various equipment items and carry the tail control cables and the tailwheel. All the 2m cubes in the inner wing were used. On the left side were a leading-edge toilet, cloak-rooms and printing press, and, in the inner end of the outer wing, a sleeping bay with (usually) three superimposed bunks. On the right were a leading-edge electrical room housing two petrol engines, one of 22kW (29.5hp) and the other of 7.3kW (9.8hp), driving generators for various DC and AC voltages up to the unprecedented level of 120V. Behind the front spar was a cloakroom, then a passageway, next a photographic processing room and behind the rear spar a baggage bay. Outboard of the centre-section joint was a second sleeping room like that on the opposite side. Natural light was provided by six roof windows in the centre section on each side (not in the sleeping rooms). The pilots' cockpit, radio room and writer/secretariat were linked by a pneumatic message-tube system similar to that used on many TB-3s.

Normal usable floor area was 109sq m (1,173sq ft), though this varied according to the reason for each flight. Depending on the requirement, other equipment could include a 'voice from the sky' loudspeaker system, an audio recording studio, a pharmacy and a leaflet dispensing system. With most of the special equipment removed, the ANT-20 could comfortably seat seventy-two passengers, with a normal operating crew of eight. In this config-uration red carpet was laid down in the passenger areas. Entry was via a large section of F-2 which hinged down with integral stairs, or by doors on each side at the trailing edge.

The tail unit followed Tupolev practice, but was enormous, the span over the tailplane being 18.3m (60ft 0½in). Most of the skin was of the 40mm by 8mm type corrugated sheet. The tailplane had a symmetric profile and was braced by upper and lower streamline-section 'wires' with which it could be driven through ±5 deg. Again Tupolev broke new ground in using rounded tips, the elevators being inset. Each elevator had an electrically-driven servo/trim tab and internal

The ANT-20 as built, with wheel spats. (Jean Alexander/RART)

leading-edge balance. So did the rudder, which again replaced the traditional horn by internal balance and an electric tab. Primary flight-controls were all operated by low-friction cables.

The main landing gears closely resembled those of the ANT-16, though the shock absorption was pro-vided by oleo-pneumatic cylinders. The side-by-side wheels on each unit were carried in a frame welded from KhMA steel. Forked lugs under the front spar were pinned to a ball joint at the top of the leg, which was braced by a forked strut at the rear pinned under the third spar and by two lateral struts pinned to the first and third spars at the side of the fuselage. The four tyres were 2,000mm by 450mm, on special wheels fitted with four-shoe air-oper-ated brakes. The original scheme was to enclose each unit in a giant 'trouser'. This was replaced by enormous spats, stiffened by three exterior beads, but these were removed before first flight and never replaced. The castoring tail-wheel was similarly held in a sprung fork of KhMA steel, and had a 900mm by 200mm tyre.

As already mentioned, Tupolev had at first thought six 900hp Mikulin M-34FRN engines would provide adequate power, but he was persuaded by the poor behaviour of the ANT-16 to add a tandem pair above the fuselage in a nacelle generally similar to that on the bomber but carried on a totally different arrangement of struts, in-clined up at 3°30′. The wing engines

were carried on anti-vibration mounts on welded KhGSA steel trusses canti-levered off the front spar, and drove 4m (13ft 1½in) two-blade wooden pro-pellers with spinners faired into the cowling. Starting was by compressed air, and each cylinder's exhaust escaped through a short upward-pointing stack. The water radiators of the wing engines were in neat short ducts under the wing similar to those of late-production TB-3s. As in the ANT-16, the nacelle engines were cooled by a double-size matrix in a dorsal duct above the oil tank serving those engines.

A total of 9,400 litres (2,068gal) of fuel could be housed in twenty-eight riveted aluminium tanks arranged in groups of four and eight in each outer wing, feeding into a collector tank on each side. All throttle linkages were connected to the three engineer sta-tions. Moving on from the 'fighter cockpit' of the ANT-14, Tupolev gave the ANT-20 three such cockpits, each with four front windows, a metal aft fairing and hinged roof providing access to the aircraft upper surface. The wing engineers were just inboard of engines 2 and 7, and the fuselage engineer was under the nacelle on the right side of the fuselage.

Bearing in mind the enormity of the task, the design and construction of the ANT-20 was remarkably quick. The completed aircraft was inspected and accepted by the special committee on 24 April 1934. Moreover, when Gromov flew it for the first time on

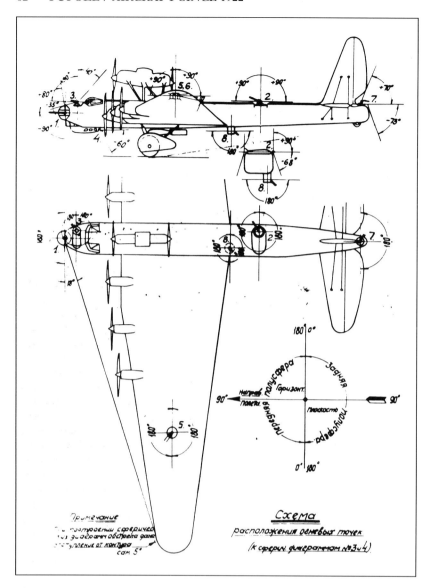

An original OKB drawing showing the planned armament of the ANT-20V. (via Yefim Gordon)

uncertain, but he began performing aerobatics. Attempting to loop around the ANT-20, he struck it from below, causing the loss of both aircraft, himself, CAHI pilot I S Zhurov, MG-squadron pilot I V Mikheyev, ten crew members and thirty-three passengers. Curiously, a recent ANTK Tupolev booklet states that in 1939 this aircraft set a record in lifting 10 tonnes to 5,000m (which would have been easily within its capabilities if it had still existed).

Early in the programme Tupolev schemed the ANT-20V heavy bomber. This would have had an unchanged wing and engines but a bulbous fully glazed nose, large chin gondola, a taller and more rounded vertical tail, internal and external carriage for up to 10 tonnes of bombs, and defensive armament of two Oerlikon 20mm cannon and six 7.62mm DA or ShKAS machine-guns. Limited to a gross weight of 41,000kg, it would have had a poor performance, the calculated maximum speed being 230km/h at 3,500m, the climb 27min to 3,000m and the range only 1,300km with a bombload of a mere 2 tonnes (4,410lb). This project was overtaken by the even bigger ANT-26, TB-6.

Span 63m (206ft 8¼in); length (tail down) 32.9m (107ft 11¼in), (flight attitude) 32.476m (106ft 6½in); wing area 486sq m (5,231sq ft).
Weight empty (typical) 28,500kg (62,831lb), including bare airframe 17,850kg and power installation 10,650kg, fuel/oil 7,150kg (15,763lb); loaded 42,000kg (92,593lb).
Maximum speed 220km/h (137mph) at sea level, about 250km/h (155mph) at 5,000m (16,400ft); cruise 195km/h (121mph) at low level; service ceiling 4,500m (14,760ft); range (maximum fuel) 2,500km (1,553 miles), (normal fuel) 1,200km (746 miles); take-off run 400m (1,312ft); landing speed 100km/h (62mph), landing run 400m (1,312ft).

ANT-20bis, PS-124

The destruction of the *Maksim Gorkii*, the embodiment of all Soviet aspirations, caused shockwaves of grief and anger throughout the USSR, and an instant determination not just to replace it but to replace it by a whole fleet of such giant aircraft. This was, of course, a purely emotional reaction; newly formed Aeroflot could not economically operate even one MG, let

17 June he expressed himself as pleasantly surprised by its ease of handling during the 35-minute workout. On the second flight, two days later, he flew low over Red Square where the return of the Chelyuskin expedition was being celebrated. During factory tests the registration L759 was applied, and the name was painted right across under the wings. There were few modifications, the most important being an increase in chord of the two inboard sections of aileron and the addition of an autopilot, for the first time in the Soviet Union.

The monster aircraft joined the MG squadron on 18 August 1934. At one time it was fitted with a complicated system of electric lamps under the huge wing with which slogans could be displayed at night, but this proved predictably erratic in reliability and was removed. Later all the upper surfaces were painted red, adding a tonne to empty weight, and the registration was changed to I-20. A routine was established whereby a crew of twenty managed all the on-board equipment when no (or few) passengers were being carried, reduced to eight when the propaganda equipment was replaced by passengers. Many of the latter were high officials, but others were people being rewarded for public service or for breaking output records.

It was common for the MG to be escorted by a small aircraft, often a fighter, to emphasise its size. On 18 May 1935 the escort was an I-5, flown by CAHI pilot N P Blagin. Whether Blagin was authorised to do so is

The ANT-20bis landing at Khodinka. (G F Petrov/RART)

alone a fleet, but so strong was the national feeling that donations to build such a fleet poured in a torrent until, according to popular belief, the total reached 35 million Rubles. More recent documents state that by 4 July 1935 the sum actually available was a staggering 68 million.

To some, both in the government and in AGOS–CAHI, this was embarrassing. Tupolev told the author he was aware of the sum said to have been collected, and equally aware that the nation did not need a fleet of ANT-20s, but the Soviet Peoples' Commissar announced that sixteen aircraft would be built, named *Vladimir Lenin, Iosif Stalin, Maksim Gorkii, Mikhail Kalinin, Vyacheslav Molotov, Kliment Voroshilov, Sergio Ordzhonikidze, Lazar Kaganovich, Stanislav Kosior, Vlas Chubar, Anastas Mikoyan, Andrei Andreev, Sergei Kirov, Mikhail Frunze, Feliks Dzerzhinskii* and *Valerian Kuib'shyev.*

What had in 1932 been reorganised as KOSOS–CAHI was actually awarded a contract to produce these sixteen aircraft, but the arrest of A N Tupolev in October 1937 caused a major upheaval. Eventually, while design changes continued at KOSOS in Moscow, preparations for manufacture of the aircraft were assigned to GAZ No.124 in distant Kazan, Tatar Republic. Tupolev's former senior aide, B A Saukke, was sent there as deputy chief designer. The arrangements could hardly have been worse, and Tupolev told the author he guessed that about two years were lost in trying to decide who did what and where responsibility lay.

Before he was removed, Tupolev had made it clear that, thanks to A A

Mikulin's ceaseless quest for more power from the M-34 engine, the tandem nacelle engines should be removed. This alone justified the new designation of ANT-20bis, but there were several further changes. Basically, the requirement for a giant propaganda aircraft had lapsed, and the ANT-20bis was from the start planned as a transport, principally for sixty passengers seated two+two.

The most important modification in the 20bis was that power was provided by six more powerful engines in completely new installations. As built, the engines were M-34FRNVs, each rated at 1,000hp for take-off. They were carried in modified 30-KhGSA mountings in low-drag cowlings with the necessary large radiator matrix mounted vertically underneath in a deep duct with inlet louvres controllable from the cockpit and an exit in-

tended to provide thrust in flight. The exhausts were grouped into a single pipe, only the end of which projected outside the cowling above the wing. The engines drove VISh-4 three-blade constant-speed metal propellers.

Most of the airframe was unchanged, the only obvious alterations being to lengthen fuselage section F-1 to form a useful ten/twelve-seat cabin with nine all-round windows, and to redesign the vertical tail. The new fin and rudder had much greater chord but reduced height, total aircraft height in flight attitude being reduced from 11.253m to 10.858m (35ft 7½in). The new rudder had its servo tab at the bottom instead of the top. Other visible changes were that the spats removed from the ANT-20 were fitted over the mainwheels, the cockpit canopy was reshaped and fitted with improved windows, the control cable

Looking back from the cockpit of the ANT-20bis. (John Stroud)

fairings along the outside of F-2 were redesigned as slim half-round covers, the elevators were given external mass-balance weights, the landing lights were moved from the leading edge of the wing to the nose, and a direction-finding loop antenna appeared under the forward fuselage. In fact the radio was completely updated, and other internal changes were that the entire structure was considerably strengthened, even though fuel capacity was for some reason significantly reduced. Surviving documents give the wing area as 486sq m, the same as for the ANT-20, and also, in several places, as 480sq m (5,167sq ft). There is no reason to believe that the wing was changed in size.

Long before the ANT-20bis was ready for flight GAZ No. 124 had known that no series aircraft were to follow it. There was no demand for them, and the factory was needed for other work, such as production of the TB-7. The single ANT-20bis was, however, completed, though by this time its ancestry had been largely forgotten and it appeared in almost all documentation at Kazan as the PS-124 (passenger aeroplane, factory 124). It was flown for the first time on 15 May 1939, the pilot being E I Shvarts (Schwartz). It was generally judged an excellent aircraft, and on 12 August 1939 it was accepted for service with the GVF (civil air fleet) with the registration L760.

It was furnished for sixty passengers, and for about a year, from the end of 1939 until December 1940, it operated 271hr 25min on scheduled services between Moscow and Mineralnye Vody. It had a crew of seven, usually captained by N I Novikov.

Span 63m (206ft 8¼in); length (flight attitude) 34.096 m (111ft 10⅜in); wing area see text.
Weight empty 32,046kg (70,648lb); fuel/oil 5,830kg + 370kg (12,853lb + 816lb); loaded weight 44,000kg (97,002lb).
Maximum speed 275km/h (171mph) at 3,500m; cruising speed 235mph (146mph); service ceiling 5,500m (18,050ft); range 1,300km (808 miles); take-off run 500m (1,640ft); landing speed 100km/h (62mph), landing run 620m (2,034ft).

In December 1940 the PS-124 underwent heavy maintenance at Moscow Vnukovo. It did not re-enter service until November 1941. By this time it was powered by AM-35 engines rated at 1,200hp. Payload had been increased from 4,800kg (10,582lb) to 6,400kg (14,109lb), and the aircraft was equipped for sixty-four passengers and a crew of eight, with special provision for heavy cargoes. It was redelivered to the Aeroflot Uzbek Flight Directorate, and operated with higher utilisation on cargo services, principally between Tashkent and Kuibyshev and Tashkent–Urgench. On 12 December 1942, whilst still at an altitude of about 500m (1,640ft) on the approach to Tashkent, the elevator electric tabs suffered an uncontrollable runaway. The pilots managed to recover from the ensuing dive but could not avoid hitting the ground short of the airfield. The aircraft was damaged beyond repair, though without fatal casualties. It had flown a further 698hr since modernisation. It was the last corrugated-skin aircraft built in the USSR.

Dimensions unchanged.
Weight empty 33,370kg (73,567lb); loaded weight 45,600kg (100,529lb).
Maximum speed 296km/h (184mph) at 3,600m; cruising speed 226km/h (140mph); service ceiling 6,750m (22,150ft); range (with 'commercial allowances') 960km (597 miles).

ANT-21, MI-3

In January 1932 the ANT-21 project was launched as a replacement for the R-6 reconnaissance aircraft, but with much higher flight performance. In typical Tupolev manner, it was hoped that as much of the original (basic ANT-7) design as possible might be retained, though it was eventually decided that almost every part had to be completely new except for the engines, cowlings and propellers. By this time Tupolev had recognised that for the fastest aircraft the extra drag of corrugated skin was becoming unacceptable. He told the author he had for many years appreciated that the flow around the aircraft 'almost never follows the corrugations', but that this had not been of significance with aircraft flying at about 200km/h (124mph), except in the special case of the RD, ANT-25. The ANT-21 was the first aircraft where he felt the time had come to switch to at least partial covering of smooth stressed skin.

The new features in the ANT-21 were a slim fuselage of oval cross-section with non-corrugated skin, a twin-finned tail and retractable under-

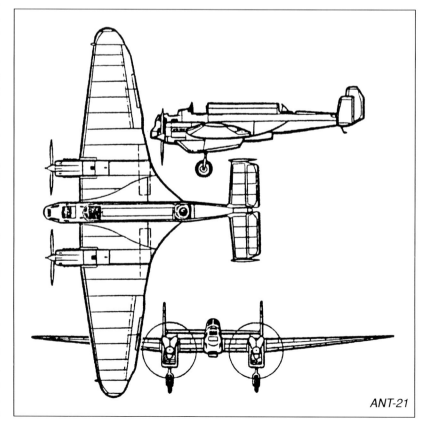

ANT-21

The ANT-21 prototype, with the entire right wing tufted to indicate airflow. (via Philip Jarrett)

carriage. During the construction of the prototype the emphasis swung towards a cruiser escort, and ultimately to a *Mnogomestnyi istrebitel* (multi-seat fighter), resulting in the VVS contract being amended to the designation MI-3. This had only a minor effect on the design, which was to some degree an exercise in creating a fast twin-engined aircraft.

The wing incorporated spar booms of ANT-7 type, and the centre section was very similar, but the outer panels were more sharply tapered to reduce span and terminate in round tips of small radius. The junction with the fuselage had an enormous trailing-edge fillet, which accentuated the impression that the wing root extended over much more than half the overall length of the aircraft! The long cable-driven ailerons were made in two sections, with nose-beak balance and a servo/trim tab in each half.

The fuselage, considerably smaller than that of the R-6, broke completely new ground for Tupolev's designers in being a semi-monocoque, with ring frames and multiple stringers all made from Dural rolled sections and lipped sheet to give a smooth oval cross-section. Dural stressed skin was used throughout, most of the area being flush-riveted. Only on the tail was extensive corrugated skin retained, the pitch being 20mm and 25mm. The almost untapered tailplane was fixed, and carried divided elevators with internal aerodynamic and mass balance. On its tips were the rectangular fins carrying traditional angular rudders with large horn-balances. There were no tabs on the rudders or elevators.

Fuel was housed in four aluminium tanks inboard and outboard of the engine nacelles. Though the engine installations were almost identical to those of the R-6, the underslung radiator ducts were extended to the rear to provide compartments to house the retracted main undercarriage. The latter were especially neat, each comprising a single oleo-pneumatic shock strut, braced laterally at the top with diagonal struts to the two attachment hinges and carrying a fork at the bottom for the 1,000mm by 250mm wheel with a drum brake. The whole unit retracted to the rear in a simple and elegant manner, unlike most retractable undercarriages of that era. Operation was hydraulic.

The ANT-21 was designed for a crew of four. In the bluff nose was an open-topped cockpit for a navigator/gunner, though he had no guns to aim. Immediately to the rear

was the open cockpit for the pilot. Far aft, behind the broad wing, was a Tur-5 with a PV or DA gun. Under the fuselage was a shallow gondola, open at the rear, providing a prone position for a ventral gunner with a pivoted DA. The main armament comprised four PV guns fixed firing ahead and aimed by the pilot, two in the lower sides of the nose and two in the wing roots. A study was made for an internal bombload, but this was not a requirement.

The prototype was built without armament, and was flown for the first time by I F Kozlov on 23 May 1933. From the start the vertical tails gave trouble, and they were repeatedly redesigned, though always retaining the rudder horns. Bearing in mind the quite normal wing loading of 97–98kg/sq m (*c*20lb/sq ft), the landing always appeared frighteningly fast at about 140km/h (87mph), and the aircraft had a strong tendency to roll to the left before touchdown. Tupolev received a host of recommendations, among them an increase in wing area and the fitting of landing flaps. Soon split flaps were fitted, an innovation for Tupolev, but their use resulted in tail buffet.

On 14 September Kozlov dived the aircraft to about 350km/h (217.5mph), when violent rudder flutter resulted in structural failure of both surfaces, one folding its upper part sideways which resulted in immediate yaw through about 180 deg and loss of control. Remarkably, Kozlov regained control at a lower speed and brought the aircraft back, though the undercarriage was

The Mi-3D (dubler). (via Philip Jarrett)

damaged on landing. This resulted in a further flurry of fixes, most urgently including redesign of the tail to have the horizontal surface mounted high up on a conventional single fin. Above the tailplane the fin was corrugated, and the all-corrugated rudder had a Flettner servo tab.

Span 19.11m (62ft 8⅜in); length, initially 10.85m, finally 10.86m (35ft 7⅞in); wing area 52.1sq m (561sq ft).

Weight empty 3,412kg (7,522lb); loaded weight, initially 5,088kg (11,217lb), finally 5,955kg (13,128lb).

Maximum speed initially 351km/h (218mph), later 350km/h (217mph); service ceiling 7,885m (25,870ft).

MI-3D Tupolev had never before encountered such difficulty in producing an acceptable aircraft, and the problems continued with the MI-3D (D for *dubler*, meaning second prototype), also called ANT-21bis. The wings were enlarged, the flaps were increased in area, the engines replaced by 770hp M-34N, the nose altered to a more streamlined enclosed form with multiple windows, the ailerons increased in area by extending their chord over the mid-section to project behind the line of the wing trailing-edge, and the fuselage lengthened to accommodate tandem cockpits for the pilot, radio operator (who could if necessary fire a PV from a floor hatch) and rear-gunner, all with sliding canopies. The wing-root PV guns were retained, but the navigator in the nose was provided with a 20mm Oerlikon cannon on a universal-joint mounting, and the upper rear gunner had a fast-firing ShKAS.

Tail trouble persisted, and a third aircraft (there is some evidence

this was the *dubler* modified) was designated as the *Etalon* (production standard). The tailplane was rigidly braced to the fin by upper and lower struts, and the fin and rudder were given smooth skin throughout. The nose was slightly shortened, multiple cooling and ventilation louvres were provided in the sides of the cowlings and nacelles, and the wings were covered in tightly stretched fabric. NII-VVS testing took place in July–December 1934, by which time (according to Shavrov) preparations were made for series production, including making the jigs and tooling. However, the MI-3D failed its acceptance testing, and was abandoned, though work continued on the derived ANT-29, DIP.

Span 20.76m (68ft 1⅜in); length 11.57m (37ft 11½in); wing area 59.2sq m (637sq ft).

Weight empty 4,058kg (8,946lb); loaded, normal 5,463kg (12,044lb), maximum 5,608kg (12,363lb).

Maximum speed 285km/h (177mph) at sea level, 350km/h (217.5mph) at 5,000m; service ceiling 8,300m (27,230ft).

ANT-22, MK-1

Though the ANT-11 was never built, Tupolev continued to study twin-hull flying-boats, and his interest was sharpened in 1930 by a similar project by R L Bartini. In July 1931 the VVS (there was then no separate naval air force) issued a requirement for a *Morskom kreiser*, literally a 'sea cruiser', with a radius of action of 1,000km (621 miles) and maximum speed of 300km/h (186mph); to these

attainable requirements was added the ambitious demand for 'a bomb load of 6,000kg' (13,228lb) and armament of two or three quick-firing cannon and four or five machine-guns. AGOS, later KOSOS, received the contract for a single ANT-22 prototype, with VVS designation MK or MK-1.

From the outset Tupolev preferred the twin-hull catamaran layout, which had looked encouraging in model tests of the ANT-11, but he agonised over the use of tandem push/pull engines. With so large an aircraft there was no need for such a high thrust-line merely in order to keep the propellers clear of waves and spray, but with engines mounted on the wing it was difficult to find room for the six engines which were the minimum needed. Drawings were produced showing five M-35 engines, but these were not available. Nobody liked engines on the trailing edge, or twin hulls so widely separated that there was room for two engines on the leading edge between them. In the end the best choice appeared to be three nacelles each with a push/pull pair of the most powerful engines available.

Other basic choices concerned the wing, hull and tail design. The wing was given the aspect ratio of 8.54 to achieve the best aerodynamic efficiency and range. The hulls had a volume at the front much greater than that needed for flotation, and at the cost of greater structure weight this was expected to enable the MK to operate in significantly higher sea states than any previous marine aircraft. In comparison, the aft part of the hulls was small, and the moment arm of the horizontal tail was short (only 2.2 times the mean aerodynamic chord), so instead of using a very broad surface the designers increased structural rigidity by adopting a braced biplane layout.

The wing was made as a 16m (52ft 6in) rectangular centre section extending just beyond the hulls, and outer panels with dihedral and straight taper. The aerofoil section was the usual CAHI-6, thickness/chord ratio tapering from just under 20 to 11 per cent. There were two main spars from tip to tip, plus a front spar and from one to three lighter rear spars, all made up from standard dural sections and tube with diagonal Warren-type bracing, both chordwise and spanwise. The booms of the main spars had a

The Mi-3 Etalong *(background censored).* (via Philip Jarrett)

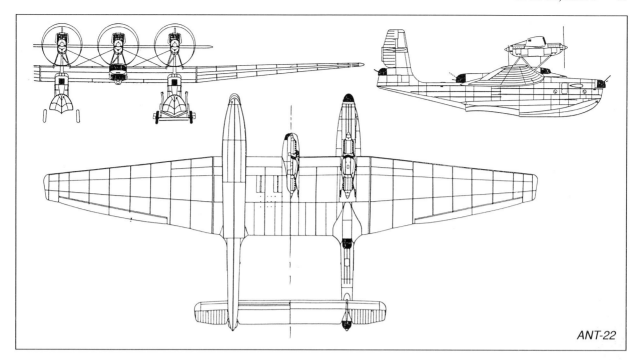

ANT-22

cross section of 125mm by 113mm as far out as the ailerons, tapering thereafter. Altogether there were nine A-type ribs in the centre section and nine in each outer panel, the corrugated skin being applied in the usual way between them. In the common Tupolev manner, the ailerons were increased in chord by curving their trailing edges behind that of the straight-tapered wing, and each was made in two parts, without tabs. There were no flaps.

The centrelines of the identical hulls were 15m (49ft 2in) apart. Each was made in two distinct parts, a massive planing bottom designed for maximum seaworthiness and a much lighter and narrower upper portion designed purely from the viewpoints of inflight strength and internal accomodation. Each lower, hydrodynamic section had a rounded top and a maximum beam of 2,592mm (102in). The planing bottom was concave on each side of the central keel, to become horizontal at the chine. There were two transverse steps made with Alclad sheet 3.5mm and 4.5mm thick, the first quite shallow, the section between them being a concave curve in side view. Downstream of the second step the bottom was a plain V. Internal structure was massive, and numerous hatches were provided for inspection. The non-corrugated skins were attached with the minimum number of rivets, all joints and holes being sealed with polymerised oil.

Above the watertight floors the upper part of each hull had a narrow rectangular section, with twenty-one principal frames and bulkheads and uncorrugated skin with three deep stiffening beads on each side (there being a discontinuity in these beads at the rear wing spar frame). At the rear of each hull a secondary structure was added to carry a gun turret and a vertical fin attached by six tapered bolts, and thus detachable. Each fin carried a rudder with a balance horn at the top, corrugated skin and a Flettner tab. Half-way up each fin was mounted the principal tailplane, while at a lower level the hulls were joined by a second

tailplane of even narrower chord. In view of the relatively short moment arm available, it is astonishing that the entire longitudinal control authority was vested in small elevators hinged to short tailplanes outboard of the fins at the upper level.

The three engine nacelles were almost identical to those of the ANT-20, though the six engines were of M-34R type, each rated at 830hp and driving a two-blade propeller of 4.2m (165.5in) diameter. Engine starting was by compressed air. Partly because this aircraft was expected to loiter at low airspeeds the water radiators were extremely large, with a

The ANT-22 on its four-part beaching chassis at Savestopol. (RART)

The ANT-22 at anchor in 1931, showing the elevators depressed.
(A Aleksandrov/RART)

matrix 1.4m (55in) square in a 'saddle duct'. All three nacelles were exactly in line. Remarkably, in view of the massive hull weight ahead of the wings, the engines had to be mounted on KhGSA struts as far forward as possible, the centre of each nacelle being above the wing leading edge.

On the aircraft centreline was a small nacelle to house the flight crew. This comprised two pilots in a fully glazed enclosed cockpit, a navigator/ bomb aimer in the nose with panoramic windows all round and a long bomb-aiming gondola underneath, and a flight mechanic in a compartment between the first and second spars, with the engine instruments and controls and a roof hatch for access to the engines. In an emergency it was possible to walk stooping through the wing to reach either hull. In each hull were three fully glazed gun turrets. In the left hull the nose turret had a ShKAS, the dorsal turret at the end of a fairing behind the wing a 20mm Oerlikon, and the tail turret a DA-2. In the right hull the nose turret had a 20mm Oerlikon, the dorsal turret a ShKAS, and the tail turret a DA-2. Altogether, the ammunition load comprised 600 rounds (ten drum magazines) of 20mm and 14,000 rounds of 7.62mm.

The total crew numbered eleven, one operating the PSK-1 radio station with a voice radius of 350km (217 miles). Electric power was supplied from batteries which could be charged by a 30kW auxiliary petrol generator similar to that fitted to the ANT-20.

Full intercom was installed, though the expected pneumatic message network was never fitted. The main entrance doors were on the outer side of each hull ahead of the wing, about 1m (39.4in) above the full-load waterline. KOSOS produced a pair of large four-wheeled beaching chassis, each of which picked up the hull along the chines.

Heavy loads of offensive weapons could be carried. Racks under the wings were to be provided for the required 6,000kg (11,023lb), or for four 1,200kg (2,640lb) torpedoes. Inside the centre section, where the thickness/chord ratio was 19.6 per cent, eight bays were provided, each covered by twin cable-operated doors, with a width and depth of 1.4m (55in). These could contain such loads as thirty-two FAB-100 bombs, or six FAB-1000, or four torpedoes.

The ANT-22 was assembled at GAZ No. 45 at Sevastopol, and began factory test on 8 August 1934. It proved to be satisfactory in stability and control, though performance was generally below expectation. Hydrodynamic behaviour was excellent.

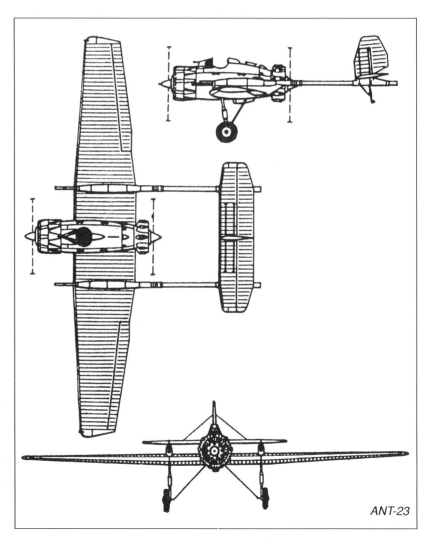

ANT-23

Take-offs and alightings at a weight of 32,000kg (70,548lb) were repeatedly made in waves 1.5m (59in) high and a wind of 12m/s (27mph), and prolonged towing or mooring at sea presented no problem. In the air the behaviour was of necessity pedestrian, a 360° circle taking 85sec; other performance figures are given in the data.

NII-VVS State testing occupied from 27 July to 15 August 1935. Few actual faults were found, but it was judged that this very large and expensive aircraft did not meet the operational demands as well as larger numbers of smaller and faster flying-boats. It was suggested that performance would be enhanced by fitting supercharged engines and improving the smoothness of the exterior, but there was little chance of series construction. The aircraft was kept airworthy, and on 8 December 1936 pilots T V Ryabenko and D Il'inskii set a world class record by lifting a payload of 10,040kg (22,134lb) to 1,942m (6,371ft). Later a different crew flew at low level with a load of 13,000kg (28,660lb). Plans were drawn for a civil transport version.

Span 51.6m (169ft 3½in); length (excluding guns) 24.1m (79ft 0¾in); wing area 304.5sq m (3,278sq ft).

Weight empty 21,663kg (47,758lb); fuel/oil 5,100kg (11,243lb) loaded weight, normal 29,450kg (64,925lb), maximum weight 33,560kg (73,986lb).

Maximum speed at sea level, normal weight 233km/h (145mph), maximum weight 205km/h (127mph); cruise 180km/h (112mph); climb 10.3min to 1,000m, 42min to 3,000m, service ceiling, normal weight 3,500m (11,480ft), maximum weight 2,250m (7,380ft); range with 5,000kg bomb load 1,330km (826 miles); alighting speed 110km/h (68mph).

ANT-23, I-12

This highly unconventional single-seat fighter was designed and built by an AGOS brigade led by Viktor Nikolayevich Chyernishov. His objective was to find a better way of using the recoilless guns developed by Kurchevskii than merely hanging the long tubes under the wings. Leonid Vasil'yevich Kurchevskii had in 1930 picked up where Prof B S Stechkin had left off the development of large-calibre guns, which had been developed (to some degree basing their design on the British Davis of 1915) from 1922. These weapons fired a large HE shell at the target, and counteracted the recoil by expelling the propellant gas and a balance weight to the rear. By the early 1930s the most immediately available gun was the APK-4 (*Avtomaticheskaya Pushka Kurchevskogo* Type 4), with a calibre of 76mm (2.99in).

The ANT-23 was planned as an all-metal low-wing monoplane in the AGOS structural tradition, but with the pilot in a central nacelle between tractor and pusher engines, and the tail carried on the end of structural tubes.

These housed the cannon and also ensured that the violent rearward blast on firing was contained until downstream of the tail. With conventional APK installations the blast had caused momentary loss of control and damage to the tail. It was also hoped that having two powerful engines with the frontal area of one would result in high flight performance.

The wing comprised a rectangular centre section with a span of 5.4m (17ft 8½in), with a thickness/chord ratio of 18 per cent, and tapered outer panels with dihedral only on the underside. Structurally there were two main spars and lattice ribs at a pitch of 1.2m, all built up from standard rolled sections and tube. For the first time corrugated skin was not used for an AGOS wing, the D6 sheet being cut to a width of 150mm (5.9in) with one edge rolled to form an external ∩-profile stiffening groove. Every eighth strip was riveted to form the flange of the rib beneath. Sheet gauge was 1mm over most of the wing, the outer strips being 0.8mm. The entire trailing edge of each outer panel was occupied by a slotted aileron, operated by cables. The ailerons were without tabs, and in the usual Tupolev fashion had a kinked trailing edge projecting behind the straight taper of the wing.

The nacelle was all-metal, welded from KhMA tube with detachable D6 panels over the central portion, which included the riveted aluminium fuel and oil tanks adjacent to each engine.

The ANT-23 prototype.

The engines, carried on extensions of the steel truss structure, were originally imported GR9Ak (Bristol Jupiter VI) but were later replaced by the locally built licensed version designated M-22, with take-off power of 570hp. They had helmeted cowlings and a Bristol-type gas starting system, and drove 2.8m (110in) wooden two-blade propellers with spinners. The pilot climbed aboard over the back of the wing into an open cockpit with an outstanding view, the windscreen incorporating an Aldis optical sight.

The tailplane and single elevator were carried on elevated platforms above the two tubes. On the centreline was the fin and rudder, braced by pairs of struts above and below. The entire tail was made like the wing, with skin applied in 150mm strips with a flange along one edge. The control commands were transmitted by tensioned cables carried in guides along the tops of the two booms, a method which in practice gave no trouble. To provide clearance for the rear propeller the undercarriage had to be tall. Each main unit had a 900mm by 200mm wheel carried on a vertical main leg incorporating a faired rubber-bungee shock absorber, braced laterally and to the rear by pin-jointed struts. A very tall bungee-sprung tailskid was provided under each tail boom.

These booms have even been described as 'water pipes'. In fact they were made from aircraft-quality steel, machined to an internal diameter of 170mm (6.7in), with a wall thickness from 1mm to 3mm. Each boom comprised three sections, each about 1.5m (59in) long, joined by threaded ends. The booms were mounted 3.7m (12ft 1⅝in) apart, attached by lugs and bolts to both spars and then covered by a light aluminium fairing. Each APK-4 was installed with the front of the barrel projecting, and with a surrounding air space. Each gun was provided with two rounds, the projectile weight being 0.56kg (20oz). Though the APK-4 could fire automatically, at a slow rate, in this aircraft each round was fired separately, though both guns could be fired together.

The ANT-23 was completed in July 1931. It was named *Baumanskiy Komsomolyets*, after the famed revolutionary who until his death in 1905 worked near the AGOS-CAHI site. I F Kozlov made the first flight on 29 August. The handling of the aircraft was generally good, but it was clear that the expected high performance would not be achieved. On 12 March 1932 Kozlov was undertaking firing trials at a height of about 1,000m (3,280ft). On being discharged, the left gun diffuser section exploded. This rendered inoperative the tail controls on that side, but Kozlov succeeded in landing the aircraft, the damaged tail boom collapsing on touchdown (he was awarded the Order of the Red Star).

The reason for the failure was found and corrected, and work on the aircraft, and on a second prototype, was continued at the end of September 1932. The second (*dubler*), designated ANT-23bis, was provided with a means for severing the rear propeller in emergency to give the pilot a much better chance of escape. Tupolev also studied severing the rear engine complete, to cause a violent nose-down pitch to help the pilot depart, but this was not accepted, and in any case the ANT-23bis was never completed.

Span 15.67m (51ft 5in); length 9.52m (31ft 2¼in); wing area 33sq m (355sq ft).
Weight empty 1,818kg (4,008lb); loaded weight 2,405kg (5,302lb).
Maximum speed 318km/h (198mph) at 5,000m (16,400ft); climb to 5,000m in 7.7min; service ceiling 9,320m (30,580ft); range 405km (252 miles); landing speed 100km/h (62mph).

ANT-25, RD

On 20 May 1931 an informal discussion was held at CAHI to study existing world records for speed, altitude and range, and consider the prospects for beating them. Tupolev pointed out how in theory an aircraft specially designed for long range, with very high aspect ratio and retractable undercarriage, ought to be able to beat the existing French record of 7,905km

The seldom-photographed ANT-25 No. 1. (G F Petrov/RART)

(4,912 miles) by a wide margin. From the outset it was recognised that by exchanging fuel for bombs or chemical weapons the same basic aircraft could become a strategic bomber.

In August 1931 A N Tupolev and the Commander of the VVS-RKKA, Gen Ya I Alksnis, made a formal proposal to the SSSR Revolutionary War Council for the creation of an aeroplane able to fly record distances. They pointed out that the records gained by France had been achieved by modified versions of existing types. Though ostensibly a civil project, the RD (*Rekord Dalnosti*) was from the outset viewed as a basis for a bomber and reconnaissance aircraft. A special committee was formed to manage this programme. Tupolev prepared an *Eskiznyi proyekt* (preliminary design), designated ANT-25, sometimes called CAHI-25.

On 7 December 1931 the Revolutionary Council authorised the construction of the RD to cover 13,000km (8,078 miles) nonstop, to fly in the summer of 1932. The meeting decided that the ANT-25 would meet the requirements. A new executive committee was formed, chaired by Kliment Ye Voroshilov, People's Commissar for Defence, to manage development of the ANT-25, its M-34 engine from the KB of A A Mikulin, various equipment items such as a gyromagnetic compass, and the construction of the first hardened runway in the Soviet Union (at Moscow Shchyolkovo, later called Chkalovskaya).

Though work on supporting items began immediately, actual design of the ANT-25 did not start until April 1932. Tupolev appointed P O Sukhoi to lead the project. Structural design of the remarkable wing was entrusted to Petlyakov and Belyaev, with control surfaces assigned to N S Nekrasov. While the engine and its reduction gear were the responsibility of Mikulin, its installation and the fuel, oil

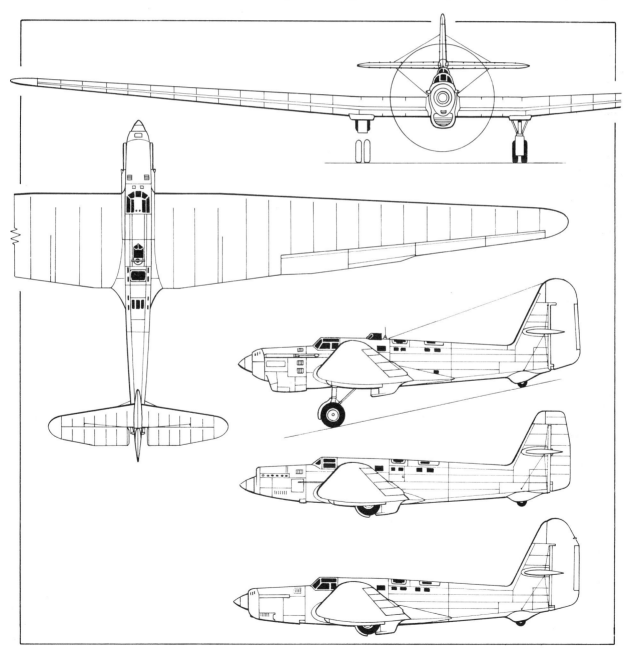

Three-view, ANT-25 No. 3: 2nd side view, ANT-25 No. 1; 3rd side view, ANT-25 No. 2.

and cooling systems were assigned to Ye I Pogosskii and K V Minkner. The retractable undercarriage was to be developed by a team led by A V Petrov. Brigade leaders on other parts of the airframe were N S Nekrasov, D A Romeiko-Gurko, G O Bertosh and N A Fomin. Aerodynamic support by CAHI was headed by Prof V P Vetchinkin, and the possibility of flutter being a problem with the high-aspect-ratio wing was to be studied by M V Keldysh. Flight test preparations were assigned to leading engineers Ye K Stoman and M A Taits.

The basis of the whole design was the remarkable cantilever monoplane wing with an aspect ratio of 13.1, and span 2.9 times the overall length. As before, it had a modified CAHI-6 aerofoil section, the thickness/chord ratio being 20 per cent at the root, 19.2 per cent at semi-span and 18.5 per cent just inboard of the pointed tip. Structurally it comprised a 3.75m (12ft 3½in) horizontal centre section and two 15.125m (49ft 7½in) tapered outer panels with 3 deg dihedral. As the chord was less than on most AGOS wings there were just two principal spars, at 18 and 44 per cent chord, plus a lighter rear spar. Most of the structure was the usual built-up diagonal and vertical lattice of D6 Dural, but the spar booms were of tube made from KhMA steel heat-treated to a u.t.s. of 140kg/sq mm. There were eighteen A-type ribs on each side projecting above the profile, with lighter intermediate ribs. Surprisingly, in view of the overriding need to reduce drag, corrugated skin was again used, the material having 40mm by 8mm corrugations riveted in strips between each pair of A-ribs.

The very long ailerons were made in four sections, each with a slotted nose and inset hinges giving 100 per cent mass balance. In the second aircraft (after modification) and third ANT-25 the inboard two sections had chord increased behind the wing trailing-edge line, the outer of this pair also having a cable-operated servo tab.

The fuselage, of basically oval-section, was constructed in two parts. The main portion was built integral with the wing centre section, and had a strong truss internal structure with closed-profile frames and tubular longerons. At the front of this section was the KhGSA firewall, ahead of which the heavy engine was carried in

a triangulated mounting of welded KhMA steel. Aft of the wing the rear fuselage was built integral with the fin as a lighter semi-monocoque, with open frames and L-section stringers. The smooth skin was riveted on in narrow longitudinal strips in 0.8mm or 1mm thickness.

The tail was all-D6, with structure built up from lattice girders or flanged sheet with lightening holes, the skin being corrugated. The fixed tailplane was mounted high and wire-braced, there being seven A-ribs on each side and the elevators having 100 per cent balance from inset hinges. At the top of the almost rectangular rudder was a horn, but the major part of the balance was provided by pivoting the surface on small welded-KhMA inset pyramid hinges.

The main undercarriage was unlike anything previously attempted by Tupolev's designers. The main shock strut was an oleo-pneumatic precision assembly in KhGSA steel, braced by a diagonal strut on each side, all three being hinged to trunnions under the front spar. At the rear the leg was braced by a pivoted strut which could be unlocked and pulled up by cables driven by an electric motor and gearbox to retract the unit to the rear. On the sprung end of the leg was an axle projecting equally on each side of the leg carrying two 900mm by 200mm disc wheels with powerful brakes. When retracted, and locked up by an electrically moved bolt, the lower halves of the wheels remained projecting under the wing, faired by an aluminium box to the rear. The tailwheel had a balloon tyre and was carried in a fork on a castoring sprung strut, the unit being faired into the rear fuselage.

The engine chosen was the M-34, with compression ratio 6, rated at 750hp. There was direct-drive to a two-blade wooden propeller of 4.5m (14ft 9¼in) diameter. To minimise drag the water radiator could be fully retracted, or cranked down by the pilot along vertical slides immediately in front of the firewall to project only as much into the slipstream as necessary. Next to it was the big 350 litre (77gal) oil tank. In the wings were multiple riveted-AMTs fuel tanks between the spars housing a total of 7,640 litres (1,681 gal). Fear of wing flutter had led to prolonged research and testing, and the distributed fuel mass was found to play a major role in moving the flutter

boundary away from normal flight conditions. On the other hand, at a time when fatigue was totally ignored in Britain and some other countries, careful fatigue tests were made, and the design factor reduced from 4.8 when new to 3 after 1,000 hours.

Accommodation was provided for a pilot, radio operator/navigator and flight engineer in tandem cockpits sealed against the weather. All three cockpits had hinged glazed roofs and side windows. Full blind-flying instruments were installed, as well as a sextant, drift sight and the first Soviet gyromagnetic compass, specially designed to operate in polar regions. LF and MF radio transmitters and receivers were installed amidships, with a voice range of 5,000km (3,100 miles), the equipment including a folding mast for use after a forced landing. Electric power was provided by a 12V battery charged by a 500W engine-driven generator. The centre fuselage was designed so that it could become a bomb bay.

The RD-1 was finished silver, without any markings, but with the wing painted red between the spars as far out as the mid-point of the ailerons. It was first flown on 22 June 1933, by Gromov accompanied by Stoman. Handling was unsatisfactory, the ailerons and rudder being insufficiently powerful, making the aircraft dangerous in severe turbulence. Moreover, the fine pitch of the propeller limited performance; all fuel would be exhausted after 48 hours, by which time the still-air distance covered would be only about 7,200km (4,475 miles).

Attention was therefore focussed on RD No. 2, which was only a few weeks behind the first. The original RD was later fitted with an engine of compression ratio 7, giving 874hp, and data are for it with this engine.

Span 34m (111ft 6⅜in); length 10.85m (35ft 7¼in); wing area 87.1sq m (937.5sq ft).
Weight empty 3,700kg (8,157lb); fuel/oil 3,890kg + 300kg (8,576lb + 661lb); loaded weight 8,000kg (17,637lb).
Maximum speed 212km/h (132mph) at sea level; cruising speed 165km/h (103mph); endurance/range, as above; take-off run 1,000m (3,280ft); landing speed 80km/h (50mph).

RD No. 2, the *dubler*, differed mainly in having a geared M-34R engine. This gave more power, and having a higher thrust-line changed the shape of the nose. The first had

The famous red-winged ANT-25 No. 3. (Hugo Hooftman)

compression ratio of 6.8, with a rating of 900hp, but to match the fuel this was reduced to 6.6, power remaining unchanged. Radiator, oil and fuel systems were little changed, but the fin and rudder were redesigned to have greater height, with a round top, the rudder also having a large Flettner tab giving greater power.

The aircraft was again silver, but with the engine cowling dark blue, and bearing registration number NO-25. It was first flown by Gromov on 10 September 1933. By this time Sukhoi had received the Order of the Red Star during CAHI's fifteenth birthday celebrations. The *dubler* proved to have somewhat better flying qualities, but range was still inadequate, the best figure achieved in the aircraft's original form being 10,880km (6,761 miles). Accordingly the corrugations over the outer wing panels and horizontal tail were filled in with balsa and then covered in Percale cambric fabric, drawn tight and sewn with curved needles, and finally coated with dope. Fairings were added at the wing roots and varnished, the propeller polished and the leading edge of the wing again cleaned and doped. This enabled fuel capacity to be increased, improving endurance to a theoretical 80.4 hours.

The *dubler* was tasked with 'exploring possibilities of a military version'. It flew the circuit Moscow/ Tula/Ryazan or Moscow/Ryazan/ Kharkov as many times as possible, each time suffering problems. The usual crew was Gromov, navigator I G Spirin and radio operator A I Filin, and on 30 June 1934 they set a national record at 4,465km (2,774 miles) in

27hr 21min. They had left Shchyolkovo to drop one tonne of scrap metal (simulating a bomb load) on the Noginsk range, and hoped to fly much further, but landed at Kacha with fuel-feed problems. On 24 July, making the first take-off from the new 4km concrete runway with a starting downhill section, they flew 6,559km (4,075 miles) in 39hr 1min. On 10–12 September 1934 they orbited for 75hr 2min to cover 12,411km (7,712 miles) (also reported as 12,101km (7,519 miles)). This was the longest flight recorded by any ANT-25.

The *dubler* was then brought up as nearly as possible to the standard of the third aircraft, differing mainly in wing structure and tankage. Gromov fell ill, and new crews were organised under S A Levanyevskii and V P Chkalov. On 3 August 1935 it took off from the 4km runway, Levanyevskii's crew intending to fly to the USA, but oil leakage over the Barents Sea forced them to return, landing at Krechevits near Novgorod. Levanyevskii then declared that such transpolar flights should be attempted only by multi-engine aircraft (a result being the ANT-37).

A recent Sukhoi history states 'Interest in the RD as a record-breaking machine declined. But Chkalov was approached with the idea of making a transpolar flight in the RD, and Stalin approved'. After further minor modifications, including uprating the engine to 950hp, the *dubler* made a second attempt on 20 July 1936, crewed by V P Chkalov, G F Baidukov and A V Belyakov. They flew a kinked route to Petropavlovsk-Kamchatka and then set course for Khabarovsk. Severe icing, especially of

the propeller, forced a dangerous night landing on the island of Udd (later renamed for Chkalov) in the mouth of the Amur. The distance was put at 9,374km (5,825 miles) in 56hr 20min.

After further small changes, including the fitting of a CAHI-developed system of de-icing, including an alcohol slinger ring on the propeller, the *dubler* was flown by the same crew to Portland, Washington, on 18–20 June 1937. Though the great-circle distance was 8,504km (5,284 miles), the actual distance flown was about 10,000km (6,214 miles), in 63hr 25min. The aircraft returned by sea and was displayed at the 15th Paris airshow.

Span unchanged; length 11.57m (37ft 11½in); wing area (after fitting larger-chord ailerons) 88.2sq m (949sq ft).
Weight empty 3,784kg (8,342lb); maximum loaded weight 10,000kg (22,046lb).
Maximum speed 244km/h (152mph) at sea level; service ceiling (max wt) 2,100m (6,890ft), (lighter) 7,850m (25,750ft); range 10,800km (6,710 miles).

Histories of the ANT-25 cover the first two aircraft in detail, but say little about RD No. 3. The paucity of information is the more surprising when it is realised that it was the third which made the final long-range flight, gaining worldwide fame. The main reason is that it was in effect the prototype of the DB-1 bomber, and was therefore secret.

The airframe was generally unchanged apart from strengthening the main undercarriage and increasing tyre pressure, and fitting a larger rudder with the navigation light below instead of above the servo tab. Ahead of the firewall the M-34R engine remained rated at 950hp for take-off, but had been subjected to small modifications resulting from prolonged endurance testing. It drove a VISh propeller, the first to be developed in the Soviet Union, with three Duralumin blades controllable in pitch by the pilot; diameter remained 4.25m (13ft 11in). The radiator matrix, of a new type, was fixed in position under the engine with controllable inlet shutters and a hinged door at the rear of the duct.

In most respects the engine installation was similar to that of the ANT-20bis, though the exhaust pipes were led through a heat exchanger on each side of the cowling before escaping through six fishtails at the rear. Fresh air from a circular inlet at the front of each heat exchanger was passed along a metal duct along each side inside the fuselage from which hot air entered the cockpits. The fuel system was completely redesigned, and – again for the first time in the Soviet Union, if not in the world – the load-carrying structure of the wing was sealed between the spars to form two integral tanks, each with a length of 7m (23ft). This made it possible to carry additional fuel, and almost eliminated flutter problems by distributing mass across the span.

Following 1/25 scale model tests, a buoyancy system of inflatable bags was installed. A total of twenty rubberised-fabric bags were fitted along the leading edge, under the cockpits and in the rear fuselage. In emergency these could be inflated from a compressed-air cylinder. Both the front two cockpits had a set of pilot controls, in the hope that the pilot might be able to rest. Full radio was at last fitted, together with a direction-finding loop. This was mounted above the fuselage at the point where the radio wire antenna was anchored. Immediately ahead of this was added an aperture covered by a 'fighter-type' canopy which facilitated rotating the loop. Instead of a single venturi beside the pilot's cockpit a group of three were placed beside this dorsal canopy, linked to the duplicated air-driven instruments in the front two cockpits.

The aircraft was finally inspected to see where drag could be reduced, and the entire surface of the fabric-covered wings and horizontal tail doped red. This aircraft was used for only one

record flight. It left Shchyolkovo on 12 July 1937, flown by Gromov, A B Yumashyev and S A Danilin. After a relatively troublefree flight they landed at San Jacinto, California, after 62hr 17min. They had over 1,500kg of fuel still available, but had no authority to cross into Mexico and were not eager to strike east across the mountains. The distance was put at 10,148km great circle (6,306 miles, a new world record), or 11,500km (7,146 miles) actual. Tupolev regretted that the ANT-25 never did demonstrate its true range potential.

The history of P O Sukhoi's work states that 'After it was recognised that a transpolar flight required a twin-engined aircraft, Sukhoi was directed to modify [the ANT-25] as a bomber'). In fact, the possibility of the same design serving as both a record-breaker and as a bomber had been envisaged in the very first discussion between Tupolev and General Alksnis. Drawings of possible bomber versions proceeded in parallel with different forms of the RD at intervals from 1932, though an instruction to proceed was not received until early 1933, and actual construction of the prototype began in August of that year. The series aircraft was the ANT-36, which see.

Two other designers were brought in to assist in improving the ANT-25. Kalinin was concerned more with the ANT-36. The other helper was light-plane entrepreneur S A Moskalyev (pronounced 'Moskalyov'), who was assigned the considerable challenge of replacing the Mikulin petrol engine by the AN-1, the first of the diesel engines of A D Charomskii to be cleared for flight in a single-engined aircraft. The AN-1 was again a water-cooled V-12, but a two-stroke burning fuel oil. Bigger than the M-34, it had a capacity of 61.04 litres, weighed 1,050kg (2,315lb) and was rated at 900hp.

Later versions of similar bulk and weight gave over 1,500hp, but even the AN-1 had the advantage that its low fuel consumption soon overcame its greater installed weight, and at maximum takeoff weight the RDD (*Rekord Dalnosti Dieselnyi*) was expected to be able to fly 25,000km (15,534 miles) non-stop. In fact, no attempt was made to fit the maximum tankage, and the undercarriage was made non-retractable. Test flying the RDD, also called ANT-25D, began on 15 June

1936. Range, cruising at 5,500m (18,050ft) with 3,500kg (7,716lb) of fuel, was found to be 10,800km (6,711 miles). A retractable undercarriage would have extended this to about 12,000km (7,457 miles). The diesel engine was usually difficult to start. In 1940 Charomskii's later M-40 diesel powered the BOK-11, one of several BOK aircraft based on the ANT-25 airframe. An AN-1 was installed in at least one ANT-36.

Dimensions unchanged.
Weight empty, about 4,300kg (9,480lb); maximum take-off weight 11,000kg (24,250lb).
Maximum speed (light) 250km/h (155mph), (max weight) about 190km/h (118mph); cruising speed (long-range flights) 165km/h (102.5mph); service ceiling (maximum weight) about 2,000m (6,560ft); take-off run about 2,200m (7,200ft).

ANT-26, TB-6

Russians are supposed to have an innate love of bigness. Whether or not this is true, in the late 1920s Tupolev, Arkhangel'skii and Petlyakov undertook basic research (mainly on paper) into the limiting sizes of structure that could safely be made in Kol'chug alloy using their established methods, with corrugated skin. In general, their answers suggested that it would be possible to build an aeroplane with a span of 200m (656ft), though of course such a giant might not fit small grass airfields.

In 1929 CAHI agreed to go ahead with two sizes of intermediate aircraft. The first was the ANT-20, which was later enlarged as described earlier. The next was the ANT-26, with a span of 95m (311ft). This was planned to be developed as a super-heavy bomber, the TB-6, and also as the ANT-28 transport, described separately. Tupolev managed this challenging project personally. He was determined that, despite the long timescale, it should never appear obsolescent, and that it should be an overall success.

The configuration was the natural one of a cantilever monoplane with a relatively small fuselage. The structure followed the established form, with bolted joints enabling the whole airframe to be dismantled into manageable parts. The huge wing was of CAHI-6 profile, with a root

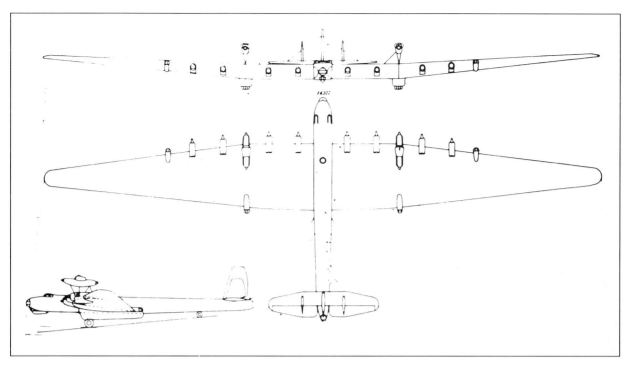

An original drawing of the ANT-26, TB-6 (via Yefim Gordon)

thickness/chord ratio of 20 per cent. Thus, the actual depth at the root was almost the same as that of the fuselage. The wing was made as an untapered horizontal centre section with a span of 26m (85ft 3½in) and equi-tapered outer panels with dihedral. For best range efficiency the aspect ratio was set at the remarkably high value of 12, exceeded only by the contemporary ANT-25.

The fuselage was similar in cross section to that of the ANT-20, though slightly longer. The cross-section was a square with rounded corners,

tapering aft of the wing in width and depth to a gun position aft of the tail. The nose was hemispherical and fully glazed with multiple flat panels. Underneath was a large gondola for the bomb aimer. A structural joint led to the next section, with two decks, the upper level having left and right pilot cockpits each with a 'bug-eye' canopy giving a view directly down on one side of the aircraft. The next section included the centre wing, with the bomb bays between the spars. Finally came the two portions of rear fuselage. Crew could walk from nose to tail, a

catwalk passing between the bomb racks, and through the wings between the fuel tanks.

The tail was surprisingly small, the horizontal surface (with a span of 18.8m (61ft 8in)) appearing tiny in relation to the wing. The monoplane tailplane was mounted at the top of the fuselage and was probably fixed in incidence. It carried tapering elevators which would have been driven by Flettner tabs. On the centreline was a fin and rudder, supplemented by smaller fins and rudders exactly 7m (23ft) apart, with single bracing-wires

The surviving model of the TB-6. (V Nemecek/RART)

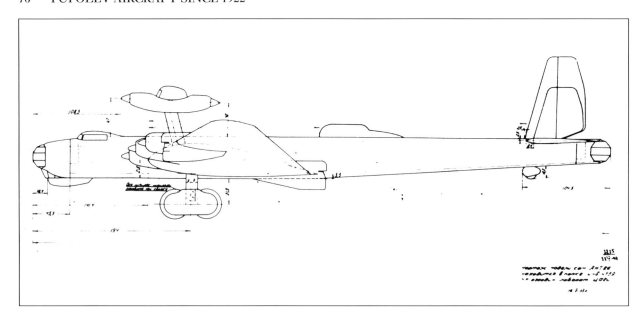

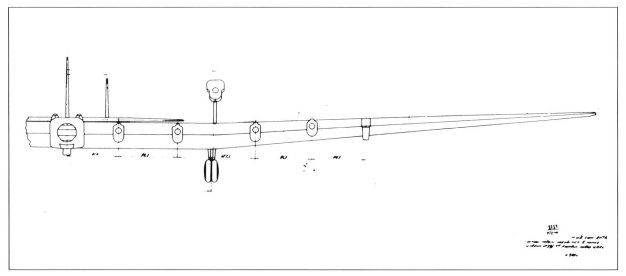

Original drawings of the ANT-26 flying scale model. (via Yefim Gordon)

from near the top of the centre fin passing through the auxiliary surfaces to meet the tailplane halfway to the tip. No information survives on the ailerons, apart from the fact they did not project beyond the outline of the wing.

Tupolev calculated that the giant bomber would need 10,000hp. He disliked tandem nacelles carried high up on struts, but after searching for engines in the 1,200–1,500hp class, and also studying pusher engines added on the trailing edge, he decided to space eight M-34 engines along the leading edge and add four more in two tandem nacelles each carried high above the joint between the centre and outer wings. By the time detail design was taking place in 1933 the 900hp

M-34FRN had become available, so the twelve engines exceeded the required power. The propellers of the upper tractor engines required the spacing of the leading-edge engines to be uneven. The inner propeller centrelines were 4.70m from the aircraft centreline, and the spacing was then 5.28m, 7m and 5m.

A three-view drawing of the ANT-26, dated December 1932, is reproduced here for the first time. The upper nacelles are seen to be carried on multiple struts broadly resembling those of the ANT-16, -20 and -27, though with a different arrangement in the side elevation. However, February 1933 drawings of the flying scale-model (mentioned later) shows each nacelle as carried on

a single built-up cantilever strut, sloping forward. If this was the intended production scheme for the bomber, each of these struts might have been wide enough to contain a staircase. There was, of course, inflight access to the leading-edge engines, and to the complete fuel system.

In the December 1932 drawing of the ANT-26 the main undercarriage can be seen to comprise four wheels with 1,700mm tyres all mounted on a common axle and able partially to retract into a giant 'bathtub' extending right across under the joint between the centre and outer wings. The tailwheel(s) can be seen no less than 9.55m ahead of the tail end of the fuselage. In the drawings of the flying scale-model each main undercarriage is shown as a faired vertical shock strut carrying a four-wheel bogie truck with tyres which, scaled to the full-size aircraft, would have been 2,000mm in diameter. The tailwheel is shown spatted and much further aft.

Bombload was to be 24.6 tonnes (54,233lb), carried in three inter-spar bays. The normal crew was to be twelve, seven of whom were to be gunners. One document lists the guns as one 37mm, four 20mm and four pairs of 7.62mm ShKAS. All drawings show gunner's cockpits with glazed cupolas on the outer leading edge 4.84m beyond the centreline of the outer engines, and identical cockpits behind the trailing edge at the centre/outer-wing junction. In the December 1932 drawing these formed the aft end of the giant 'bathtub', whereas in the scale-model they were completely separate structures. The 1932 drawing shows a Tur gun ring above the centre fuselage and in the tail, plus waist guns just behind the trailing edge. The later scale-model aircraft had a large dorsal 'doghouse', and the tail had an enclosed turret.

CAHI tunnel testing was satisfactory, and in 1935 B N Kudrin flew the scale-model, though no photographs of it have been found. It was a two-seat glider of 17.27m (56ft 8in) span, which faithfully duplicated the aerodynamics and flight-controls of the giant TB-6 bomber. We are left pondering on where the pilot and observer sat in this model's 7m (23ft) fuselage; to get the centre of gravity in the right place, the 'doghouse' could have been a fairing for the head of the observer, balancing the weight of the pilot in the nose (who of course would have been relatively like a giant in comparison with the bugeye canopies). Be that as it may, in 1935, the whole programme was discontinued, mainly on grounds of cost/effectiveness and vulnerability. Tupolev concurred with this decision.

Span 95m (311ft 8in); length 38.8m (127ft 3⅜in); wing area 754sq m (8,116sq ft).

Weight empty 48,600kg (107,143lb); loaded weight 70,000kg (154,321lb).

Maximum speed (est) 300km/h (186mph); service ceiling 5,500m (18,050ft); range with maximum bomb load 2,000km (1,242 miles).

ANT-27, MDR-4

This long-range flying-boat had one of the most chequered careers of any Soviet aircraft. It began life in 1931 as the MDR-3, the designation meaning *Morskoi Dalnyii Razvyedchik*, marine long-range reconnaissance. The Soviet Union urgently needed such an aircraft, and it was the first major project of I V Chyetverikov, who at the age of 27 was appointed head of the Marine Brigade of the Central Construction [ie, design] Bureau. His inexperience led him wisely to use as much existing design as possible.

For example, the hull was almost a direct scale of the Grigorovich ROM-2, though with the overall depth increased so that the monoplane wing could be mounted directly on it, and with the planing bottom modified to have a curved instead of a vertical step, upstream of which the central keel had an auxiliary keel on each side. The complete wing was constructed from the drawings of another Grigorovich design, the TB-5 heavy bomber, though with a corrugated-skin centre section carrying engines above instead of below and with stabilising floats further outboard. The tail was also based on that of the TB-5, though with the twin fins and rudders redesigned.

Altogether the MDR-3 was a workmanlike all-metal flying-boat, powered by two tandem pairs of 680hp BMW VI engines and carrying a crew of six, with up to eight machine-guns and 500kg of bombs. The single MDR-3 was first flown at Sevastopol on 14 January 1932. In many ways it was highly successful, but – even without armament, and despite the considerable installed power – its performance was unacceptably poor. For example, the take-off time on smooth water was 36 seconds, the time to reach 2,000m (6,560ft) was 40min, and the service ceiling was only just above this level. GUAP (the central management of the aviation industry) lacked confidence in Chyetverikov, and accordingly, in late 1932 the MDR-3 was transferred to KOSOS, Tupolev being instructed to improve it.

Tupolev studied the drawings carefully, but left the assignment to the head of the marine-aircraft brigade,

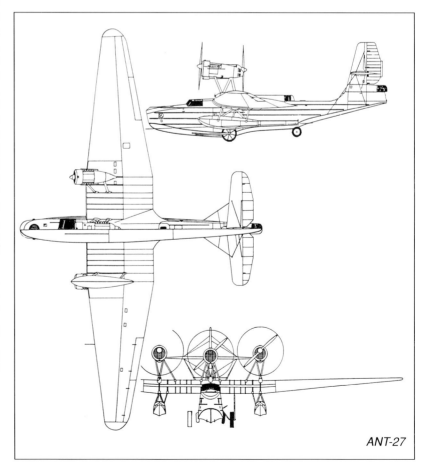

ANT-27

The prototype ANT-27, MTB-1. (G F Petrov/RART)

I I Pogosskii, who had led the design of such aircraft as the ANT-8 and -22. Under Pogosskii a totally different aircraft emerged, with little of the original retained apart from the planing bottom and most of the inner structure of the hull. Even the hull was modified, the second of the tandem bow cockpits being skinned over, the dual pilot cockpit being altered, and the upper aft portion being rebuilt to have a fairing behind the wing leading to a Tur-9 enclosed gun turret, with a second turret at the extreme tail. The new hull was reskinned with sheet stiffened by longitudinal half-round beads, in the Pogosskii manner. The stabilising floats were redesigned, with the step further forward, but carried on the same struts. KOSOS designed a beaching chassis comprising four separate units, each with a single wheel with a flat-tread solid-rubber tyre attached to sockets in the hull, the main units having a strut pinned under the third spar of the wing and a fore/aft bracing-strut attached to the keel under the cockpit.

Apart from having unchanged root dimensions, to bolt on the original centre-section root ribs, the wing was entirely new, and considerably greater in span and area. For the rectangular centre section traditional structure was adhered to, with projecting A-profile ribs and 40mm by 8mm corrugated skin applied in strips between

them, but the strange decision was taken to use fabric covering over the tapered outer wings. The single vertical tail was almost identical to those used on the ANT-22, and at the same high level the fin carried the variable-incidence tailplane, with wire bracing in line with the pivot and altered in angle by the rear wires. The rudder had fine corrugations and inset hinges, being driven by a Flettner tab added behind the trailing edge. Its movement was allowed by cutouts at the root of each elevator, with six projecting A-type ribs on each side.

The propulsion was completely redesigned. Instead of four 680hp engines the other brother, Ye I Pogosskii, selected three 830hp Mikulin AM-34R geared V-12s, two tractor and one pusher. This arrangement involved moving the outer engines further out from the centreline, and they were installed in streamlined nacelles on forward-raked struts attached to the same A-type rib as the stabilising floats. Each nacelle was braced laterally by two struts on the inner side. The central nacelle was carried on three sets of ∧-struts. the middle pair sloping to provide fore/aft bracing. Unlike the tractor engines, which were set at an angle of +5 deg, the centre engine was horizontal. All three were geared to a two-blade fixed-pitch propeller of 4m diameter, and were cooled by a circular radiator on the

front of the nacelle. Each nacelle had a self-contained oil system complete with tank and radiator. Fuel was housed in multiple tanks between the ribs of the centre section.

The ANT-27 was designated as the MDR-4, and was eagerly awaited by the *Morskaya Aviatsiya* (naval air force). The sole prototype was completed in March 1934, and on 16 (often reported as the 15th) April it suffered a catastrophic failure during take-off in a choppy sea. The struts supporting the central nacelle failed, and the nacelle rotated down on to the cockpit, killing pilot Ivanov and also I I Pogosskii who was beside him.

Span 39.4m (129ft 3⅛in); length 21.9m (71ft 10in); wing area 177.5sq m (1,911sq ft).
Weight empty 10,500kg (23,148lb); fuel/oil 2,450kg + 196kg (5,401lb + 432lb); loaded weight 14,382kg (31,706lb), maximum weight 14,660kg (32,319lb).
Maximum speed at sea level 233km/h (145mph); cruising speed 170km/h (106mph); climb to 3,000m (10,000ft) 13.25min (18min at 14,660kg); service ceiling 5,450m (17,880ft); range 2,130km (1,324 miles) [2,215km at 14,660kg].

So important was the programme that a second prototype, the ANT-27bis, was quickly ordered. Designated MTB-1 (*Morskoi Torpedonosyets-Bombardirovshchik*), it differed in numerous details, most notably in having 930hp M-34RN engines, increased fuel capacity and full

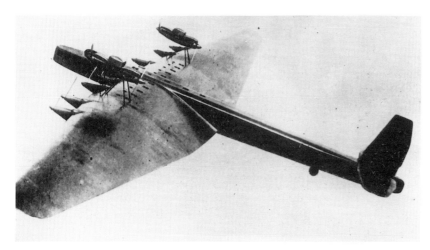

The surviving model of the ANT-28. (M Passingham/RART)

armament. Defensive armament was intended to be twin ShKAS in the nose, an Oerlikon 20mm cannon in the Tur-9 dorsal turret and twin ShKAS in the tail. Thanks to the elimination of the previous underwing bracing struts the wing racks could now carry a torpedo on each side (with reduced fuel). Alternatively the load could comprise bombs or mines up to a maximum of 2,000kg (4,409lb).

The ANT-27bis was completed at GAZ No. 31 at Taganrog in January 1935. Generally judged to be a satisfactory aircraft, a series of fifteen were ordered from GAZ No. 31 in summer 1935, but the ANT-27bis suffered an accident as serious as its predecessor. The fabric separated from part of the wing, and in the emergency forced alighting it hit a rock, crashed and burned. Despite this, the production order stood, five being delivered in 1936 and the other ten in 1937. They had an uneventful service career, used mainly in the reconnaissance role, some continuing in use during the Great Patriotic War. Shavrov states that a transport version was projected as the ANT-29, but nothing on this has been discovered and the number was used for a completely different aircraft.

Dimensions, unchanged.
Weight empty, as built 10,521kg (23,194lb); fuel and oil 3,746kg+370kg (8,258lb+816lb); loaded weight 16,250kg (35,825lb).
Maximum speed 225km/h (140mph) at sea level, 266km/h (165mph) at 3,000m; cruising speed 185km/h (115mph); climb not stated (VB Shavrov cites 25min to 3,000m with M-34R engines, but these were not installed); service ceiling 6,550m (21,490ft); range 3,480km (2,160 miles).

ANT-28

Never built, this would have been the transport version of the ANT-26. It was always principally a military project, though if suitable airfields were available it could also have transformed the opening up of Siberia and other unexploited regions of the USSR. Like the bomber, the project was led by Petlyakov. The airframe was to be based on that of the bomber, with a derived wing and almost the same gross weight, but the fuselage would have had much in common with the ANT-20, with a similar nose, two rows

of five cabins inside each wing, each providing accomodation at two levels, and an unobstructed interior accessed by four stairways to the ground. The bulkiest or heaviest loads would have been slung externally. Loads would have included 250 troops or 25 tonnes of cargo 'including tanks'.

Unlike the bomber the four-wheeled main undercarriages would have been non-retractable, and the tail would have had a single fin. No armament was specified. Several basic decisions had not been taken when in 1936 the ANT-28 was discontinued together with the ANT-26. The surviving model shows that the eight wing engines would have been mounted internally, driving via long shafts. Basic data would have been broadly similar to those of the ANT-26, though maximum weight and fuel capacity would been less. Range with a 25-tonne load was estimated at 3,600km (2,237 miles). Tupolev later considered that, whereas it was correct to cancel the ANT-26, the ANT-28 could have been a worthwhile programme.

ANT-29, DIP

The obvious outcome of the ANT-21 programme was the DIP (*Dvukhmestnyi Istrebitel Pushechnyi*), two-seat

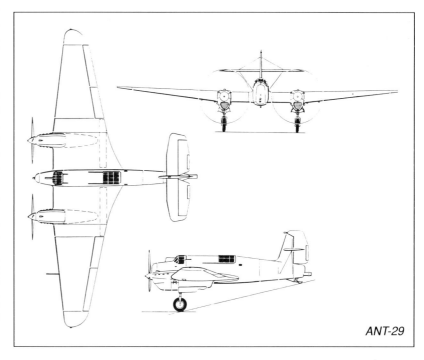

ANT-29

One of the few photographs of the ANT-29. (V Nemecek/RART)

fighter with cannon. The design was entrusted to P O Sukhoi, who produced an outwardly excellent aircraft which for the first time was entirely covered with smooth stressed skin, most of the exterior being flush-riveted. The other new feature was that the main armament was based on Kurchevskii's recoilless cannon, which had been tried in the ANT-23 but in the ANT-29 were less-riskily installed in the fuselage.

The wing was aerodynamically based on that of the ANT-21 *dubler*, though on a slightly reduced scale and with totally different structure. The wing was again made as a horizontal centre section, with the trailing edge curved back at the root into a large fillet, and outer panels with dihedral (mainly due to taper in thickness) and tips of small radius. There were three main spars from tip to tip, the first and third being at 16 and 66 per cent chord and forming the structural box. Like the ribs the spars were a mixture of riveted trusses built up from standard profiles and sheet pressings with flanged lightening holes, the load-bearing skin being flush-riveted in tapered strips. As before, there were split flaps under the centre section and long-span ailerons divided into inner and outer sections, but the ailerons were now fully balanced, did not project behind the wing and had no tabs.

The fuselage was a semi-monocoque of oval section, firmly establishing Sukhoi as a master of the new form of stressed-skin construction. As before, the pilot sat over the leading edge and the gunner over the trailing edge. The latter would have worked radio had it been fitted, but his additional duty in the DIP was to ensure proper functioning of the big guns. The pilot's enclosed cockpit, which precluded any rear view, had a small glazed roof which slid to the rear for access. The gunner's much larger canopy slid forwards on side rails as a single unit.

The tail was of typically severe angular outline. The fin carried a rudder with a large upper horn with a projecting mass balance, driven by a Flettner tab. High on the fin was pivoted the tailplane, with incidence adjusted by the single upper and vee lower bracing wires. Inset into it, stopping well short of the tips, were the rectangular elevators with inner-end cutouts for the rudder and driven by trailing-edge Flettner tabs.

The engines, completely unrelated to those of the ANT-21s, were Hispano-Suiza 12Ybrs V-12s each rated at 760hp, and later to be made under licence as the M-100. These were quite light for their power, and were to be adopted for the very important SB (ANT-40). Their installation drew heavily on that of the third RD, with the coolant radiator directly underneath in a duct with adjustable shutters at the sloping inlet and a hinged flap at the exit. Carburettor air was rammed in through a small inlet on each side under the leading edge. The simple stub exhausts followed the usual Hispano 1–2–2–1 style. No suitable propellers were available, so French Chauvière 350 series were purchased, each with three Dural variable-pitch blades with a diameter of 3.5m (138in).

The undercarriage was little changed from that of the ANT-21 *dubler*, even a tailskid being retained. Operation of the single-strut main units was hydraulic, a fork holding the wheel with 900mm by 280mm tyre, part of which projected below the two bay doors to reduce damage in the event of a wheels-up landing. The two doors were pulled shut by simple links to the legs. The shock-absorbing tailskid was pivoted under the tail end of the fuselage, with a small fairing upstream and a steel shoe. The pitot head was well outboard on the port wing.

The unusual armament was not installed during early test flights. It comprised two APK-8 recoilless guns, each of 102mm (4in) calibre, mounted one above the other on the lower centreline of the fuselage. The barrel of each gun was about 4m (157in) long, so that the backseater could check on the reloading of successive rounds. Each gun was arranged so that, before firing, it was linked by a gas-tight seal to the recoil tube which extended beyond the tail of the fuselage to a point at which the violent blast could not interfere with the tail or with control of the aircraft.

The pilot sighted the guns with an optical sight faired in a prominent narrow box ahead of the windscreen. Sighting could also be assisted by firing tracer from two ShKAS fixed in the wing roots. A third ShKAS could be aimed defensively by the gunner.

There was no provision for a bomb load.

Priority on the SB programme delayed the DIP severely. Not until winter 1935 could it be prepared for flight, and the very first test flight, by S A Korzinshchikov, showed that the flight controls were unacceptably ineffective. He reported that the rudder and ailerons were particularly inadequate, and the result was a return of the prototype to ZOK to have all control surfaces reskinned. Flight testing was then briefly continued, while the whole concept of Kurchevskii's big recoilless guns was debated. Though it was suggested they should be replaced by ordinary 20mm guns, there was one further KOSOS prototype with APK armament, the ANT-46. The ANT-29 was abandoned in March 1936.

Span 19.19m (62ft 11½in); length (overall) 11.1m (36ft 5in), (excl APK) 10.2m (33ft 5⅜in); wing area 56.88sq m (612sq ft).
Weight empty 3,876kg (8,545lb); normal loaded weight 5,300kg (11,684lb).
Maximum speed 296km/h (184mph) at sea level, 352km/h (219mph) at 4,000m (13,125ft); climb 5.6min to 3,000m (10,000ft), 9.6min to 5,000m (16,400ft); no other reliable data.

ANT-30, SK-1

This unbuilt project dating from January 1930 was for a long-range fighter, to be powered by two M-38 engines. The meaning of its designation might have been 'fast cruiser'. An obvious reason for its non-appearance is that the engine, a 575/600hp nine-cylinder radial unrelated to Mikulin's later AM-38, was never qualified for use.

ANT-31, I-14

With this experimental fighter, designed to VVS-RKKA order from mid-1932, P O Sukhoi's brigade went all-out to create the most modern and innovative fighter in the world. Among its new features were streamlined aerodynamics, a cantilever monoplane wing, flush-riveted stressed-skin construction without corrugations (except on the tail control surfaces), fully retractable mainwheels with brakes, an

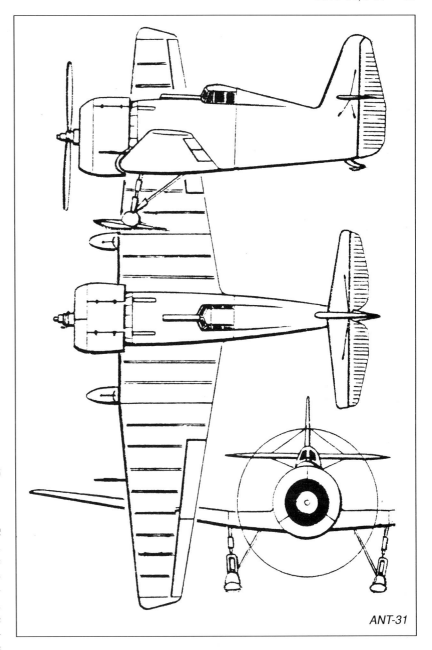

ANT-31

enclosed cockpit, an engine enclosed in a long-chord cowling and advanced armament. At the outset this appeared likely to be the superior single-seat fighter the Soviet Union so urgently needed.

The wing had relatively long span, almost equal taper and a NACA 16 per cent profile. Structurally it comprised a wide centre section with slight dihedral to which were bolted outer panels with more pronounced dihedral. The two spars from tip to tip had tubular KhGSA booms and riveted KhMA truss webs. The centre section had three A-type ribs on each side and each outer panel ten, all riveted from standard sections. Smooth skin 0.8mm or 1mm thick was riveted in strips

between the ribs. The long-span slotted ailerons were each made in inner and outer sections, without tabs.

In comparison, the fuselage appeared stumpy in the extreme, and tapered sharply in width from the circular section at the engine to the width of the fin at the stern frame carrying the rudder. It was made in front and rear sections, the joint being at the rear of the cockpit, the structure being a modern semi-monocoque similar to the ANT-29. All skin was 0.8mm flush-riveted, the tall fin being integral. On the front was the steel firewall, through which extended the four main KhMA longerons to carry the imported Bristol Mercury IVS2 engine, rated at 500hp. Exhaust was

The ANT-31. (A Aleksandrov/RART)

collected by a ring on the front and discharged at the bottom, and the propeller was a wooden two-blade unit of 2.8m (110in) diameter. Immediately behind the firewall were the oil tank and 260 litre (57gal) fuel tank.

The fin carried the tailplane well above the level of the fuselage, its incidence being controllable by the usual arrangement of bracing wires, one above and a vee below on each side. The elevators and rudder, of typical angular shape, were skinned in 20mm-pitch corrugated sheet, and though the hinges were inset there were no horns or tabs. The neat undercarriage comprised a bungee-sprung tailskid and single-leg main units pivoted to the outermost rib of the centre section and braced by a diagonal strut to the rear, the whole unit being retracted directly inwards by cables from a cockpit hand-wheel so that the wheels with 700mm by 150mm tyres were housed almost touching each other inside the wing.

The cockpit was above the trailing edge, with a glazed hinged canopy. In front of the windscreen was an optical sight tube in a narrow fairing similar to that of the ANT-29. Armament was to be one 7.62mm PV-1 machine-gun above the nose, used to assist sighting, and two APK-11 recoilless cannon hung under the wings outboard of the propeller disc, each fed with fifty rounds of 45mm ammunition. An alternative fit was to be two 20mm ShVAK guns, or two synchronised PV-1 and four D-1 light-bomb containers.

The prototype was completed at ZOK in May 1933 without armament and finished in unpainted polished metal, riding on three fixed skis. It was first flown on 27 May by K A Popov, the test programme being directed by leading engineer A Kozlov. A 1985 history of Sukhoi states that the first flight was made by B L Bukhgol'ts, 'in the presence of the Directors of CAHI and ZOK and of Tupolev', while another source gives the first-flight date as 17 June. In general, handling was excellent, though V B Shavrov reported 'some aspects of control were difficult', another source commenting that in a tight turn the horizontal tail was in the wake of the wing (which is unlikely to have been true).

Span 11.2m (36ft 9in); length 6.095m (19ft 11⅞in); wing area 16.85sq m (181.4sq ft).
Weight empty 1,101kg (2,427lb); fuel/oil 187kg (412lb); loaded weight 1,455kg (3,208lb).
Maximum speed 316km/h (196mph) at sea level, 389km/h (242mph) at 4,550m (14,930ft); climb to 3,000m in 4.43min and to 5,000m in 8.03min, service ceiling 9,400m (30,840ft); take-off run 120m (394ft), landing speed 110km/h (68mph), landing run 260m (853ft).

ANT-31bis Series production appeared likely, and in August 1933 work began on a second prototype. This was powered by an imported Wright Cyclone F-2, driving a Hamilton Standard metal propeller. Though this engine was larger in diameter than the original, it was more powerful at 640/710hp, and in any case the real reason for its choice was that the same engine was to be available from licence production as the M-25. An important change in the second aircraft was that the pilot was seated in an open cockpit, though with the windscreen extended back to form the front half of a very narrow enclosing canopy. Pilots

The ANT-31bis. (via Philip Jarrett)

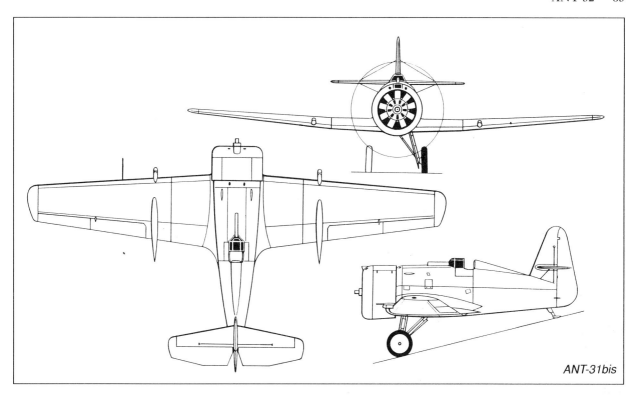

ANT-31bis

preferred to be able to lean out and, especially, to look to the rear. Other changes were the fitting of elevators with horns, and a rudder with a curved outline giving slightly greater area (as had been retrofitted to the ANT-31) and complete redesign of the under-carriage. Now the legs and bracing struts were pivoted to the main lower fuselage longerons and retracted out-wards instead of inwards. Even though the legs were splayed apart, the track was narrow. The bis was first flown in mid-February 1934. From March to May 1934 it was subjected to factory testing by K K Kokkinaki, A P Chernyavsky and I P Belozerov.

The ANT-31bis was later fitted with armament, two PV-1s above the fuse-lage and two APK-11 recoilless cannon of 45mm calibre, plus racks for four AO-10 bombs of 10kg each. Testing of the guns was handled by P M Ste-fanovskii. In autumn 1934 the APK-11s were replaced by two APK-37s, of 75mm calibre (obviously with a much smaller number of rounds), and the engine was changed for a Cyclone F-3 of similar power. NII-VVS trials were handled by A I Filin, whose report stated that the 31bis was 'equal to the best foreign aircraft' in speed at 5,000m, and superior at 1,000–3,000m, with greater firepower, but 'slightly inferior in ceiling and rate of climb'. Production was recommended.

Span unchanged (11.2m); length 6.1m (20ft 0⅛in); wing area 16.93sq m (182sq ft).
Weight empty 1,169kg (2,577lb); fuel/oil 195kg; loaded weight (PV-1) 1,483kg (3,269lb), (F-3 and APK-11) 1,527kg (3,366lb).
Maximum speed 358km/h (222.5mph) at sea level (352km/h with F-3 engine), 372km/h (231mph) at 1,050m (with F-3 engine, 402km/h [250mph] at 3,000m); climb 3.9min to 3,000m, 7.9min (8.7min with F-3 engine) to 5,000m service ceiling 8,800m (28,870ft) (8,500m with F-3 engine); range 600km (373miles); take-off run 175m (582ft); landing run 320m (1,065ft).

I-14 In late 1934 a series of fifty-five aircraft based on the ANT-31 *dubler* were ordered from the newly built GAZ No. 125 at Irkutsk. Again there are confused reports, various arma-ment schemes being cited and the engines said to be F-3 Cyclones. In fact the standard armament was either two or, normally, four ShKAS, the recoil-less cannon having by this time been abandoned. The engines were licensed M-25s, each similar to the F-3 and with the same 710hp take-off rating, driving a two-blade VISh (variable-pitch propeller) produced under Hamilton licence. The expected large orders went to Polikarpov's I-16, almost entirely because that aircraft used a much smaller amount of aluminium alloys.

Even the modest production run of I-14s was never completed. GAZ No. 125 delivered four in 1936 and

fourteen in 1937, plus a further four from another new factory, GAZ No. 153 at Novosibirsk. Acceptance tests at Irkutsk were made by E Yu Preman. A total of twenty I-14s was issued to a VVS *Polk* (regiment) in 1937, and except for the narrow track proved generally popular. Their time to fly a 360 deg circle of 6.5 seconds compared with 6.1sec for an I-15 biplane, 6.2–9.2sec for I-16 versions, 11.4sec for the two-seat biplane DI-6 and 11.4sec for the APK-armed I-Z monoplane. So far as is known, none saw action.

Dimensions, as 31bis.
Weight empty 1,170kg (2,579lb); fuel/oil 200kg (441lb); loaded weight 1,540kg (3,395lb).
Maximum speed 375km/h (233mph) at sea level, 449km/h (279mph) at 3,400m; climb 3min to 3,000m, 6m to 5,000m; service ceiling 9,420m (30,900ft); range 615km (382 miles); take-off run 220m (722ft); landing speed 110km/h (68mph), landing run 320m (1,065ft).

ANT-32

Dating from 1934, this project was for a single-seat fighter. It has been suggested that it would have had con-ventional cannon armament.

ANT-33, ANT-34

No reliable information has been found on these numbers, though in 1980 it was reported (not by the Tupolev archivist) that the ANT-34 was to have been a smooth-skinned development of the ANT-29, powered by Wright Cyclone F-2 engines and armed with conventional guns.

ANT-35

In May 1934 Aviavnito, the aviation section of Vnito, the USSR's national scientific and technical organisation, organised a competition for two sizes of civil airliner. By this time KOSOS had abandoned traditional construction using non-load-carrying corrugated skins, and had become a world-class exponent of streamlined stressed-skin structures. Tupolev had already assigned to AA Arkhangel'-skii's brigade the ANT-40 (SB) high-speed bomber, which was to take precedence over all other projects. He additionally charged Arkhangel'skii with producing a civil passenger version to meet the larger of the Avia-vnito requirements.

The payload of bombers occupies only a fraction of the volume required by the same weight of passengers. Douglas showed, in the DC-2/3 and B-18 bomber, that it is possible to derive a satisfactory mass-produced

Interior of a PS-35 cabin. (G F Petrov/ RART)

bomber from an established passenger transport. The reverse operation tends to result in a transport aircraft with high performance but extremely cramped passenger accommodation, and so it proved in this case. At the same time, while the Soviet Union had very little fare-paying air traffic, either internally or internationally, it was keenly interested to show off its ability to build modern high-speed aircraft.

In the event, apart from the undercarriage, few parts of the ANT-40 were actually used in the transport, though the wings were closely related. They shared the same CAHI-6mod profile, tapering from a thickness/chord ratio of 16 per cent at the end of the centre section to 12.5 per cent at the tip.

Structurally the wing comprised a horizontal centre section and tapered outer panels with 7 deg dihedral. The two main spars had KhGSA tubular booms, the webs and ribs all being built up from standard light-alloy sections. Skin thickness was 0.6–1mm on the centre section, extending just beyond the nacelles, and 0.5–0.6mm on the outer wings, the riveting round the leading edge to the front spar being flush. On the trailing edges out to the ailerons were split flaps, driven electrically to 65 deg for landing. The slotted and balanced ailerons were inset and constructed in two parts, the slightly larger inner sections having tabs.

The fuselage was a semi-monocoque structure of oval section, the width and height being only 1.5m and 2m (59in by 79in). The cockpit was fully equipped for two pilots, no separate navigator or radio operator being carried. The main cabin could accommodate ten single seats, each next to a window. The low headroom of 1.68m (66in) did not make it any easier to step over the main spars passing under the second and third seat rows. Each passenger had an electric light and an outlet for fresh air, which could be heated by passing over the engine exhausts, though there was no airframe de-icing. At the rear was the main door on the left, a toilet, and a baggage compartment accessed either from the cabin or by a separate hatch on the left side of the fuselage.

The ANT-35 prototype. (G F Petrov/ RART)

The tail showed no trace of the severe angularity that had been a 'Tupolev trademark'. The fin was built integral with the fuselage, and carried a large rudder with a small horn-balance, inset hinges and a Flettner tab. The horizontal tail was of very low aspect ratio and had a sharp taper. The tailplane was mounted at mid-level on the fuselage, with fixed incidence, and carried elevators (entirely below the level of the rudder) with inset hinges and tabs. Again, leading-edge riveting was flush.

The engines were among the first M-85s to be produced in the Soviet Union, under Gnome-Rhône K14 Mistral Major licence. Rated at 800hp, they were installed in long-chord NACA cowls with adjustable gills, driving propellers of 3.2m (10ft 6in) diameter with three ground-adjustable Dural blades. In winter spinners could be fitted, together with a baffle at the front of each engine with pilot-adjustable radial apertures. Fuel and oil capacity totalled 690kg (1,521lb), indicating a fuel capacity of about 850 litres (186gal).

The main undercarriage was almost identical to that of the ANT-40. Each unit comprised an air/oil shock strut carrying at its lower end a welded KhMA fork for a wheel with drum brake and a 1,000mm by 300mm tyre. As in the ANT-25, the leg was laterally braced by diagonal side struts, all three being pivoted under the front spar. An electrically signalled hydraulic jack, forming the upper part of the rear drag strut, pulled the unit up into the nacelle, mechanical links from the leg closing the two doors. Part of the retracted tyre remained visible. The castoring tailwheel had a balloon tyre and was non-retractable. Equipment included one of the first autopilots in the Soviet Union, linked into the rod-driven elevator circuit and rod/cable ailerons. The pitot head was on the nose, the landing light in the leading edge of each wing, and a direction-finding loop aerial was in front of the fin.

Construction of the ANT-35 prototype went ahead as soon as the ANT-40 had been ordered into production. Few problems were encountered, and M M Gromov made a successful first flight on 20 August 1936. On 15 September, bearing registration SSSR-NO35, it flew to Leningrad and back, 1,266km (787 miles) in 3hr 36min. In November/December it was displayed at the Paris Salon, where John Stroud felt he 'was not alone in considering the workmanship of poor quality'. It was, however, with the Polikarpov I-17, the first modern stressed-skin Soviet aircraft to be seen by the rest of the world.

Span 20.8m (68ft 2⅞in); length 14.95m (49ft 5⅞in); wing area 57.8sq m (622 sq ft).
Weight empty 4,710kg (10,384lb); fuel/oil 690kg (1,521lb), loaded weight 6,620kg (14,594lb).
Maximum speed, [CAHI figure] 390km/h (242mph), [Shavrov figures] 350km/h (217.5mph) at sea level, 376km/h (234mph) at 4,000m; cruising speed [CAHI] 350km/h (217.5mph); service ceiling, 8,500m (27,890ft); range 1,300km (808 miles); take-off run 200m (656ft); landing speed 95km/h (59mph), landing run 250m (820ft).

ANT-35bis, PS-35 Clearly a fast and successful flying machine, the ANT-35 was flawed on account of its cramped cabin. Authorisation was given to build a second (*dubler*) aircraft with a fuselage cross section 0.15m (6in) deeper, and also slightly longer. This was a great improvement, the headroom being adequate and the wing spars making only small ridges in the floor. Passenger capacity remained ten, but the payload was increased to 1,050kg (2,315lb).

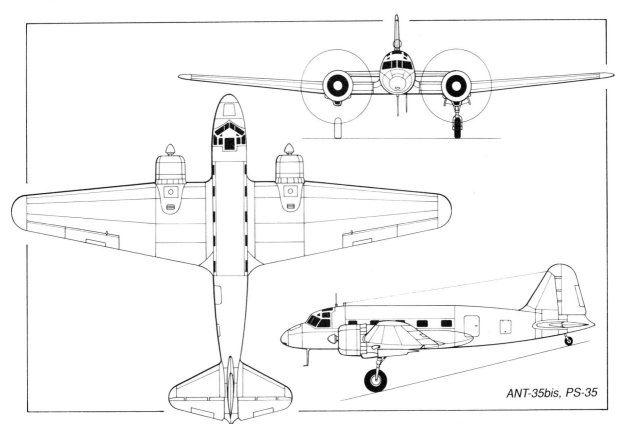

ANT-35bis, PS-35

A series PS-35. (B Vdozenko [photographer] via G F Petrov/RART)

The engines were changed to the M-62IR, the licensed version of the Wright Cyclone, rated at 1,000hp for take-off. This engine required a cowling of slightly increased diameter. As before, the carburettor inlets were recessed into the top of the cowling, but now they were larger. The propellers were changed to VISh-2 (licensed Hamilton Standard) two-position bracket type, with three blades of unchanged diameter.

The capacity of the pair of tanks in the centre section was increased to 990 litres (218gal). Other changes included the provision of an emergency escape hatch in the roof of the cockpit, relocation of the D/F loop immediately aft of this hatch, just in front of the lead-in of the HF radio wire antenna strung to the top of the fin, and of the pitot head to the tip of a long post under the nose.

Flight testing enabled the improved aircraft to be cleared to higher weights, with the main undercarriage strengthened, and for the centre of gravity range to be opened out to 24.5–28.5 per cent of mean aerodynamic chord. The ANT-35bis was then accepted for service with Aeroflot as the PS-35 (passenger aeroplane 35), and nine series aircraft were delivered in 1939. By this time it had been agreed that the Douglas DC-3, with the same engines, would be a much more useful aircraft, built in large numbers as the Li-2.

The PS-35s served on routes to Riga/Stockholm and later to Prague, Leningrad, Lvov, Simferopol and Odessa. After June 1941 survivors were used to support the battlefront. Data are for the series aircraft.

Span and wing area unchanged; length 15.4m (50ft 6¼in).

Weight empty 5,012kg (11,049lb); fuel/oil 710kg+90kg (1,565lb + 198lb); loaded weight 7,000kg (15,432lb).

Maximum speed, 350km/h (217mph) at sea level, 372km/h (231mph) at 1,500m; cruising speed 346km/h (215mph); range 920km (572 miles); climb 6.1min to 3,000m, 13min to 5,000m; service ceiling 7,200m (23,620ft); take-off run 10sec, 225m (738ft); landing speed 105km/h (65mph), landing run 300m (984ft).

ANT-36, RD-VV, DB-1

Production of the military variant of the ANT-25 was ordered in May 1933. ANT-36 was the KOSOS designation; RD-VV signified 'record distance, military variant', and DB-1 indicated 'long-range bomber Type 1'. Building these bombers was the first task assigned to the newly built GAZ No. 18 at Voronezh. The initial order was for twenty-four aircraft, but manufacture was stopped in May 1936 at the twentieth series aircraft, mainly because the DB-2 was expected to take the DB-1's place.

A major design change was that the ANT-36 was covered in smooth stressed skin throughout. As noted in the ANT-25 entry the bomber had a crew of three, a bomb bay amidships for ten FAB-100 bombs carried nose-up (total weight in practice about 2,400lb), though four was a more usual load, and two 7.62mm DA machine-guns, one firing up to the rear and the

other down and to the rear. Various camera fits were possible, filling many pages of the handbook, in most cases replacing the bomb load. In every configuration a single AFA-14 was retained in the rear cockpit. Fuel capacity was reduced to 4,900 litres (1,078 gal), gross weight being 7,806kg (17,209lb).

As early as 1933 it was recognised that this aircraft would be vulnerable to modern defences, and it was intended to be used – if possible – where an enemy did not expect to encounter hostile aircraft. A rival constructor, K A Kalinin, was assigned the task of modernizing the basic aircraft, while S A Moskalyov was charged with changing the engine to the AN-1 diesel (*see* ANT-25, RDD). V A Chizhevskii's specialist high-altitude BOK-1, BOK-7 and BOK-11 were all constructed separately, their design being based on that of the RD.

The prototype was flown by M M Gromov in 1934. Of the twenty series aircraft, two were retained for technical development, two were scrapped unused, three were assigned to CAHI, one to Leningrad and two to NII-VVS. The remaining ten became operational with a front-line *polk* (regiment) near the Voronezh factory. Despite their long endurance they flew an average of only 25–30hr in 1936–37, though some exceeded 60 hours. Documents suggest that the DB-1 was regarded as a failure, not capable of being sufficiently updated to survive in warfare. DB-1s in service had the M-34R engine, data being very similar to later ANT-25s.

Though Moskalyov stated that the airframe selected to be powered by the first flight-cleared Charomskii diesel (AN-1) was an ANT-25, documents recently discovered show that it was in fact the last of seven ANT-36 bombers refused by the VVS, factory number 1818 (GAZ No. 18, eighteenth aircraft). Some details of the installation are given in the ANT-25 entry. The newly unearthed papers describe the aircraft as riddled with problems, some caused by the shift in centre of gravity. The undercarriage had to be made non-retractable. It first flew on 15 June 1936, as noted under ANT-25.

After being stored unused, two of the best DB-1 bombers, Nos. 1813 and 1814, were selected on 20 December 1938 for a record flight by an all-woman crew. While Nesterenko, Berezhnaya (replaced by Mikhalyova) and Rusakova made training flights, Voroshilov reported adversely on the condition of the aircraft. Eventually it was decided that the ladies should fly an Ilyushin DB-3. No photographs of the ANT-36 have yet been discovered.

ANT-37, DB-2

In autumn 1934 Tupolev's design collective received an order to build a single prototype DB-2 long-range bomber. The new aircraft, the ANT-37, was as far as possible based on the ANT-36 DB-1 bomber, but it was obvious – even without Levanyevskii's insistence that transpolar flights demanded more than one engine – that the DB-1 was severely underpowered. The ANT-37 was therefore a twin-engined project from the outset, even though it was not designed to break long-range records.

Again the task was assigned to P O Sukhoi's brigade. Though the ANT-37 superficially had much in common with the ANT-25/-36, very few parts were common, and even the undercarriage was modified. The exterior was smooth stressed-skin throughout. With hindsight, the chief importance of the ANT-37 was as a bridge between the older technology of the ANT-25 and the ANT-40 high-speed bomber which, as the SB, was to be built in greater numbers than any

other aircraft bearing the Tupolev name.

The wing bore only a superficial aerodynamic resemblance, the structure having three tubular-boom spars and simplified ribs, none being of the projecting A-type. The horizontal centre section, which extended to the outer edge of the nacelles, had no taper on the leading edge. Outboard of the nacelles the very long tapered outer panels, with 6 deg dihedral, featured two small sections of split flap, followed by three large sections of aileron projecting behind the trailing edge like many earlier Tu aircraft, the middle section having a tab.

The fuselage was again based on that of the ANT-25, but was slightly longer and instead of housing an engine the nose was fully glazed for the navigator/bomb aimer, with additional side windows. The pilot's cockpit above the leading edge was almost the same as before, though it had a vee windscreen and a fully-glazed canopy opened by sliding to the rear. Next came the bomb bay, similar to that of the DB-1 with vertical cells for ten FAB-100 bombs, though alternative loads were possible. Finally came the cockpit for the radio operator/gunner, which resembled that of the ANT-40 (SB), with a large glazed canopy sliding forward on rails. Defensive armament of three pin-mounted ShKAS was never fitted.

The tail differed only in detail from that of the final ANT-25. All parts were smooth-skinned, the rudder having a diagonal horn-balance at the top, deeply recessed hinges and a

large Flettner tab, while the elevators had small balance areas at the tips and were hinged to a tailplane whose incidence was controlled by the usual arrangement of one bracing wire above and a vee below. The engines were imported fourteen-cylinder Gnome-Rhône K14 Mistral Majors, each rated at 800hp. The oil-cooler inlets were above and the carburettor inlets below the long-chord cowls, which were made in four sections curving round the front of the engine and fitted with controllable gills. The two-blade fixed-pitch wooden propellers had a diameter of 3.25m (10ft 8in). Fuel was contained in six tanks in the outer wings between spars 1 and 2.

Each main undercarriage was a slightly taller version of that of the ANT-25, with twin braked wheels retracting to the rear. Of course, instead of needing a special box fairing each unit was housed in the nacelle, with twin doors cut away so that part of each tyre remained exposed. The castoring tailwheel was fixed (though most drawings show it as retractable, pulled up by cables into a compartment with twin doors). Like the flaps, undercarriage operation was effected electromechanically, the pilot selectors being pushbuttons, then novel in Russian cockpits. Like some of the latest American aircraft, Sukhoi made the ANT-37 almost all-electric, with batteries charged by a generator on each engine.

The prototype was completed at GAZ No. 18 on 15 June 1935 and flown on the following day by N S Rybko. Tail buffet was noticed on early test flights, and on 20 July a shallow dive suddenly produced such violent oscillations that the tail came off. K K Popov and leading engineer M M Yegorov escaped by parachute, but an electrical engineer in the rear fuselage was killed. Shavrov states the cause was 'vibration of the horizontal tailplane', while the CAHI history cites 'vibration of the fuselage and vertical tail'.

Span 33.2m (108ft 11in); length 15m (49ft 2½in); wing area 85sq m (915sq ft).

Weight empty 5,800kg (12,787lb); fuel/oil 4,050kg + 380kg (8,929lb + 837lb); loaded weight 9,450–11,200kg (24,691lb).

Maximum speed 301km/h (187mph) at sea level, 342km/h (212.5mph) at 4,000m; service ceiling 8,000m (26,250ft); range with bomb load 5,000km (3,110 miles); take-off run 970m (3,280ft).

The ANT-37 DB-2 nearing completion in the Tupolev assembly hangar at Khodinka on 9 January 1936. Note the slogans, portraits and statue of Lenin. (G F Petrov/RART)

The ANT-37 probably before first flight. (via Philip Jarrett)

ANT-37bis, DB-2D With a designation meaning 'long-range bomber 2
dubler', Sukhoi's brigade produced the
second prototype almost in parallel
with several others, of which only one
more is known to have been completed. The *dubler* had a redesigned
elevator system incorporating modified hinges and anti-flutter masses
carried on short arms ahead of the
hinge axis above and below the surface.
Similar masses were added to the rudder, which was cut down at the top but
extended in chord. The fuselage was
strengthened, the wingtips shortened,
the navigator was given extra windows
on the right, the tailwheel was again
fixed, the engines were M–85s in
improved cowls with twin exhaust
pipes at the top, and the propellers
were changed to three-blade VISh-2
bracket-type variable-pitch units,

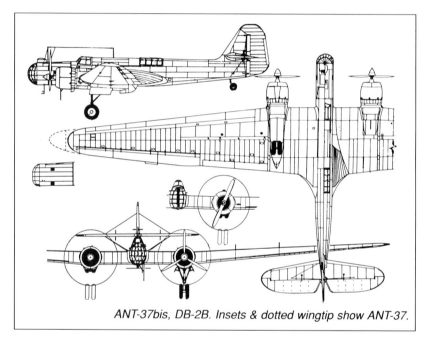

ANT-37bis, DB-2B. Insets & dotted wingtip show ANT-37.

The ANT-37bis, DB-2B, showing the wing/fuselage fillet. (via Philip Jarrett)

the diameter being unchanged. A photograph taken in GAZ No. 18 on 9 January 1936 shows that the flaps were extended inboard to the roots, and that at that time the wing-root fillets were as on the first aircraft.

The second aircraft was first flown on 18 February 1936, with the design bureau experimental tail letters IYe. At once it was discovered that tail flutter was manifest at 140–150km/h (87–93mph). Following urgent further fixes, which included enormously enlarging the wing-root fillets, this ANT-37 got as far as NII-VVS testing, when on 20 August 1936 it was flown by M Yu Alekseev and N S Rybko with a simulated 1,000kg bomb load Moscow–Omsk–Moscow, a distance of 4,955km (3,079 miles) in 23hr 20min, an average of 212km/h (132mph).

Span 31m (101ft 8½in); length 15m (49ft 2½in); wing area 84sq m (904sq ft).
Weight empty 6,025kg (13,283lb); loaded weight 11,500kg (25,353lb).

ANT-37bis, DB-2B The DB-2B was planned as the production long-range bomber, an order for thirty of which had been placed in 1935. It superficially showed few changes beyond addition of a cutaway portion just behind the trailing edge for a ventral gun, and revised engine installations with the carburettor inlet extended forward under the cowling and the exhaust pipes grouped to discharge under the leading edge. This was not built, the production order

for thirty being cancelled, in favour of the Ilyushin DB-3.

However, one of the additional aircraft in GAZ No. 18 was completed. This ANT-37bis became a long-range civil aircraft, named *Rodina* (Motherland), and featured further modifications. The engines were M-86s, developed from the M-85 by SK Tumanskii to give 950hp. Their installations were little changed from the DB-2D apart from adding spinners to the propellers, but fuel capacity was increased, there being twelve wing tanks and a tank in place of the bomb bay. All armament and purely military equipment was removed, the airframe was refined to reduce drag, with additional flush riveting, the tail was further altered in detail, the nose was made hemispherical and glazed with curved Plexiglas, the pitots were mounted on a pillar under the nose, a D/F loop was housed in a fairing above the nose, the pilot was given a flat front windscreen, dual venturis were fitted beside the cockpit, a wire radio antenna was strung between two masts above the fuselage and additional navigation equipment was installed.

The ANT-37bis appeared capable of breaking the world record for distance with an all-women crew. After painstaking preparation and crew training it left Moscow on 24 September 1938 flown by V S Grizodubova, P D Osipenko and M M Raskova. After 26hr 29min they made a wheels-up landing at Kerbi settlement, 5,947km (3,695 miles) to the east

[other sources state the submitted great-circle distance of 5,908.7km]. The ANT-37bis was recovered and subsequently used for exploratory and survey work by Aeroflot. In 1940 it was transferred to a Moscow aircraft factory and used as a research and liaison aircraft until 1943.

Construction of at least one more ANT-37bis, a bomber with additional wing tankage, was started in 1936, but it was never completed.

Data for *Rodina*, dimensions essentially as DB-2D.
Weight empty 5,855kg (12,908lb); fuel/oil 5,525kg + 430kg (12,181lb + 948lb); loaded weight 12,500kg (27,557lb).
Maximum speed 340km/h (211mph) at 4,000m; range 7,300km (4,536 miles); endurance 30hr; take-off run 1,000m (3,280ft).

ANT-38

This unbuilt project of 1933–34 was for a high-speed bomber. It has been speculated that it was a bomber from which the ANT-41 was derived, with a similar airframe and engines.

ANT-39

No information.

ANT-40, SB

The SB (*Skorostnoi Bombardirovshchik*, fast bomber) was built in larger numbers than any other Tupolev aircraft. At the start of the Great Patriotic War on 22 June 1941 it formed 94 per cent of the Soviet front-line bomber force. By 25 June over 1,000 had been destroyed, mainly on the ground, and in subsequent operations attrition was very severe because, like the British Fairey Battle and Bristol Blenheim, it was not fast enough to escape from Messerschmitt Bf 109s and had little defence. It was nevertheless a type of the greatest importance, used in many versions for many purposes.

The official NII-VVS requirement was issued in October 1934. It called for a modern bomber to co-operate

Preparing the DB-2B, Rodina. (RART)

with front-line land and naval forces, able to reach 330km/h and carry a 500kg bomb load to a target 400km away. A seemingly obvious answer would have been a short-span version of Sukhoi's ANT-37, but Tupolev wisely saw that what was needed was a fresh design of the simplest possible nature. When the SB became known outside the USSR it was instantly called 'a copy of the Martin bomber'. Tupolev was irritated by this. He told the author 'Of course we were aware of the Martin, but it did not influence the SB in the slightest. Our design was the obvious extrapolation of the ANT-21, –29 and –38'. Another common Western misapprehension was to call it the SB-2. This arose from the fact that the full designation included the number and type of engines, written in the form 'SB-2M-100A' and 'SB-2M-103', just as the full designation of an earlier bomber was 'TB-3-4M-17' (often written 'TB3-4M17').

Like its predecessors (21/29/38), the SB was the work of the KOSOS brigade led by Arkhangel'skii. The overall layout was the natural one of a twin-engined cantilever monoplane with the main undercarriage retracting into the nacelles. The fuselage was made as slim as possible, the bomb bay and the spars of the mid-mounted wing isolating the radio-operator/gunner from the pilot above the leading edge and the navigator/bomb-aimer in the nose. All three crew positions could be enclosed, and the structure was light-alloy stressed-skin throughout, with flush riveting used as widely as possible.

The wing resembled that of the ANT-29 but was of simpler structure, with just two main girder spars. The centre section, 5m (16ft 5in) wide, was horizontal and untapered on the leading edge, the aerofoil profile being CAHI-6mod with thickness/chord ratio 16 per cent. Steel bolted joints at the spars attached the outer panels, tapered almost entirely on the trailing edge, with 2 deg dihedral and a thickness/chord ratio reduced to 12.5 per cent at the rounded tip. Each spar comprised upper and lower booms of 30-KhGSA stainless steel joined by diagonal (Warren-truss) bracing and vertical members at every third or fourth rib. The latter were ∩-section with diagonal bracing, spaced every 200–250mm (8–10in). Skin thickness was 0.6–1mm over the centre section

and 0.5–0.6mm over the outer panels, in the first prototype flush riveted throughout.

Four sections of split flap were fitted, two on the centre section and two on the outer panels. These were driven to a maximum of 60 deg by pull rods, bellcranks and finally push rods, all operated by a transverse hydraulic ram in the fuselage. From the flaps to the tips were two-section ailerons with recessed hinges, the chord extending slightly aft of the trailing edge inboard. Operation was by push/pull rods. The right inboard section had a trimmer, operated by cable and screwjack.

The relatively small fuselage, only just roomy enough for crew of normal stature, and (in the first aircraft) extremely short, was a simple stressed-skin semi-monocoque structure of pear-shaped section, with two strong tubular upper longerons, two strong lower longerons of a section built up from flanged sheet, tubular and top-hat stringers, light transverse rings from pressed sheet and flush-riveted strips of skin. In the usual Tupolev style it was made in sections which could be unbolted. Section F-1 was the tiny compartment for the navigator/bomb-aimer, with an almost hemispherical glazed nose with twin vertical slots for two machine-guns. Entry was via a floor hatch, and in a belly landing a small man could escape through a roof hatch. F-2 was by far the largest, made integral with the centre wing, which was set at +2 deg incidence. At the front was a sheet bulkhead with apertures through which items such as a piece of paper

could be passed between the navigator and pilot. The latter sat in line with the leading edge, with a deep vee windscreen and aft-sliding glazed canopy. Immediately behind him was the bomb bay. F-3 contained the radio-operator/gunner with a large forward-sliding canopy and prone ventral position. Armament details are given later.

The internal structure of the tail was almost entirely pressed sheet, with multiple lightening holes, with flush-riveted skin. The fin and rudder had a typical severe angular shape, the unbalanced rudder having a long Flettner tab. The horizontal tail was of so-called butterfly shape, the tailplane being fixed and wire-braced and the elevators being unbalanced but having large tabs.

Each main landing gear comprised a single air/oil shock strut of KhMA steel with a diagonal brace on each side and a fork carrying the drum-braked wheel with 950mm by 250mm tyre. A hydraulic jack acting on the rear breaker strut retracted the unit back into the nacelle, the side braces having pin-jointed ties which closed the two doors, leaving part of the tyre exposed. The castoring tailwheel was faired by a spat.

Arkhangel'skii was eager to use the Hispano-Suiza V-12 engine, which was to be made under licence, 'Sovietised' by Vladimir Klimov, as the M-100. At a conference in February 1934 with the director of CAHI, by then I M Kharlamov, Tupolev insisted that it was necessary to build two prototypes, with different engines. For reasons of timing, the second would

The SB 2RTs after being lengthened. No photograph has been discovered showing the original short fuselage. (RART)

have the French engine, which he agreed was the most promising.

SB, ANT-40 2RTs

For the first prototype, however, he recommended an engine with which his team were familiar: the Wright Cyclone F-3 air-cooled radial, rated at 710hp and driving the Hamilton three-blade bracket-type propeller of 10ft 6in diameter. The engines were installed in short-chord cowlings centred on the wing, without gills. Exhaust was piped under the wing, and the rear end of the nacelle above the wing was left open to extract cooling air, a feature of all subsequent versions. Provision was made for 1,670 litres (367gal) of fuel in four wing tanks, but only two 470 litre inner tanks were installed.

The first prototype, with KOSOS number ANT-40, was known simply as the SB. It was first flown on 7 October (not, as in most Western accounts, 25 April) 1934. The test pilot was K K Popov, and the wheels had been replaced by non-retractable skis measuring 2,800mm by 820mm, the tail ski being 800mm by 320mm. Early testing showed the need for numerous modifications to all the flight controls, and for other changes. A particular problem was unacceptable elevator control, leading to 'spontaneous soaring or diving with little pressure on the control wheel'.

Span 19m (62ft 4in); length 10.485m (34ft 4¼in); wing area 47.6sq m (512sq ft).

Weights and performance, not known.

No photograph of the SB in its original form is known to survive. Landing after the ninth flight, on 31 October 1934, it was seriously damaged and substantially rebuilt. The fuselage was considerably lengthened, the outer wings were replaced by new panels with increased taper (which as span was unchanged resulted in the tips being almost pointed) and the engines were replaced by 800hp M-85 two-row radials, installed as in the ANT-37bis. The rebuilt aircraft was tested from 5 February to 31 July 1935, but faded from the scene because of its remarkably poor performance, and the much better handling of the second prototype. It was not submitted to NII-VVS testing.

Span 19m (62ft 4in); length 12.3m (40ft 4¼in); wing area 46.3sq m (498sq ft).

Weight empty 3,132kg (6,905lb); loaded weight 4,717kg (10,399lb).

Maximum speed 325km/h (202mph) at 4,000m; service ceiling 6,800m (22,310ft); range 700km (435 miles).

ANT-40 No 1 SB 2HS

The second prototype, which was not designated ANT-40$_2$ as has appeared in Western articles, had Hispano-Suiza 12Ybrs water-cooled V-12 engines, each rated at 780hp at 3,300m, driving simple fixed-pitch two-blade metal propellers of 3.3m (10ft 10in) diameter, fitted with small spinners. Each engine was carried on a triangulated truss of welded steel tube bolted to the booms of the front spar. The radiators were of the flat vertical type, forming the entire front of the flat-sided cowling. In front of each matrix was a shutter formed from sixteen vertical 'Venetian blinds' opened and closed under thermostat control. The two exhaust pipes ran inside the cowling to the open rear end above the wing.

To reduce wing loading (from 102kg/sq m to 93.4kg/sq m), and thus improve ceiling and field length, the outer wings were again redesigned, with greater span and chord, despite leading-edge taper being increased from 4.5 deg to 9 deg. A landing light was added in the port wing leading edge, the tailwheel spat was removed, and numerous changes were made to systems and equipment, including fitting all four fuel tanks.

Armament was also installed. The bomb bay, fitted with two doors pulled open against spring force by two cables, normally accomodated six FAB-100s, four in a group carried nose-up between the wing spars and two lying horizontally at the rear. As an alternative, two FAB-250s or one FAB-500 (weight, about 1,260lb) could be carried. In the nose were two ShKAS mounted on a frame able to pivot them together from +48 deg to –50 deg but offering very limited lateral movement. These had 960 rounds each. The rear cockpit had a single ShKAS on a Tur-9 mount able to fire in all directions (with nothing to protect one's own tail or wings) extending to 10 deg below horizontal around the sides and to 65 deg from the vertical directly ahead. This gun had 1,000 rounds. The same man could lie prone to fire the 'dagger' ventral ShKAS, with a maximum depression of 70 deg; this gun had 500 rounds. Many other weapons were carried later.

This prototype was completed in December 1934 and first flown on 16 February 1935. Though numerous faults remained, Arkhangel'skii received NII-VVS instructions to organise mass production, and this was quickly done at GAZ No. 22 at Fili and the new GAZ No. 125 at Irkutsk, which had previously built only a small number of I-14s.

Span 20.3m (66ft 7¼in); length 12.17m (39ft 11¼in); wing area 51.95sq m (559sq ft).

Weight empty 3,499kg (7,714lb); fuel/oil 530kg + 60kg (1,168lb +132lb); loaded weight (normal) 4,850kg (10,692lb), (maximum) 5,350kg (11,794lb).

Maximum speed 332km/h (206mph) at sea level, 404km/h (251mph) at 5,000m; climb to 5,000m (16,400ft) 9.4min; service ceiling 9,400m (30,840ft).

ANT-40 No 2

The third prototype incorporated further major aero-

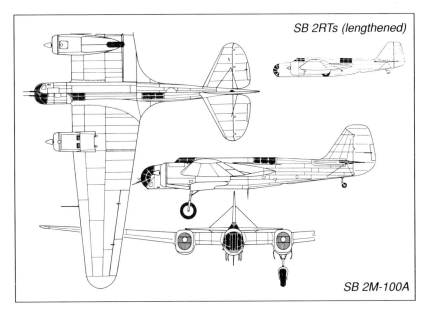

SB 2RTs (lengthened)

SB 2M-100A

dynamic changes to improve stability and control. Equally important, the entire flight-control system was re-designed with the assistance of the CAHI experimental aerodynamics section, led by Ye P Grossman and M V Keldysh. To reduce flutter, aileron balance was increased to 91–93 per cent. The horizontal tail was enlarged, and elevator balance increased from zero to 50–80 per cent across the span of each surface. The engines were moved forward 100mm (4in) and the propellers changed to the production ground-adjustable VFSh type, with Hucks starter dogs exposed by removing the spinners.

This aircraft was tested from September 1935 to April 1936. During this period a famous incident occurred in which, hearing that Commissar S Ordzhonikidze was investigating the bomber's problems, NII-VVS engineers festooned the aircraft with placards each pointing to a fault. Tupolev began tearing them down, shouting 'Hooligans!', but this led to an equally famous further meeting, mentioned in the Introduction, at which Stalin proclaimed 'There are no trivialities in aviation!'. Considerable further modification took place, though those most externally obvious had to wait for the first series aircraft. Stalin was displeased that major changes should be needed at a time when over thirty aircraft were on the assembly line in GAZ No. 22, and GAZ No. 25 was fully tooled.

Dimensions unchanged.

Weight empty 3,900kg (8,598lb); loaded weight 5,468kg (12,055lb).

Maximum speed 356km/h (221mph) at sea level, 418km/h (260mph) at 5,300m; climb 2.2min to 1,000m (3,280ft), 19.8min to 8,000m (26,250ft); service ceiling 9,560m (31,365ft); range 980km (610 miles); take-off/landing runs both c300m (984ft).

SB 2M-100A in wartime summer camoflague. (G F Petrov/RART)

SB 2M-100 Aircraft began to come off the GAZ No. 22 line in June 1936, when the SB was far from fully developed. Subsequently there were many versions, and it is impossible to describe which sub-types were produced in which month, but the annual total deliveries from the Fili factory from 1936 to 1941 inclusive were 268, 853, 1,250, 1,435, 1,820 (fifty per day including Sundays), and sixty-nine. The corresponding figures for Irkutsk, starting in May 1937, were seventy-three, 177, 343, 375 and 168. The combined total is thus 6,831, the greatest for any of Tupolev's aircraft.

The M-100 was the first of the Soviet licence-built Hispano-Suiza 12Y engines, rated at 750hp at 4,100m, but otherwise with the installation and propellers unchanged. Standard fuel capacity was 1,670 litres (367gal) in four aluminium tanks. The most important changes, however, were aerodynamic. The outer wings were again redesigned, with a further increase in area achieved by restoring the 4.5 deg leading-edge taper whilst at the same time increasing the span slightly, despite increasing dihedral from 2 deg to 5 deg. The ailerons were further improved to eliminate flutter, which had been manifest on the third prototype. The wing-root trailing-edge fairings were enlarged and made horizontal instead of curving upwards. The horizontal tail was yet again improved, and the vertical tail was also redesigned, the rudder being increased in area and rounded, and fitted with a large diagonal-edge balance horn at the top terminating in a projecting lead mass balance. The Flettner tab was made much smaller. The entire structure was modified in detail to ease mass-production, one change being to restrict flush-riveting to the leading edges.

Span 20.33m (66ft 8⅜in); length 12.24m (40ft 1⅞in); wing area 56.7sq m (610sq ft).

Weight empty 4,060kg (8,951lb); fuel/oil 530kg+60kg (1,168lb + 132lb) normal, 1,325kg +120kg maximum; loaded weight (normal fuel) 5,628kg (12,407lb).

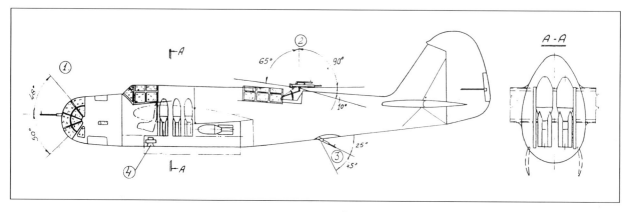

SB armament: 1, twin ShKAS (1,920 rounds); 2, Tur-9, ShKAS (1,000 rounds); 3, ShKAS (500 rounds); 4, camera; in bomb bay, eight FAB-100.

SB 2M-100A on skis with external bombs. (G F Petrov/RART)

In 1937 one M-100A aircraft was fitted with the MV-3 dorsal turret developed by G M Mozharovskii and I V Venevidov, giving a single ShKAS with the OP-2 optical sight 360 deg coverage, and the MV-2 improved glazed ventral blister, again with a single ShKAS. This aircraft weighed 5,810kg, and had a maximum speed of 412km/h (256mph).

Dimensions unchanged except length 12.273m (40ft 3¼in).

Weight empty 4,138kg (9,123lb); fuel/oil 1,130kg+120kg (2,491lb + 265lb); loaded weight (normal) 5,748kg (12,672lb), (maximum) 6,462kg (14,246lb).

Maximum speed 372km/h (231mph) at sea level, 423.5km/h (263mph) at 4,000m; climb to 1,000m in 1.6min, to 5,000m in 7.4min; service ceiling 9,560m (31,365ft); range (max fuel) 2,150km (1,336 miles); take-off/landing run both 300m (984ft).

Maximum speed 326km/h (203mph) at sea level, 393km/h (244mph) at 5,200m; climb to 1,000m 2.8min, to 5,000m 11.7min; service ceiling 9,000m (29,530ft); range (normal fuel) 980km (609 miles), (max) 2,187km (1,360 miles); take-off/landing run, both 300–350m (984–1,148ft).

SB 2M-100A The M-100A was the first production engine incorporating improvements by Klimov. These increased the maximum rating to 860hp at 3,300m. There were few changes to the installation, apart from fitting electric starters from late 1937, the Hucks dogs being retained. Also from late 1937 the VFSh propeller was increasingly replaced by the three-blade variable-pitch VISh-2, diameter remaining the same. Fuel capacity was slightly reduced by fitting protected tanks, the total being 1,520 litres (334gal), comprising 400 litre inner cells and 360 litre outers.

Thanks to the greater power, weight was permitted to increase, and external armament was added. A rack under each inboard wing was added to carry various bombs or an AM–300 aerial mine. When the full internal bomb load was also carried the fuel load had to be reduced. In spring 1938 the wing racks were used to test the RO-132 gun firing 132mm (5.2in) rocket projectiles. In 1939–40 six aircraft were in service with this weapon, later replaced by the RS-132 free-flight rocket.

The SB 2M-100A was the first version made in large numbers. A few with the M-100 were supplied to Republican Spain, doing very well in combat (for example, against the slower Fiat C.R.32) from 26 October 1936. By 1937 the M-100A version was maintaining this good reputation, being popularly dubbed *Katyusha* after a character in a musical. However, by 1938 it was found to be vulnerable to faster fighters, notably the Messerschmitt Bf 109, and also to catch fire easily in combat, despite its protected tanks.

Considerable numbers (including the later M-103 version, a total of 292) were sold to China, plus sixty designated B 71 to Czechoslovakia. The B 71 was fitted with Avia-built HS12Ydrs engines, and sixty-six were later built under licence by Avia plus forty-five by Aero, all survivors soon finding their way to the Luftwaffe or puppet forces.

SB 2M-103 In 1938 an increasing proportion of production was fitted with Klimov's improved M-103 engine, rated at 960hp at 4,000m. The structure was strengthened, and the maximum bomb load was increased to 1,500kg (3,307lb), enabling external bombs, mines or 368 litre (81gal) drop tanks to be carried as well as an internal bomb load. Other new features included a rudimentary set of duplicate flight controls for the navigator, communications radio using a wire strung from a post above the windscreen to the top of the fin, and semi-retractable main skis able to swing aft on parallel links.

The first M-103 aircraft was tested from 19 October 1936, and on

SBbis-3. (G F Petrov/RART)

1 November reached an altitude of 12,695m (41,650ft). The first pre-series aircraft incorporating all other planned modifications (apart from the MV-2/MV-3 armament) was flown by M Yu Alekseyev on 2 September 1937 to the record height of 12,247m (40,180ft) whilst carrying a load of 1,000kg. It underwent NII-VVS testing from 27 July to 19 September 1938, being cleared for production as the SB bis.

SB bis 2M-103 This was a single aircraft tested in March 1938 with a polished wing surface (this is believed to mean flush-riveted throughout). It reached 428km/h (266mph) with 1,240kg fuel, at a gross weight of 5,905kg.

Dimensions, unchanged.
Weight empty 4,427kg (9,760lb); fuel/oil 530kg + 60kg (1,168lb + 132lb)/or 1,130kg + 120kg (2,491lb + 265lb); loaded weight (normal) 6,175kg (13,613lb), (maximum) 7,750kg (17,086lb).
Maximum speed 358km/h (222mph) at sea level, 419km/h (260mph) at 4,000m; climb 7.45min to 4,000m, 8.4min to 5,000m; service ceiling 9,600m (31,500ft); range 1,800km (1,118 miles); take-off run 310m (1,017ft); landing run 400m (1,312ft).

SBbis-3 2M-103 Introduced at the fourteenth series at both GAZ, this was built in larger numbers than any other. The main new feature was a redesigned engine installation, with the frontal radiator replaced by a smaller matrix beneath the engine in a profiled duct, with a 'chin' inlet and an exit flap at the rear controlled by the pilot or by a thermostat. Standard propellers were three-blade fully variable VISh-2, later VISh-22, with large spinners faired into the front of the redesigned cowling. The exhaust pipes were now outside the cowling, taken back into the wing to exit almost vertically on each side of the open upper end of the cowling. The oil, previously cooled by a block matrix under the engine inside the cowling, was now piped through twin drum coolers in the wing outboard of the engine, with two circular ram inlets in the leading edge and rectangular exits in the upper skin.

The NII-VVS bis-3 was tested from 1 November 1937 to 17 January 1938. During 1938 the simple aft ventral hatch was progressively replaced by the MV-2 glazed blister. This enabled

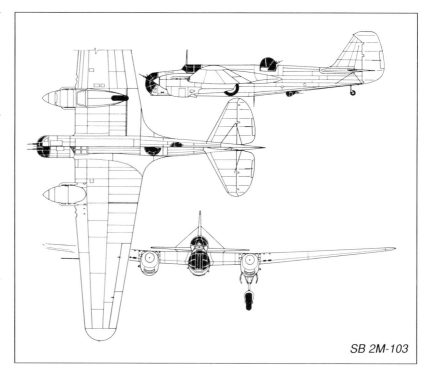

SB 2M-103

the backseater to stay in his seat and man the radio and both upper and lower rear guns, the MV-2 having a periscopic sight. Data are for the original armament.

Dimensions unchanged.
Weight empty 4,303kg (9,486lb); fuel/oil 530kg + 60kg (1,168lb + 132lb); loaded weight 6,013kg (13,256lb).
Maximum speed 375km/h (233mph) at sea level, 445km/h (276.5mph) at 4,500m; service ceiling 9,800m (32,150ft); range 1,600km (994 miles).

SB 1938–39 In the eighteenth and later series the weight was allowed to rise considerably. The MV-2/MV-3 combination was made standard, leading edges were polished, and equipment fit was enhanced. These versions saw extensive action in the Khalkin Gol, Nomonhan, Winter and Great Patriotic wars.

Dimensions unchanged.
Weight empty (typical) 4,768kg (10,511lb); fuel/oil 1,130kg + 120kg (2,491lb + 265lb);

SB with 2M-103 on test with VAP-1000 poison-gas dispensers on the external weapon racks. The test programme was from 27 July to 19 September 1938. (G F Petrov/RART)

Late production SB with MV-2/MV-3 rear armament. (Philip Jarrett)

loaded weight (normal) 6,352kg (14,026lb), (maximum) 8,050kg (17,747lb).

Maximum speed 375km/h at sea level, 420km/h (261mph) at 4,000m; climb 9.5min to 5,000m; service ceiling 9,300m (30,500ft); range 1,350km (839 miles); take-off run 370m (1,213ft) landing 397m (1,302ft)

SB The final production series were powered by the Klimov M-105, a further development of the M-100 rated at 1,050hp at 4,000m. This enabled maximum bomb load to be increased to 1,800kg (3,968lb). There were few other changes, and the reduced ceiling and range are not explained.

Dimensions unchanged.
Weight empty (typical) 4,800kg (10,582lb); fuel/oil 1,130kg+120kg (2,491lb + 256lb); loaded weight (normal) 6,700kg (14,771lb), (maximum) 8,050kg (17,747lb).

Maximum speed 395km/h (245mph) at sea level, 445km/h (276.5mph) at 4,000m; service ceiling 9,000m (29,530ft); range 1,200km (746 miles).

SB 2M-104 In the 1939 production about thirty aircraft were delivered with the M-104 engine, rated at a little over 1,000hp but with a two-stage supercharger maintaining high-altitude performance. No data.

SB 2M-106 At least two aircraft were fitted with the M-106, a further development rated at 1,200hp. This engine remained experimental.

PS-40 2M-100A In January 1938 one aircraft was stripped of military equipment and equipped as a cargo transport. Attachments were provided for three aluminium bins with a [presumed total] capacity of 2.58cu m (91cu ft). After testing, it was used by

Aeroflot. Loaded weight was 6,400kg (14,109lb), and maximum speed with fixed skis 308km/h (191mph) at sea level, 341km/h (212mph) at 3,800m.

PS-40U This second civil aircraft was fitted with a completely new all-metal nose containing a second pilot cockpit for training purposes. It was tested on 11–16 March 1938.

PS-41 This aircraft was modified in 1938 from an SB 2M-103 for evaluation by Aeroflot as a mail, cargo and passenger aircraft.

PS-41 2M-103U Also known as the PS-40bis, these were series-built civil transports derived from the SBbis-3. The interior was completely re-designed for efficient carriage of 970kg (2,138lb) of cargo and mails, though seats could be put in the rear-fuselage

compartment (which retained the sliding glazed canopy). The nose was a metal version of that of the bomber, with a loading hatch. Each aircraft was delivered with wheels and retractable skis (main 2,800mm by 910mm, tail 800mm by 320mm). Data for aircraft with skis.

Dimensions as SBbis-3.
Weight empty 4,380kg (9,656lb); fuel/oil 1,200kg+100kg (2,646lb+220lb), loaded weight 7,000kg (15,432lb).

Maximum speed 428km/h (266mph) at 4,000m; range 1,180km (733 miles); take-off run 660m (1,429ft) landing run 220m (121ft).

PS-41bis These aircraft were strengthened and further modified, the main difference being provision for two 340 litre (270kg) auxiliary tanks hung under the inner wing. Gross weight was still 7,000kg, of which fuel accounted for 1,730kg. In postal use the payload is given as only 180kg.

Loading a PS-41 (PS-40bis), with an Li-2 in the background. (G F Petrov/ RART)

The USB, or SB-3 or SB-UT, came in many forms. This one has M-100A engines, which was unusual. (G F Petrov/ RART)

SB-3/2M-103 Also called the USB this was a military dual-control trainer, considered necessary because the SB initially suffered high attrition from pupils converting from much slower aircraft. The pupil occupied the original pilot cockpit. In the nose was an open cockpit for the instructor, with a small windscreen. He could crank his seat higher to make the take-off and landing. About 120 were delivered in 1938.

SB-3B/2M-103 These could double as pilot trainers and also as tugs for A-7 gliders.

SB-3bis Intended new standard cooling system with rear-mounted radiator tunnels. Tested in 1938 by P M Stefanovskii and M A Lipkin, but considered defective.

SB modifications One aircraft was tested at CAHI in 1939-40 with the electrically-powered Tur-DU dorsal turret able to slew at 90 deg/sec. The SB/TSH also known as the SB/3K, was used to investigate nosewheel 'tricycle' landing gear, which had not then been tried in the USSR. The non-retractable strut-braced installation was designed by I P Tolstikh, and flight testing was handled by Mark L Gallai. The NII-VVS photographs were taken on 12 September 1940.

MMN With Tupolev incarcerated, further development was handled by Arkhangel'skii himself. This work led to the MMN, SB-RK (Ar-2) and B (SBB). A more distant relative was the ANT-41. Of these, the MMN requires inclusion here, because it was specifically an improved SB. Features

included a new wing of reduced span with a NACA-22 high-lift profile and improved slotted flaps, new streamlined nacelles, and a longer fuselage with a more capacious navigator compartment. Powered by two 1,050hp M-105 engines, the MMN, first flown in 1939, weighed 6,420kg (14,153lb) loaded and reached 458km/h (285 mph). The Ar-2 and SBB were even further removed from the Tupolev aircraft.

ANT-41, T-1

One of Tupolev's lesser-known aircraft, the ANT-41 was in principle a bigger, heavier and more powerful version of the SB. It could have been a valuable military and naval attack aircraft, had it not encountered aeroelastic problems similar to those afflicting other KOSOS prototypes. It was ordered in April 1934 as the T-1 (*Torpedonosyets 1*), the first Soviet aircraft explicitly designed as a torpedo bomber.

The project was launched in KOSOS in August 1934 in Myasishchev's brigade, the leading engineer on the T-1 being I P Mosolov. Though obviously similar in conception to the SB, it was substantially more massive, some of the material thicknesses being more than 50 per cent greater, and the chosen engines, Mikulin's water-cooled M-34, were much more powerful. The airframe was smooth stressed-skin throughout.

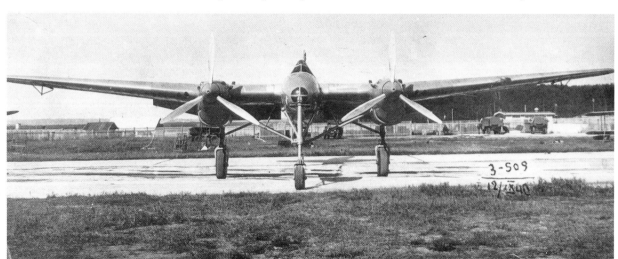

The SB/TsN photographed with flaps down on 12 September 1940. (G F Petrov/RART)

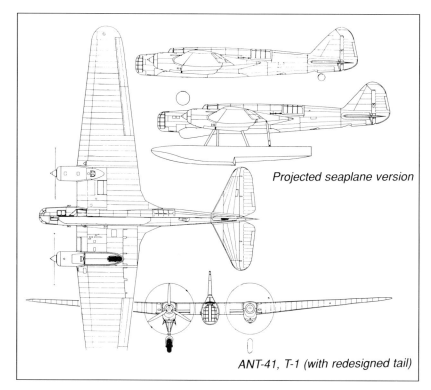

Projected seaplane version

ANT-41, T-1 (with redesigned tail)

The wing had two main and three secondary spars, with traditional built-up truss construction. The horizontal centre section was fitted with four large sections of split flap, outboard of which were large balanced ailerons with a tab on the left inboard section. The fuselage was again basically a strengthened version of the SB, but with a glazed nose resembling the ANT-37 and with a remarkable internal weapon bay with twin doors measuring 6.5m (21ft 4in) long, approaching half the overall length of the aircraft.

This bay was designed to accomodate two torpedoes of 820kg (1,808lb) each, or a single heavy torpedo weighing 1,700kg (3,748lb), or one of the new TAFN pattern weighing 920kg (2,028lb). Bombs and mines up to an individual *kalibr* of 1,000kg (2,205lb) were alternative loads. The crew numbered four, though for defence there were just three ShKAS (one fewer than in the SB), in the nose, rear cockpit and 'dagger' aft ventral positions.

The large water-cooled engines were installed in the style Myasishchev was to make his trademark, with the cowlings fitted tightly and the radiators located in the wing immediately inboard of the engines, with 'letterbox' inlets in the leading edge and flush exits with controllable flaps in the upper surface. Each was to drive a

3.55m (11ft 7¾in) two-blade fixed-pitch wooden propeller. The exhaust pipes were taken back inside the cowling to exit close together in the top of the nacelle, which was not left open at the rear. Approximately 2,600 litres (572gal) of fuel could be housed in tanks between the spars inboard of the radiators and above the bomb bay.

The three units of the landing gear were strengthened SB-type, the main legs being slightly longer and the castoring tailwheel being retractable. The tail was again very similar to that of the SB, the unbalanced rudder

having a very large Flettner tab, and the elevators having inset hinges. Unlike the SB, the tailplane could pivot from 0 deg to +3 deg.

By 1935 the improved M-34FRN, rated at 900hp, became available, and this allowed maximum weight to be increased from 7,500kg to the figure given in the data. By this time the KOSOS project designation was LK, and there was every expectancy of series manufacture. Production drawings were sent to GAZ No. 84 at Moscow Khimki. Drawings were also produced of a planned twin-float seaplane version, and of engines with individual ejector exhausts.

ZOK completed the parts for the prototype in January 1936, and these were taken to Moscow Central Aerodrome for assembly. By this time the problems with the SB had made Myasishchev's engineers, and Ye P Grossman of CAHI, very uneasy about the ANT-41 tail design. Without being flown, the aircraft was returned to ZOK where a completely redesigned tail was fitted, the chord of the fin and rudder being increased, and the rudder provided with deeply inset hinges and a large balance horn, the tab being made smaller. Changes were also made to the horizontal surface. Another change was to replace the wooden propellers by ground-adjustable three-blade metal propellers of the same diameter.

With the CAHI experimental insignia IYe on the tail, the ANT-41 was returned to the airfield on 28 May 1936, where on 2 June it was flown for 25 minutes by a three-

Assembling the ANT-41, with the original tail, on 5 January 1936. (G F Petrov/RART)

man crew headed by pilot A P Chernyavskii. Unfortunately, during high-speed diving trials on the fourteenth flight on 3 July it suddenly encountered aileron flutter violent enough to cause major wing failure. The crew escaped by parachute. Urgent design fixes were put in hand, whilst other modifications would have resulted in the series aircraft having a span of 26.02m (85ft 4in), wing area reduced to 86sq m (927sq ft) and the fuselage significantly lengthened to 16.84m (55ft 3in). However, later in 1936 it was decided that Ilyushin's DB-3 and twin-float DB-3T would be selected as torpedo-bombers instead.

Span 25.73m (84ft 5in); length 15.543m (50ft 1¼in); wing area 87.78sq m (945sq ft).

Weight empty (bare) 5,630kg, (equipped) 5,846kg (12,888lb); fuel/oil 1,900kg + 150kg (4,188lb + 331lb); loaded weight (normal) 8,390kg (18,496lb), (maximum) 9,150kg (20,172lb).

Maximum speed 358km/h (222mph) at sea level, 400km/h (248.5mph) at 2,100m; service ceiling (normal weight) 9,500m (31,170ft); range (with auxiliary fuel) 4,200km (2,610 miles); take-off run (normal weight) 370m (1,213ft).

ANT-42, TB-7, Pe-8

Having achieved such success with the ANT-6/TB-3, it is remarkable that its successor, the TB-7, should have become a halting and troubled

programme, with different versions built in trivial numbers. This is sad since the basic design was as good as all the others to bear Tupolev's initials. A contributory factor to the problems was that during most of the programme Tupolev and many of his senior aides were imprisoned.

In 1931 the NII-VVS issued a specification for a heavy bomber to replace the TB-3, to carry a 10-tonne bomb load 1,500–2,000km at 250km/h at a height of 7,000m. This concept gradually gave way to a faster, higher-flying bomber carrying a smaller load, the bomb load even falling to 500kg in 1934. On 29 July 1934 Tupolev authorised start of work on the ANT-42, confirmed by official order No. 7342 two days later. Funds were made available on 27 December 1934 by decision of the Council for Work and Defence. A revised specification called for a 2-tonne (4,410lb) bomb load to be carried 4,500km (2,796 miles) at up to 440km/h (273mph), at no less than 10,000–11,000km (32,800–36,090ft). Because of the high speed the requirement for an escort version, carrying guns instead of bombs, was dropped, but there remained a demand for a transport version to carry fifty equipped troops (this was not built).

By this time Tupolev had already decided that the ANT-42 would be a four-engined cantilever monoplane with smooth stressed-skin. The most difficult decision concerned the need for the engines to retain high power at high altitude. The eventual decision was to instal a fifth engine in an

ATsN (*Agregat Tsentralnovo Nadduva*) driving a giant supercharger piping air to the four propulsion engines. A recent article states that this idea (which unknown to Tupolev had been used in German bombers of 1916–18) was inspired by 'air-conditioning in the mining industry', but Tupolev told the author they had adopted it after reading 'an article, with calculations, by your own Frank Whittle'.

Following the general reorganisation of KOSOS, a special design bureau was formed in 1935 to manage this aircraft from start to finish, and as it was the first such group it was called KB-1. Petlyakov was put in charge, with I F Nezval as deputy. While Tupolev himself, assisted by B M Kondorski, decided the approximate overall design, the following team leaders were appointed within KB-1: aerodynamics, V N Matveyev; stressing, V N Belyayev; centre wing and engine mounts, K I Popov; outer wings, B A Saukke and M M Glinnikov; fuselage and tailplane, V M M Myasishchev and I F Nezval; flight controls, M M Sokolov; engine installations, P S Kotenko and B S Ivanov; landing gear, A G Agladze; armament, S M Meyerson and M S Sviridonov; equipment, B L Kerber.

The ruling airframe material was D16 Dural, and aerodynamics and stressing resembled an ANT-40 with the linear scale multiplied by 2. The main exception was the use of 30KhGSA high-strength steel for the two main spars of the wing. The horizontal centre section, extending to the

The incomplete ANT-42 prototype at Khodinka. (via Philip Jarrett)

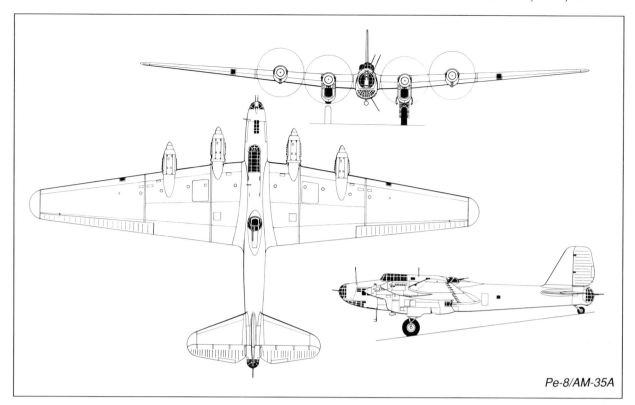

Pe-8/AM-35A

outer side of the outer nacelles, had a span of 7.61m (24ft 11½in), the tapered outer panels having 6 deg dihedral measured on the underside. Aerofoil profile was CAHI–40, thickness/chord ratio being 19 per cent at the root and 15.5 per cent at the tip. The skin was flush-riveted in over sixty long but narrow spanwise strips on each wing. Mean chord was 5.35m (17ft 7in), and design centre of gravity varied from 28.4 per cent empty to 34 per cent at full load. The long but narrow ailerons, pivoted on their undersides to external brackets, comprised an outer section, a middle section with tab and an inner section with slightly extended chord, all fabric-covered. The flaps on the outer wing and centre section were of the split type, driven hydraulically to 45 deg.

The complex fuselage had tubular main longerons, including four principal ones of 30KhGSA, the rest of the structure being D16 assembled by round- or pan-head rivets. Under the wing was the large bomb bay, 5.87m (231in) long. There were two nose-fuselage sections, F-1 and -2 ending at the bomb bay well ahead of the leading edge. Next came the mid-fuselage, ending at the rear spar. Finally came the complete rear fuselage and tail, incorporating the rear portion of the bomb bay, 2.03m (80in) long, though

the entire bomb bay was closed by just two hydraulically driven doors.

A strange decision was to adopt a pear-shaped fuselage cross section. This meant the two pilots had to be seated in tandem in narrow cockpits, an unusual arrangement, and it made it impossible to use five similar 950hp M-34FRN engines, the ATsN engine having to be the more compact 750hp M-100. Thus, as planned in January 1935, the fuselage comprised a nose housing the navigator and bomb aimer, with a chin gondola (called the 'beard') and a powered turret covering a 120 deg cone ahead, the tandem pilot cockpits with sliding canopies, a radio operator on the left and engineer on the right, an upper turret firing from horizontal to +70 deg with 360 deg traverse, a ventral position covering a downwards cone of 130 deg and a powered tail turret covering a rear cone of 145 deg if small 7.62mm ShKAS guns were used and 100 deg in the case of a 20mm ShVAK. This armament was to change.

The tail was conventional. The fin was integral with the fuselage and carried a rudder with finely corrugated skin, a large Flettner tab and a small balance horn. The tailplane, mounted at the top of the fuselage with wire bracing, carried elevators which again had corrugated skin, Flettner tabs

and two large external arc-type mass balances on each surface.

The propulsion engines were carried on pinned and welded trusses of 30KhGSA tube, driving 3.9m (153.5in) VISh-2 three-blade Dural propellers. Each was cooled by a water radiator in a profiled duct directly below, with a controllable side exit door. A total of 11,500 litres (2,530gal) of fuel was housed in protected AMTs aluminium tanks in the leading edge and between the spars. Each main undercarriage had a single wheel with a 1,600mm by 500mm tyre, with pneumatic expanding-drum brakes and hydraulic retraction to the rear into the inner nacelle, part of the tyre protruding to reduce damage in a wheels-up landing. The tailwheel had a 700mm by 300mm tyre, and could castor but did not retract.

Despite such distractions as the ANT-20bis, the failure of the Lenin works to supply KhGSA steel and repeated changes to the technical requirements, the unarmed prototype was rolled out on 9 November 1936 from GAZ No. 156, the special factory at the NKVD secure site at Moscow Khodinka where aircraft designers were imprisoned. It was without either its engines or the ATsN, but the main engines arrived on 23 December, and on 27 December the first flight was

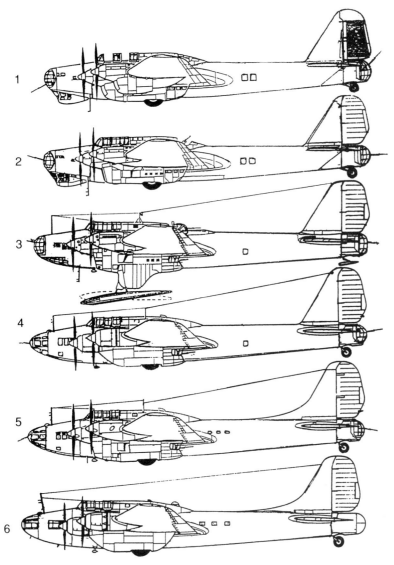

Pe-8 variants: 1, prototype ANT-42 March 1937; 2, prototype May 1940; 3, first series No. 4211; 4, Pe-8/ASh-82; 5, Pe-8ON/ACh-30B; 6, Pe-80N Polyarnyi.

made by M M Gromov, with N S Rybko as co-pilot, A S Rakhmanin as engineer and M F Zhilin as mechanic. After a total of fourteen flights Gromov reported generally favourably, but asked for a more effective rudder and improved engine installations.

Span 39.01m (127ft 11¼in); length 23.4m (76ft 9¼in); wing area 188.4sq m (2,028sq ft).
Weight empty 17,885kg (39,429lb); fuel 8,250kg (18,188lb); loaded weight 23,860kg (52,602lb) normal, 30,000kg (66,138lb) maximum.
Maximum speed 320 km/h (199mph) at sea level, 403km/h (250mph) at 8,000m (26,250ft); climb to 5,000m in 18.8min; service ceiling 10,800m (35,430ft); range 3,000km (1,864 miles).

In late March 1937 GAZ No. 156 received the ATsN, the M-100 engine being installed in a fireproof compartment with the exhaust pipe along the right side of the fuselage. It drove through a step-up gearbox to a 25,000rpm centrifugal blower sending the compressed air through aluminium pipes along the rear spar to the propulsion engines. The rudder was redesigned with slightly greater area and smooth skin, the tab being completely revised and the horn replaced by deeply recessed hinges. A more surprising change, the effectiveness of which was obvious from tunnel testing, was to place all four engine radiators in deep ducts under the inner engines. Airflow was controlled by left/right curved inlet doors pivoted about a vertical axis at top and bottom to open clamshell-fashion.

Flight testing resumed in July, but construction of runways at Khodinka

forced relocation to Podlipki, where work was again interrupted by a landing heavy enough to cause damage. Testing resumed on skis in November, followed by NII-VVS testing at Eupatoria, in the warmer Crimea, from March 1938.

Dimensions unchanged.
Weight empty 18,000kg (39,683lb); loaded weight 24,000kg (52,910lb) normal, 32,000kg (70,547lb) maximum.
Maximum speed 315km/h (196mph) at sea level, 439km/h (273mph) at 8,600m; climb to 5,000m in 16.3min; service ceiling 11,250m (36,910ft); range 3,500km (2,175 miles).

The ANT-42 *Dubler* (2nd prototype) was powered by AM-34FRNB engines each rated at 1,200hp, fed with compressed air from an ATsN-2 installation powered by an 850hp M-100A, with the M-103 (ATsN-2) planned in production. Because of the supposed near-invulnerability of the supercharged aircraft at high altitude, defensive armament (now fitted) was reduced to: nose turret, twin ShKAS; dorsal, a powered turret with one ShVAK; nacelle stations, completely revised with twin ShKAS; tail, twin ShKAS, other positions being deleted. The 'beard' was enlarged, fin and rudder enlarged, tailplane located lower and fixed without bracing, fuselage and centre section 100mm (3.9in) wider, main tyres enlarged, controls/instruments modified, improved AP-42 autopilot fitted, electrical system redesigned, two extra tanks installed, armour improved, engine cooling revised with new inlets and separate superimposed ducts each with its own controllable exit flap, and the bomb-carrying provisions finalised.

The latter were very comprehensive, and included: forty FAB-100s (twenty-four internal and sixteen under the inner wings); twelve FAB-250s (eight internal); six FAB-500s (four internal); four FAB-1000s (two internal); or a single FAB-2000 or FAB-5000, the latter weighing over 11,500lb. A variety of mines, fragmentation and armour-piercing bombs, photoflashes, target markers and cluster dispensers could be carried, as well as VAP-500 or VAP-1000 poison-gas dispensers on special racks further forward than the underwing bomb carriers.

The arrest of Tupolev and Petlyakov in October 1937 disrupted the pro-

A diesel-engined Pe-8/ACh-30B. Note the captured Ju 52/3m. (A Aleksandrov)

gramme, which was then further delayed by the execution in the first half of 1938 of over 40,000 senior Army and Air Force officers, and fifty leaders of the aviation industry, including the Director and Deputy Director of CAHI, the Ministry of Aviation Industry department heads and the Director of the V M Frunze engine factory where the AM-34 was built. On top of this, neither the AM-34FRNB nor M-103 had been cleared for production.

The second aircraft made its first flight on 26 July 1938, and between 11 August and 28 December 1938 it completed NII testing in the hands of Stefanovskii, with V V Dazko as co-pilot, A M Bryandinski as navigator and I M Markov as engineer. Numerous changes followed, but lack of engines held up production, and this problem was to bedevil the entire history of the ANT-42. There was even a

proposal from the Kharkov Aviation Institute to fit two 2,600hp PT-1 condensing steam turbines. This particular aircraft, however, flew 113 bombing operations in 1941–45. It was popular, because it was lighter and handled better than series machine.

Dimensions unchanged.
Weight empty 18,755kg (41,348lb); fuel 10,800kg; loaded weight 24,594kg (54,220lb) normal, 32,000kg (70,547lb) maximum.
Maximum speed 310km/h (193mph) at sea level, 365km/h (227mph) at 3,500m, 414km/h (257mph) at 7,500m; climb 20min to 5,000m; service ceiling 10,400m (34,120ft); take-off run 450m (1,476ft); landing run 550m (1,804ft).

On 20 April 1938 the ANT-42 was accepted for service as the TB-7, Nezval being sent to GAZ No. 124 at Kazan to supervise production. At this time the vast factory complex was incomplete, and numerous problems resulted in further ANT-42 redesign. Most notably, the ATsN-2

was abandoned, and the main engines changed to the 1,200hp AM-35 driving VISh-24 4.1m propellers with 20 deg pitch range. Other changes included deletion of the 'beard', addition of a nineteenth fuel tank, tandem radio masts canted to the left, and comprehensive radio and navigation aids.

The first series aircraft, 4211, flew on skis in late 1939, and was damaged on landing. Subsequently it was found that the large tyres were safe on ice and snow, and skis were not used again (almost unique for Soviet aircraft). By 1940 production was halted for lack of engines, deliveries comprising four aircraft with the AM-34FRNV (later replaced by the AM-35A) and ATsN-2 and two with the AM-35 and the ATsN-2 replaced by a station for a commander and radio operator, increasing the operational crew to eleven. On production aircraft armament was changed to a TAT (MV-6) dorsal turret with a retractable 20mm ShVAK, a similar gun in the KEB tail turret and a 12.7mm UBT in the ShU barbette behind each inner nacelle.

In May 1940 it was decreed that Nezval should be replaced by Commissar Kaganovich, who would reinstate production using the Charomskii M-30 or M-40 diesel, for which Nezval's OKB would design the installation. This version is described next. Meanwhile, in 1940 all completed aircraft (eighteen by early 1941) were re-engined with the AM-35A, rated at 1,350hp, which was to be the most important engine in VVS service. The first aircraft delivered reached 14 TBAP (heavy bomber regiment) at

Pe-8/ASh-82, with new nose and dorsal fin, carrying a test 14X cruise missile. (GFI photo, courtesy RART)

N396 at an Arctic base with Polar Aviation Il-12, Tu-2 and Li-2 in the background. (G F Petrov/RART)

Kiev–Borispol in September 1939, but this regiment's flying fell away to zero in 1940 through absence of spares. Not until April 1941 was this situation realised and rectified, and the unit was then reorganised as 412 TBAP (Brig. M V Vodopyanov), attacking Berlin from a base near Leningrad on 9 August 1941. Later the TB-7s were grouped into 81 DBAD (long-range aviation division).

Altogether forty-two aircraft were flown with the AM-35A, some having been built with diesel engines. In May/June 1942 No. 42066 flew Foreign Minister Molotov and others to Washington via Scotland, this previously unknown aircraft astonishing Western observers. Later in 1942 the aircraft was designated Pe-8 in honour of Petlyakov, who had been killed on 12 January of that year.

Dimensions unchanged.
Weights (typical) empty 19,986kg (44,061lb); fuel 13,025kg; loaded 27,000kg (59,524lb) normal, 35,000kg (77,160lb) maximum.

Maximum speed 347km/h (215.6mph) at sea level, 443km/h (275mph) at 6,360m; climb 14.6min to 5,000m; service ceiling 9,300m (30,500ft); range with 4-tonne bomb load 3,600km (2,237 miles).

The original AN-1 supercharged two-stroke diesel by A D Charomskii ran in 1932, and it led to a series of large V-12 water-cooled engines. By 1940 the M-30 had been cleared for production. Equipped with four turbochargers, it weighed 1,150kg and was rated at 1,400hp. It fitted the ANT-42 well, without significant changes in the propellers or cooling system, and outwardly distinguished only by the six ejector exhausts on each side being replaced by a single upward-curving pipe. Aircraft 4225 was the first to fly with experimental M-40 engines, in spring 1941, the test pilot being G F Baidukov.

Altogether, five aircraft flew with the M-40 diesel, rated at 1,500hp, followed by eleven powered by the generally similar M-30. Their range

was considerably increased, but most other aspects of performance were poorer. Reliability was poor, power and general behaviour were unacceptable at high altitude, starting was often difficult, and lubricating-oil consumption enormous. Nine aircraft were re-engined with the AM-35A. Charomskii cured most of the faults, and six later aircraft were powered by the ACh-30B.

Dimensions unchanged.
Weight empty 19,790kg (43,629lb); fuel/oil 13,025kg + 670kg (28,715lb + 1,477lb); loaded weight 26,000kg (57,319lb) normal, 33,500kg (73,854lb) maximum.
Maximum speed 345km/h (214mph) at sea level, 393km/h (244mph) at 5,600m; climb 16.2min to 5,000m; service ceiling 9,200m (30,200ft); range 5,460km (3,393 miles).

To try to resolve the engine situation Deputy Commissar P V Dementyev suggested fitting the ASh-82, Shvetsov's excellent fourteen-cylinder air-cooled radial, rated at 1,600hp. Nezval's KB designed the installation,

A diesel-engined testbed (test engine not known). (RART)

with the exhaust discharged through seven short pipes, each served by a front and a rear cylinder. The lower front of the inner nacelles, where the radiators had been, was skinned over. Later aircraft had exhaust collected in a ring and discharged by four large flame-damping fishtails, two on each side.

The first radial-engined TB-7 was 42047, though this was overtaken by 42058, the trials aircraft, flown by factory pilot B G Govorov in September 1942. From this point the TB-7/Pe-8 was a successful programme, though still available only in trivial numbers, thirty-four aircraft being built with the ASh-82 engine. GAZ No. 124 ceased production with twenty-three aircraft in 1941, manufacture of all later Pe-8 aircraft being at Moscow Fili (GAZ No. 22), which delivered twenty-two in 1942, twenty-nine in 1943 and a final five in 1944. Some sources state that total production was ninety-three (including the two prototypes) but the combined figure from the two factories appears more likely to have been 149.

Among other updates, the nose turret was replaced by a hand-aimed ShKAS in the tip of a simple pointed nose, reducing weight, drag and cost. With a trivial increase in tankage, the range of later ASh-82 aircraft actually exceeded that with diesel engines.

Dimensions, unchanged except pointed nose increased length to 23.59m (77ft 4¼in).
Weight empty 18,570kg (40,939lb); fuel/oil 13,120kg + 670kg (28,715lb + 1,477lb); loaded weight 27,200kg (59,965lb) normal, 36,000kg (79,365lb) maximum.
Maximum speed 362km/h (225mph) at sea level, 422km/h (262mph) at 5,600m; climb 15min to 5,000m; service ceiling 9,500m (31,168ft); range 5,800km (3,605 miles).

As noted earlier, a handful of later Pe-8s had the ACh-30B, the refined and more reliable two-stroke diesel, rated at 1,500hp. Apart from the VISh-61V1 propellers, the installation was the same as for the M-30/M-40. This engine was fitted to 42038 (the 45th aircraft) and to 42029 and 42039 (the 54th and 55th). Extensive additional equipment increased empty weight. Nezval commented 'No operational improvement is apparent'. The ACh-30B also powered the final four aircraft, which some sources designate Pe-8ON.

Dimensions as previous (revised nose).
Weight empty 22,864kg (50,406lb); fuel/oil c13,000+800kg; loaded weight 30,000kg (66,140lb) normal, 35,500kg (78,263lb) maximum.
Maximum speed 342 km/h (212.5mph) at sea level, 390km/h (242mph) at 6,000m; climb 19.5min to 5,000m; service ceiling 8,200m (26,900ft); range 5,600m (3,480 miles).

The final four aircraft, powered by the ACh-30B engine, were designated Pe-8ON (*Osobogo Naznachyeniya*, special assignment). They were long-range VIP transports, instantly identified by the addition of a curved dorsal fin. A cabin in the rear fuselage had three small windows on the left and three plus a door on the right. The first two aircraft, 42612 and 42712, carried full armament and had eight or twelve seats, plus provisions for cargo and auxiliary tanks in what in previous aircraft was the bomb bay. Delivered in 1944, these aircraft had a range of 7,000km (4,350 miles).

The dorsal fin and transport conversion were applied postwar to several Pe-8 bombers, both for transport by Aeroflot and, especially, for *Polyarni* (Polar) operations. All military equipment was removed. Early Polar aircraft, such as N395, had the ASh-82 engine, were painted orange and had special radio and navigation equipment. Later examples were powered by the 1,900hp ASh-82FN, driving propellers with four rectangular hollow-steel blades. Examples included N396 and N419. An exception was N562, which in 1949–57 operated unpainted, powered by the

ASh-71 (2,200hp) and later the ASh-73 (2,600hp). These larger-diameter eighteen-cylinder engines turned 4.1m AV-9Ye four-blade propellers. In 1952 this aircraft brought to a polar station an Mi-1 helicopter, slung underneath, and two years later landed at the geographical pole.

Numerous Pe-8s were used to test nine types of engine, mounted on the wings, in the nose or under the fuselage. One was the parent of the 5–1 and 5–2 high-speed air-launched research aircraft, while another, with many modifications including a new nose, was parent of the 10X, 14X and 16X cruise missiles based on the German Fi 103 'V 1'.

ANT-43

Assigned to Sukhoi's brigade in 1936, this high-speed transport was a stressed-skin design powered by an 800hp Gnome-Rhône 14Krsd engine. It would have seated a pilot and seven passengers, but in Tupolev's absence work was abandoned.

ANT-44, MTB-2

This major KOSOS programme was initiated in spring 1935 to meet an MA (navy aviation) requirement for a heavy bomber flying-boat, which of

The ANT-44 after having its landing gear fitted. (RART)

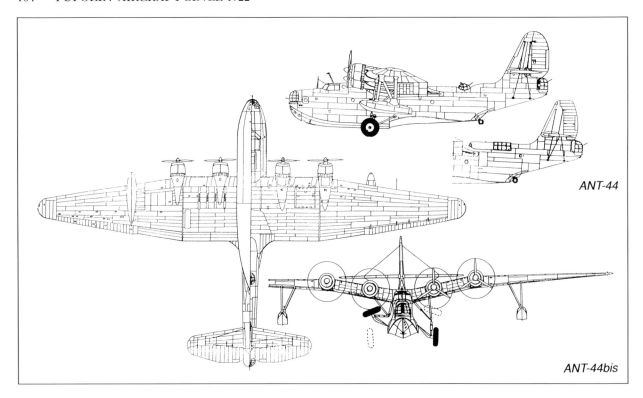

ANT-44

ANT-44bis

course could also fly oceanic reconnaissance and other duties. Tupolev
sketched one of the first flying-boats in
the world to have a completely modern
layout with the wing attached to the
hull and carrying the four engines on
the leading edge. The engines were
French GR14Krsd, of 810hp.

The wing had a profile of CAHI-
6mod type, with a thickness/chord
ratio of 15 per cent over the centre
section, which was untapered but had
a dihedral of no less than 10 deg. The
outer wings had equal taper to pointed
tips, the upper surfaces being horizontal. Structurally there were two
principal spars of built-up truss type,
but many ribs were assembled from
sections of pressed sheet. Hydraulically driven slotted-flaps were fitted in
six sections, two identical portions on
each centre section and one on each
outer wing. Each aileron was made in
two balanced but untabbed sections,
extending inboard beyond the wing
trailing edge in the usual KOSOS
manner, though not to the extent
shown in some drawings.

The deep but well-profiled hull had
an excellent planing bottom, the result
of prolonged testing in the CAHI
towing tank. The cross-section followed the ANT-22 in having a broad
lower part for flotation and a narrow
upper part for accommodation. The
bottom was a broad vee with a narrow

vertical ridge projecting down along
each side at the chine. Aft of the 90 deg
transverse step the bottom curved
gracefully up, with no chine ridge,
tapering to vanish at a point beyond
which the stern swept up to the tail.
The single-fin tail carried the fixed
tailplane quite high on the fin, with
two bracing wires above on each side
and an N below. The elevators and
rudder were tabbed and fabric-
covered, and fitted with external arc-
type mass balances, the rudder also
having a horn.

The engines, driving three-blade
two-pitch propellers with spinners,
were almost centred on the leading
edge, the nacelles being larger above
the wing than below. Each tight-fitting
long-chord cowl had adjustable rear
cooling gills, behind which a single
exhaust pipe emerged just above the
leading edge on each side. Fuel was
housed in metal tanks ahead of and
behind the front spar. On each side of
each nacelle the leading edge could
hinge down to form a work platform.
The underwing stabilising floats had
no step but were set at a positive angle,
and worked well.

The crew numbered seven, or eight
for some duties, the door being behind
the wing on the right, with small
hatches on each side of the nose.
Armament comprised single ShKAS
in nose and tail turrets, and in the rear

dorsal cockpit whose sliding canopy
was elevated above the main part of
the hull. Up to 2,500kg (5,511lb) of
bombs, mines or other stores could be
carried externally under the centre
section. Drawings show four racks on
each side, while a photograph of the
mock-up shows just two but provided
with vertical pylons carrying bombs
superimposed.

The ANT-44, MTB-2 (sea heavy
bomber type 2) was first flown by T V
Ryabenko and engineer Il'inskii at
Sevastopol on 19 April 1937. NII
testing followed. Though the OKB
later claimed 'perfect combination of
wing area, aircraft weight and installed
power, giving high flight and nautical
performance' only one further prototype was constructed.

Span 36.45m (119ft 7in); length 22.42m (73ft
6⅜in); wing area 144.7sq m (1,558sq ft).
Weight empty 12,000kg (26,455lb); loaded
weight 18,500kg (40,785lb) normal, 21,500kg
(47,400lb) maximum.
Maximum speed 330km/h (205mph) at sea
level, 345km/h (214mph) at 3,000m; climb to
1,000m in 3.5min, to 5,000m in 26.5 min; service
ceiling 6,600m (21,650ft); range 4,500km (2,796
miles); endurance 16hr; alighting speed
125km/h (77mph)

The second aircraft was the ANT-
44bis or ANT-44D (*Dubler*). It was fitted with 950hp M-87 engines, driving
VISh-3 propellers of unchanged 3.5m
diameter. Another obvious change was

The ANT-44 with bombs. (G F Petrov/ RART)

that the elevated aft dorsal cockpit was replaced by an MV-type dorsal turret, again with a single ShKAS. Other changes included redesign of the rudder to have a rounded shape of greater height and area, the horn being removed; modification of the elevators and elimination of the diagonal in the tailplane N-bracing; removal of the D/F loop antenna above the nose but addition of a pitot head on a mast above the cockpit; and addition of a water-deflecting dam round the bow below the turret.

According to ANTK Tupolev the ANT-44 was a flying-boat and the ANT-44bis an amphibian. The landing gear, which was eventually also fitted to the first aircraft, comprised single-wheel main units and a castoring semi-retractable tailwheel. Each main unit carried the large wheel and low-pressure tyre on the end of an axle which curved up to a pivot on the side of the hull, braced at the rear by a

strut pivoted further aft. The aircraft's weight was taken by a vertical strut pivoted to the underside of the front spar inboard of the inner engine. This strut consisted of two parts, a lower portion incorporating a faired shock-absorber and a rigid upper section. Where the two were joined, a long-stroke hydraulic jack was connected which, on being retracted, pulled the support strut inwards, rotating the wheel unit up clear of the water but remaining in the slipstream. Drawings show both main and tail gears fitted with skis.

The ANT-44bis was first flown by Ryabenko on 7 June 1938, and was subsequently subjected to factory and NII-VVS/MA testing at Sevastopol and Moscow reservoir. Flown by I M Sukhomlin, with landing gear removed, it set world class records: 17 June 1940, 1,000kg load lifted to 7,134m; 19 June, 2,000kg lifted to 6,284m and 5,000kg lifted to 5,219m;

28 September, with underwing floats jettisoned, 1,000km circuit Kerch – Kherson – Taganrog flown at 277.45 km/h; and 7 October, the same circuit flown with 2,000kg payload at 241.9km/h.

During the Great Patriotic War the ANT-44bis, often flown by Sukhomlin, operated as a naval bomber and priority cargo transport. No photographs have been found showing either aircraft bearing any markings, but in the Zhukovskii museum a beautiful model exists with red stars and 09 in black on the rear of the hull.

Dimensions as ANT-44.

Weight empty 13,000kg (28,660lb); loaded weight 19,000kg (41,887lb) normal, 22,000kg (48,500lb) maximum.

Maximum speed 355km/h (221mph) at sea level, probably more at height; climb to 1,000m in 3min, to 5,000m in 18min; service ceiling 7,100m (23,300ft); range (normal gross weight) 2,500km (1,553 miles); endurance 14hr; alighting speed 130km/h (80mph).

ANT-45

This 1936 project for a two-seat fighter was abandoned after the imprisonment of the KOSOS leaders.

ANT-46, DI-8

Though unrelated to the ANT-45, this also was a two-seat fighter (DI-8, two-seat fighter, type 8). The single prototype was ordered in November

ANT-44bis as it was originally built, without landing gear. (Jean Alexander/RART)

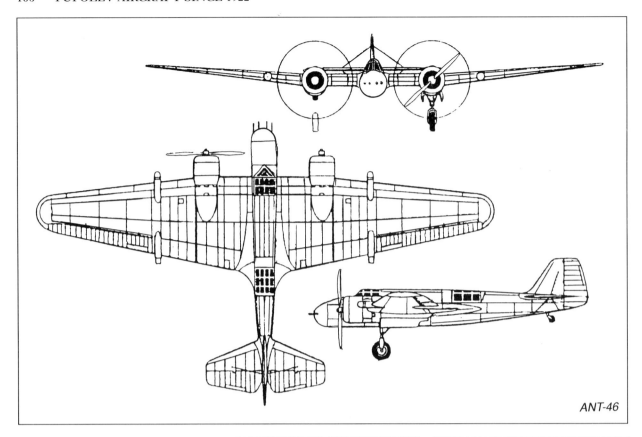

ANT-46

The only known photograph of the ANT-46. (via Philip Jarrett)

Span 20.3m (66ft 7¼in); length 12.24m (40ft 1⅜in); wing area 55.7sq m (600sq ft).

Weight empty 3,487kg (7,687lb); loaded weight 5,291kg (11,664lb) normal, 5,553kg (12,242lb) maximum.

Maximum speed 344km/h (214mph) at sea level, 388km/h (241mph) at 4,250m; climb to 3,000m in 6.8min, and to 5,000m in 11.4min; service ceiling 8,570m (28,120ft); range 1,780km (1,100 miles).

ANT-47

This high-speed fighter was abandoned after Tupolev's imprisonment.

ANT-48

Another abandoned project, this was to have been a high-speed bomber.

ANT-49

A projected reconnaissance version of the SB-2, with M-100A engines, carrying cameras in a heated bay replacing the bomb bay, plus additional fuel.

1934, and assigned to Arkhangel'skii on condition it did not in any way delay the SB. It had virtually the same airframe, but with a simple metal-skinned nose housing four ShKAS machine-guns, each with 500 rounds, and with the main armament of two APK-11 (DRP type) recoilless guns buried in the wings immediately inboard of the ailerons. These guns had an automatically fed supply of 45mm ammunition, the rear blast tube projecting behind the wing.

The only other significant differences from the SB were that the engines were 810hp Gnome-Rhône 14Krsd in installations resembling the ANT-44, but driving two-blade VFSh propellers, the bomb bay housed an auxiliary tank (and of course had no doors) and the tail was redesigned with a straight-edged angular outline. In CAHI markings, the aircraft flew as early as 9 August 1935, and continued factory testing to June 1936, but the closure of Kurchevskii's gun bureau and Tupolev's arrest halted the programme.

The model of the ANT-50. (G F Petrov/ RART)

ANT-50

This 1937 project was for a high-speed passenger aircraft, to be powered by two AM-34 (another document states AM-37) engines.

ANT-51

In February 1936 Stalin is reputed to have demanded a multirole tactical aircraft suitable for ground attack, level

bombing, reconnaissance and escort of heavy bombers, saying 'It must be simple to make, so that we can make as many as there are people called Ivanov [Stalin's personal telegraphic address] in our country'. Submissions came from Grigorovich, Nyeman (at KhAI), Polikarpov and Sukhoi.

P O Sukhoi collected a strong team led by D A Romeiko-Gurko, Ye S Fel'sner and V A Alybin, and produced a low-wing monoplane of the simplest construction, with wide use of pressings and castings. Features included an 820hp M-62 (Cyclone-derived) engine driving a two-blade variable-pitch propeller, continuous glazing joining the cockpits for the

pilot and navigator/gunner (the latter in an MV-5 turret with one ShKAS but able to aim a second ShKAS firing aft from the bottom of the fuselage), four ShKAS fixed in the wings, an internal bay under the pilot's cockpit for four FAB-100 bombs, and hydraulic drive for the bomb doors, flaps and inward-retracting main undercarriage.

Three prototype ANT-51s were built at ZOK. The first was flown on 25 August 1937 by Gromov, who reported favourably. Subsequently the engine was changed to an M-87A and ultimately to an M-88B of 1,000hp driving a VISh-23-7 propeller, while shortage of metal required the fuselage

The ANT-51 prototype. (Jean Alexander/RART)

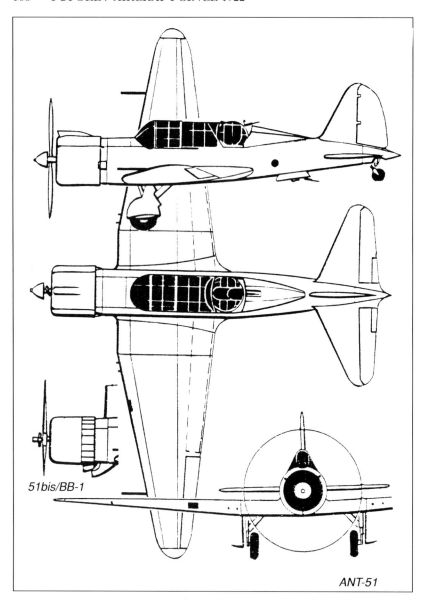

51bis/BB-1

ANT-51

to be redesigned in wood. Following NII-VVS testing in January/March 1938 by Yu G Makarov the type was accepted for production as the BB-1. With Tupolev imprisoned, it became effectively a Sukhoi aircraft. In September 1939 Sukhoi and team were relocated as an independent KB (construction bureau) in an hotel room in Kharkov, in order to be near the chosen series-production factory. In 1940 the BB-1 was redesignated Su-2. Data are for the ANT-51 as originally built.

Span 14.3m (46ft 11in); length 9.92m (32ft 6½in); wing area 29sq m (312sq ft).

Weight empty 2,604kg (5,741lb); fuel/oil 550kg (1,213lb); loaded weight 3,937kg (8,679lb).

Maximum speed 360km/h (224mph) at sea level, 403km/h (250.5mph) at 4,700m; climb to 4,000m in 8.3min and to 5,000m in 16.6min; service ceiling 7,440m (24,410ft); range 1,200km (746 miles); take-off run 380m (1,246ft); landing run 240m (787ft).

ANT-53

The ANT-53 was a 1936 project in Petlyakov's brigade for a passenger airliner powered by four AM-34FRNV engines. Perversely, the only known illustration shows a model with radial engines (possibly M-62 or M-85) and a remarkable B-29 type nose.

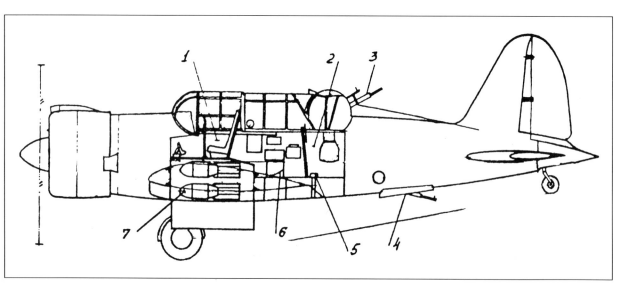

ANT-51 internal arrangement: 1, pilot's seat; 2, navigator/gunner's space; 3, ShKAS in MV-5 turret; 4, ShKAS fired through kinzhal *(dagger) hatch; 5, bombsight; 6, AFA-13 camera; 7, four FAB-100 bombs.*

The ANT-51bis prototype. (A Aleksandrov/RART)

56

High-speed bomber/reconnaissance project studied in the CCB-29 prison in 1940.

57, PB

This was the first task assigned to the detainees at CCB-29 in 1939. It was for a fast and powerful *Pikiryushchia* *Bombardirovshchik* (dive-bomber) with four M-105 engines. A wooden mock-up was being constructed when it was cancelled. The drawing entitled 'Aeroplane A', shows the side elevation against the ANT-42 (dotted outline) to the same scale.

Tu-2

58, Aeroplane 103 Though much smaller than many of Tupolev's earlier creations, Aeroplane 103 was to be perhaps the most important of his career. As outlined in the Introduction, it was created under prison conditions by Tupolev and some of his design team, who worked round the clock despite being 'enemies of the people', regarded as criminals and who for a year had suffered a mixture of hard labour and enforced idleness. From a single prototype, designed and built under almost impossible conditions, came an outstanding front-line attack bomber produced in time to play a role in the Second World War. This in turn led to a diverse family of derived aircraft.

The OKB preserved this model of Type 53. (RART)

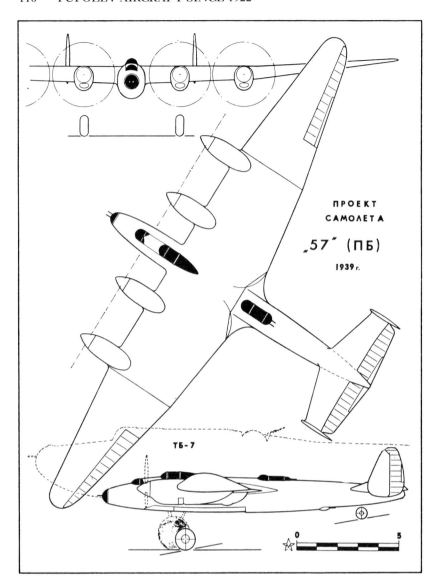

ПРОЕКТ
САМОЛЕТА
„57˝ (ПБ)
1939 г.

ТБ-7

0   5

A recent drawing of the 57, incorrectly labelled ANT-57 (dotted outline, ANT-42 to the same scale).

Soviet histories skated over the true history, and some suggested Tupolev was told 'Design a dive-bomber better than the Ju 88'. No evidence has been found for this, and today it appears the suggestion that the detainees should create a modern front-line dive-bomber was Tupolev's own, supported by the UVVS-RKKA. In fact, Tupolev's former brigade leader Petlyakov, organised into Special Technical Department 100, produced just such an aircraft in the Pe-2. Tupolev, later organised into group 103, did not duplicate Petlyakov's work but set out to create a heavier and more powerful aircraft with a similar layout. Tupolev assigned his own number 58 though of course he could no longer use the prefix ANT.

Work was able to begin in October/November 1938, soon after the NKVD formed CCB-29 in a fenced enclosure at GAZ No. 156 near the Khodinka airfield. The project was later designated Samolyot (Aeroplane) 103, from the number of the Special Technical Department. It was a clean stressed-skin aircraft, to have two engines in the 1,500hp class. It was to be as fast as contemporary fighters, and capable of adaptation to level or dive-bombing, reconnaissance, torpedo attack, air combat, long-range escort, *shturmovik* (armoured attack), transport and crew training. Features included a large weapon bay under a mid wing, a crew of three and a twin-finned tail.

The wing had a horizontal rectangular centre section with a front spar at 17.7 per cent chord and a main spar at 31.5 per cent, plus secondary spars nearer the leading edge and at the rear to carry the split flaps, hung on piano

hinges and depressed to a maximum of 45 deg. Most of the structure was assembled from rolled or extruded Dural sections, but the main strength of the wing was ahead of the second spar, where the flush-riveted outer skin was attached on top of a double inner skin with spanwise corrugations. The profile was CAHI-40, with a thickness/chord ratio of 13.75 per cent at the root and 9.9 per cent near the tip. The outer panels, attached by nineteen bolts, had 5 deg dihedral and carried the outer flaps and three-section balanced Frise ailerons, with fabric covering, the outer two sections having extra trailing-edge strips and the innermost a tab. Steel Venetian-blind dive-brakes were pivoted to the main spar, normally recessed under the wing. Their operation, like the flaps, landing gear and bomb doors, was hydraulic. Outboard of the starboard brake was a retractable landing light.

The fuselage was entirely a semi-monocoque, of the smallest possible oval section, with bolted structural breaks ahead of the wing and at the aft weapon-bay bulkhead. The mid-section was structurally dominated by the enormous weapon bay, which filled the entire space under the wing. Ahead of the wing the pilot sat on the centre-line above the front of the weapon bay. His canopy folded up and to the right, and he had a PBP-1A reflector sight for aiming two ShVAK 20mm cannon in the wing roots and twin 7.62mm ShKAS in the upper part of the nose. Behind the wing was a compartment for the navigator/bomb-aimer and radio-operator/gunner. The former had a small chart table, and his optical bombsight looked through flat glass windows in the floor. In an emergency he could fly the aircraft using a folding control wheel. If necessary he could turn round to man a single ShKAS on a TSS mount firing to the rear of the large bulged Plexiglas canopy. He also could assist the third man, whose radio included a direction/finding loop under the front of the canopy but whose main task was to lie facing aft to man the ventral ShKAS. This could be swung aft through a sliding hatch, the gunner also having a small porthole on each side. Ammunition for both rear guns was fed by two belts from a large box on the starboard side.

The weapon bay was completely unobstructed, and was normally closed

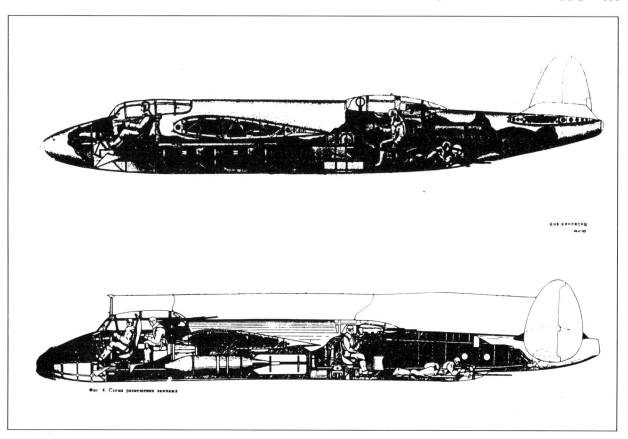

Two original drawings from the Tupolev OKB: upper, the original scheme for Aeroplane 103; lower, the scheme actually adopted (FAB-2000 on board).

by twin doors. It could carry nine FAB-100s or four FAB-500s or a single FAB-2000. As an overload two FAB-1000s could be carried under the wing roots, braced against the bomb doors, for a total load of 3 tonnes (6,614lb). An FAB-2000 is shown in the accompanying section drawing from STO-103, which also shows a completely unglazed nose. In fact, during review of the mock-up (built by Alimov in the yard with the help of Tupolev himself, who was a skilled carpenter) it was decided to improve the pilot's field of view by making the instrument panel a shallow arc under the windscreen and glazing the underside of the nose. This misled Western reporters into thinking the bomb-aimer lay in the nose. All crew positions were armoured. The radio antenna wires ran from the fins to a mast (which also carried the pitot heads) above the mid-fuselage.

The mock-up was approved on 21 April 1940. It featured M-120 engines with TK-2 turbosuperchargers. These engines were only in preliminary development by Klimov; they had three banks of M-103 type cylinders in Y-formation. On 1 July the GKO

issued detail requirements for three prototype 103 aircraft, the first with AM-37 engines rated at 1,380hp and the others with the M-120/TK-2. Both engine installations were finely streamlined, drag being minimised by installing the main coolant and oil radiators between the spars of the centre section, fed by three oval ram inlets in each leading edge inboard of the nacelles and with exit louvres in the upper surface behind the main spar. Small carburettor inlets were

provided immediately below the spinners of the VISh-61T three-blade 3.4m (11ft 1⅞in) propellers. A total of 2,000kg (4,409lb) of fuel could be housed in eight self-sealing wing cells with inert-gas protection.

The tail had stressed-skin, with fabric-covered control surfaces. The tailplane was fixed in incidence at the top of the fuselage, with 7 deg dihedral. Each main undercarriage had a single oleo shock strut with a steel fork fitting carrying a wheel with

The vital 58, Aeroplane 103. (via Philip Jarrett)

drum brake and 1,142mm by 432mm tyre, retracting completely into a deep nacelle bay with twin doors. The steerable tailwheel, with a 470mm by 210mm tyre, also retracted to the rear into a compartment with twin doors.

On 16 July 1940 the NKAP issued the construction schedule for the three aircraft, and instructed CAHI to do all related research, tunnel-testing and static/dynamic testing. The first aircraft was completed, put through systems testing, dismantled and taken to the NII test airfield, where M Nyukhtikov made the first flight on 29 January 1941. Bearing in mind the circumstances, had anything serious gone wrong, Tupolev and the STO-103 team would almost certainly have been executed, shortening this book considerably!

In fact, from the first flight it was clear that Aeroplane 103 was outstanding. Preparations for production were put in hand immediately, together with a second prototype incorporating major changes to improve fighting efficiency, described next. GAZ No. 156 testing was completed on 28 April 1941, and NII-VVS testing was completed in June. Centre of gravity range was fixed at 25.6–30.6 per cent. At an unknown later date the 103 was lost following fire in the starboard engine. Nyukhtikov escaped, but engineer A Akopyan was killed when his parachute caught on the tail.

Span 18.86m (61ft 10⅛in); length 13.2m (43ft 3⅝in); wing area 48.8sq m (525sq ft).

Weight empty 7,626kg (16,812lb); fuel/oil 2,147kg (4,733lb); loaded weight 9,950kg (21,936lb) normal, 10,992kg (24,233lb) maximum.

Maximum speed 482km/h (300mph) at sea level, 635km/h (395mph) at 8,000m; climb to 5,000m in 8.6min; service ceiling 10,600m (34,780ft); range 1,900km (1,180 miles); take-off run 440m (1,444ft); landing speed 155km/h (96mph), landing run 730m (2,395ft).

59, 103U Construction of this second prototype (U = *Uluchenyii*, improved) proceeded almost in parallel with the first. It was developed to meet new demands of the VVS, the fuselage being completely redesigned, and made slimmer in side elevation despite a new layout for a crew of four. The forward fuselage was slightly lengthened, the nose being more downsloping, with increased side glazing, but without the 103's ventral windows under the pilot's cockpit. The pilot was moved 0.65m (25in) forward, the

canopy being modified with a slightly higher roof hinged to the left, the starboard window unit hingeing to the right, and the windscreen sloping less acutely. The navigator/bomb-aimer was moved close behind the pilot. He retained his emergency flight control, and could aim bombs through the flat panels under the nose. As before, he could swivel round to man a ShKAS firing aft on a BUSh-1 mount.

The mid-fuselage was cut down in height to give the navigator a better field of fire. Aft of the wing the upper line continued straight into the canopy of the dorsal gunner, without a bulge, with twin ShKAS on a TSS mount which could fire under the opened rear part of the canopy. The two guns drew belts filling the original ammunition box. Access was provided by hingeing the front roof section to the right and the left window down. This gunner also served as radio operator, the antenna wire being attached to the port fin only, and the mast being left of centre at the rear of the navigator's canopy. The ventral gunner had an improved installation which broke the bottom line of the fuselage and provided small

extra side windows. His ShKAS now had an ammunition box of its own. A reconnaissance camera was added on the centreline immediately behind the bomb bay.

Overall length was slightly increased, span was slightly reduced and the bomb doors modified with curved fronts (where the auxiliary window panel had been). Provision was made for ten RS-132 rockets under the outer wings. The M-120/TK-2 still being unavailable, the 103U was fitted with AM-37 engines, uprated to 1,400hp. VISh-61P propellers were fitted, with diameter increased to 3.8m (12ft 5⅝in). They had distinctive spinners with a 150mm (5.9in) hole to admit cooling air to the cowling. The main radiators remained in the wing as before, with three oval leading-edge inlets on each side. Outboard of the engine was a further oval inlet in the leading edge, ducted to the carburettor and supercharger. The oil cooler was relocated under the engine, served by a larger inlet behind the spinner, and the nacelle was lengthened to project further behind the trailing edge. Fuel capacity was slightly increased.

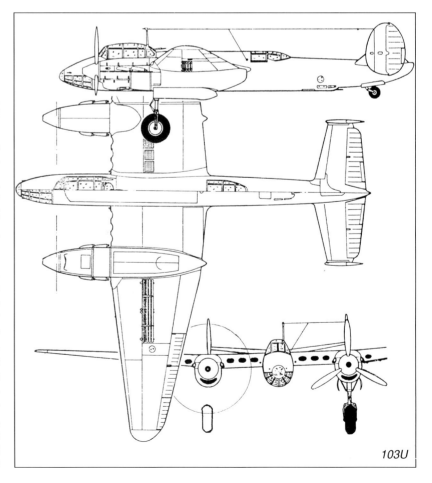

103U

The 59, 103U, prototype. (via Philip Jarrett)

The 103U was completed on 9 April 1941, and Nyukhtikov and engineer V A Miryuts made the first flight on 18 May. Again, this proved to be an outstanding aircraft, not least of its achievements being the remarkable centre of gravity range of 16.3–32.25 per cent. It later flew with VISh–61Ye propellers and other minor modifications.

Span 18.8m (61ft 8¼in); length 13.8m (45ft ¾in); wing area 48.52sq m (522sq ft).
Weight empty 7,823kg (17,246lb); fuel/oil 2,456kg (5,414lb); loaded weight 10,435kg (23,005lb) normal, 11,477kg (25,302lb) maximum.
Maximum speed 469km/h (291mph) at sea level, 610km/h (379mph) at 7,800m; climb to 5,000m in 9½min; service ceiling 10,500m (34,450ft); range 1,900km (1,180 miles); take-off run 435m (1,427ft); landing speed 155km/h (96mph), landing run 765m (2,500ft).

60, 103V Despite this excellent performance, the future of the whole programme was thrown into jeopardy by lack of an available engine. The trou-

bled M-120/TK-2 was cancelled, as was the AM-37 in order to increase output of the AM-38F, which was desperately needed for Il-2 produc-

tion. It appeared that the only alternative was to switch to the M-82 (later ASh-82) fourteen-cylinder air-cooled radial. This task needed 1,500 new drawings.

On 22 June 1941 Hitler struck at the Soviet Union, and before long Tupolev and his team were toiling round the clock digging air-raid shelters. The shelters were completed on 20 July, on the 21st Tupolev was released from detention, on the 22nd the shelters were occupied during the first raid on Moscow, and on the 25th the first evacuation train left CCB-29 on its way to the new location at Omsk. Many of the drawings needed for the M-82 installation were produced during the evacuation. Most of the other detainees were released soon after arrival, which facilitated their heroic efforts arranging accommodation whilst building

Two views of 60, the 103V. (G F Petrov/RART)

and equipping a new factory, GAZ No. 166.

No prototype flew with M-120/TK-2 engines, and one airframe almost identical to the original 103 languished incomplete, later to fly as the ANT-63, SDB No. 1. On arrival at Omsk work went on round the clock completing the third flying prototype, the 103V with ASh-82 engines, and in tooling up for production.

The 103V was modified in detail to facilitate series production. The airframe was subdivided into numerous subassemblies, as far as possible equipped with its electric and hydraulic systems before final erection, which also facilitated damage repair. Machining was minimised by widespread use of casting and fabrication from welded sheet and strip. The corrugated sheets inside the wing skins were modified to be of simpler sawtooth profile instead of top-hat rectangular form. The radial engines were installed in tight cowlings well ahead of the wings, driving 3.8m AV-5-167 three-blade constant-speed propellers. The nacelle incorporated the oil radiator underneath and a dorsal ram inlet to the downdraught injection-type carburettor. Other changes included a significant increase in the height and area of the vertical tail surfaces.

Despite the almost impossible conditions the 103V was built at Omsk between 1 August and 13 November 1941, and it was flown by M P Vasyakin on 15 December. Despite the fact that maximum speed at high altitude was reduced by almost 100km/h, the all-round performance at lower levels, where the type was expected to operate, was comparable with those of the predecessor aircraft, and in some respects – such as under negative-g – it was markedly improved. The 103V arrived at the NII test airfield at the end of February 1942. After rectification of engine-installation faults the 103V completed State NII-VVS testing, by which time production was building up and several aircraft were in action. In early 1942, with A N Tupolev no longer 'an enemy of the people', the 103V was redesignated Tu-2.

Dimensions unchanged except length 13.71m (44ft 11¾in).

Weight empty 7,335kg (16,171lb); fuel/oil 2,411kg (5,315lb); loaded weight 10,343kg (22,802lb) normal, 11,773kg (25,955lb) maximum.

Maximum speed 460km/h (286mph) at sea level, 528km/h (328mph) at 3,800m; climb to 5,000m in 10min; service ceiling 9,000m (29,530ft); range 2,000km (1,242 miles); take-off run 516m (1,693ft); landing speed 152km/h (94mph), landing run 640m (2,100ft).

60, 103VS Aircraft 100308, designated 103VS (S = *seriinyi*, series), is regarded as the first production example. Significant differences included fitting ASh-82A engines, rated at 1,600hp, and replacing all the ShKAS machine-guns by heavy 12.7mm UBT guns. The three UBT were each aimed by hand by the three aft-facing crew, and drew ammunition from redesigned magazine boxes. The navigator's mounting was designated BUSh-1, the upper rear position was designated VUB-2 and the lower rear mount was of Pe-2 type, known as the Lu/Pe-2. The dive-brakes were removed, the main tyres were 1,100mm by 425mm, the fins were more pointed at the top, and several systems were modified, mainly to reduce complexity. Operation of the main and tail landing gears was made pneumatic. The VS designation applied to a small batch, the first of which was taken to the Omsk airfield on 27 February 1942.

Factory testing took place in March and April, and in late April the first VS and two others flew to the Kalinin front for operational evaluation with the 2nd Air Army commanded by Gromov. The aircraft were supported by an OKB team led by Dmitri Markov. Initial reaction to the new aircraft was extremely favourable.

Span 18.86m (61ft 10½in); length 13.8m (45ft 3¼in); wing area 48.8sq m (525sq ft)

Weight empty 7,415kg (16,347lb); loaded weight 10,538kg (23,232lb) normal, 11,768kg (25,944lb) maximum.

Maximum speed 521km/h (324mph) at 3,200m; climb to 5,000m in 5.2min; service ceiling 9,000m (29,530ft); range 2,020km (1,255 miles); take-off run 450m (1,476ft), landing speed 152km/h (94.5mph), landing run 545m (1,788ft).

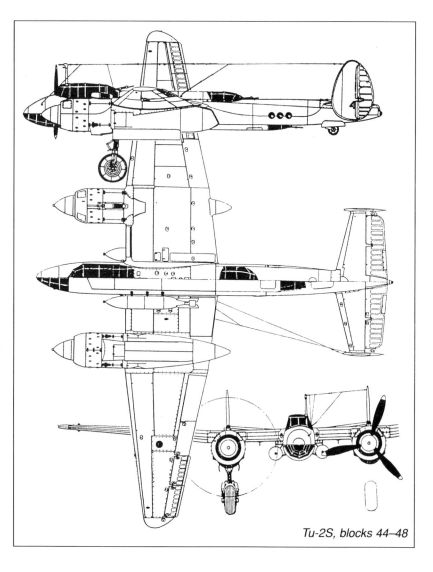

Tu-2S, blocks 44–48

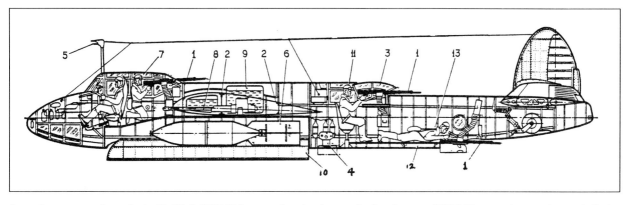

Internal arrangement of a production Tu-2S: 1, UBT 12.7mm gun; 2, main wing spar; 3, pivoted canopy of VUB-68 mount; 4, camera (see text); 5, pitot heads; 6, FAB-2000; 7, navigator/bomb aimer; 8, fixed ShVAK 20mm cannon; 9, fuselage tank; 10, bomb doors; 11, radio operator; 12, floor hatch; 13, ventral gunner.

61, Tu-2S Throughout the war the designation Tu-2S applied, despite numerous modifications. The initial standard of build for both GAZ No. 166 and, running a few months later, GAZ No. 156, was set by aircraft 716. This introduced the standard engine, the ASh-82FN, rated at 1,630hp on 91-octane fuel and 1,830hp or 1,850hp on 100-octane. It differed principally in having direct fuel injection. The initial standard propeller was the three-blade AV-5V-167, of unchanged 3.8m diameter. The engines could be started by Hucks dogs on the front of the propeller shaft, but were usually started pneumatically. They were carried on welded KhGSA trusses attached to the front two spars. In the front of the tight cowling was an iris-type ring of shutters to control cooling airflow, with automatic or pilot control. Immediately behind the cowling was an upper inlet feeding the supercharger and a shallow but broad lower duct for the oil cooler. Exhaust pipes from the fourteen cylinders were grouped into two discharge stacks, one projecting a short way through the nacelle below mid-level on each side. The 50 litre (11gal) oil tank was behind the firewall.

The main fuel tank was above the bomb bay, and the wing housed two inner and two outer tanks between the spars, total capacity usually being 2,800 litres (616gal), with self-sealing layers and NG (neutral gas) protection. There were no apertures in the leading edge, apart from those for the wing-root cannon. The hydraulic split flaps were driven to 15 deg for take-off and 45 deg for landing. The three-part Frise ailerons, with fabric skin over the top and the outer part of the under-surface, were driven to +25 deg/–15 deg, the inboard sections having trimmers with limits ±15 deg and the centre and outer sections fixed bendable tabs behind the trailing edge. A landing light could be folded out from under the outer starboard wing. Wingtips were wood, those of GAZ No. 156 being round and from GAZ No. 166 being more pointed.

The fuselage nose was of multi-ply wood, giving an almost conical outline with limited double-curvature. The canopy over the pilot, still seated 120mm (4¾in) left of the centreline, and navigator remained bulged, ending in the BUSh-1 gun mount. The navigator's retractable bombsight occupied the right side of the nose. Above the left side of the bulletproof windscreen was the radio mast, carrying the RSB-3bis/AD wire to the port fin and an RPKO-2B wire to the left side of the nose. A diagonally inclined mast on the left of the mid-fuselage carried the RSIU-3 or RSI-6 VHF wire to the port fin. Yet another wire, from high on the starboard side of the rear fuselage to the bottom of the starboard fin, served the SCh-3M uhf. The radio suite was more comprehensive than in equivalent Western aircraft. The radio operator was enclosed by the VUB-2B canopy, the rear portion folding forward for use of his UBT. The ventral gunner's Lu/Pe-2 was unchanged, but he had a larger circular window on each side. Access to the rear fuselage was via a ventral trapdoor, or via the hinged canopy roof. Tail controls remained fabric-skinned, elevator movement being +35 deg/–20 deg (trimmer +6 deg/0 deg) and the rudders ±2 deg (trimmers ±15 deg).

Production built up steadily at both factories. From May 1942 BAP-132 (bomber aviation regiment 132) con-

Tu-2S, block 20–44. (via Philip Jarrett)

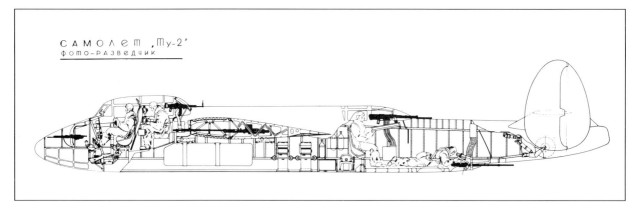

Late-series Tu-2S converted as a TU-2R: unlike the Tu-6 this had ShKAS 7.62mm machine-guns in the nose and forward dorsal positions. (G F Petrov/RART)

verted from the SB-2, assisted by GAZ No. 166 pilot Ya Paul' and two Omsk flying-school instructors, V Tyreshchyenko (who remained with the regiment to 1945) and A D Perelyot (who became a senior Tupolev test pilot). This initial regiment, with aircraft, left for the front in September 1942. A second regiment was operational by the end of 1942.

Early in production, at Block 20, the cowling diameter was reduced, with small blisters over the valve gear, and the nose was redesigned in metal, with compound double-curvature. At Block

44 the ventral gunner was given a row of three circular windows on each side, the radio operator was given the simpler VUB-68 gun mount open at the back, and a second retractable landing light was fitted beside the first. At Block 46 the ventral gunner was given the improved Lu-68 gun installation. At Block 48 the nose glazing was extended 50cm (20in) to the rear, bringing it back under the pilot's legs, and the pilot's canopy was given a straight top, leading to a VUS-1 mount for the navigator's gun (Savelyev's turret design). At Block 50 the radio

operator was given a fixed canopy of improved form, and the starboard wing root was given a hand-hold and non-slip walkway. From the twenty-first aircraft of Block 52 the wingtips were of metal. At Block 59 the leading edges of the wings and all tail surfaces were fitted with pulsating-rubber de-icers, the fins and rudders were increased in area and the propellers were changed to the AV-9VF-21K, with four square-tipped 'paddle' blades with diameter reduced to 3.6m (142in). At Block 61 the engine air inlets were enlarged to incorporate

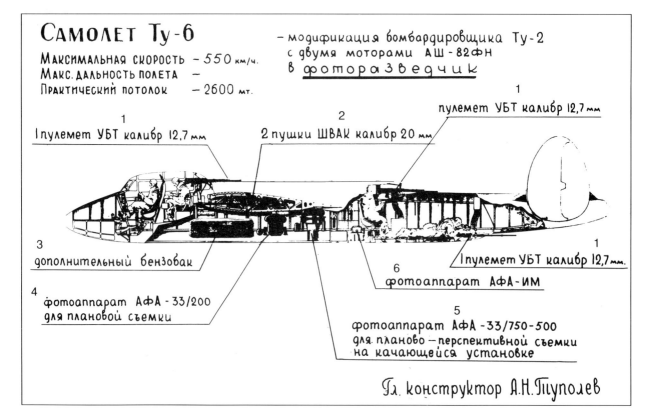

Internal arrangement of Tu-6: 1, UBT 12.7mm gun; 2, two ShVAK 20mm; 3, supplementary tank; 4, AFA-33/200 camera; 5, AFA-33/750-500 panoramic camera; 6, AFA-1M. (G F Petrov/RART)

Nose of Tu-2S, block 48 onwards; note AV-9VF-21K propellers. (M Passingham/RART)

Maximum speed (Block 10) 482km/h (300 mph) at sea level, 547km/h (340mph) at 5,400m, (Block 60) 551km/h (342mph) at 570m; climb to 5,000m (Block 10) in 9½min, (Block 60) 10.8min; service ceiling (Block 10) 9,500m (31,170ft), (Block 60) 9,000m (29,530 ft); range (Block 10) 2,100km (1,300 miles), (Block 60) 2,180km (1,355 miles); take-off run (Block 10) 485m (1,541ft), (Block 60) 540m (1,771ft); landing speed (Block 10) 158km/h (98mph), landing run, 67m, (2,216ft), (Block 60) 500m (1,640ft).

Tu-2T Though a torpedo-carrying version had been part of the original requirement, design effort for this did not become available until 1944. Two series aircraft were then modified. The first, in one document called Tu-2NT (*Nizhkii*, low), merely had a TD-44 pylon and Der-4-44U safety system under each wing root in place of the FAB-1000 rack, and a pilot torpedo sight. The inboard flaps had part of the trailing edge cut away. Drop tests with 45-36-AN torpedoes in February 1945 were excellent; data below are for this aircraft.

The second aircraft had 1,020 litres (224gal) of fuel in an auxiliary tank in the sealed bomb bay, strengthened landing gear and other changes. First flown by Opadchii and V P Marunov on 2 August 1946, it demonstrated the ability to fly 3,800km (2,361 miles), and to reach 490km/h (304.5mph) with two torpedoes or just over 500km/h (310mph) with one. It was also flown with three 800kg (1,764lb) torpedoes, but with reduced fuel. A slightly modified version was built at

dust filters. From early 1943 a proportion of Tu-2 production was equipped to carry cameras in the bomb bay, a typical fit being three or four AFA-33, AFA-3c/50 or AFA-33/50. In 1947, with the introduction of the large AK/AFU-156L in the rear of the bomb bay, causing a bulge, reconnaissance versions were redesignated Tu-6.

Total deliveries from GAZ Nos. 156/166 amounted to a little over 3,000, completed in 1948. To replace attrition a further 218 were built in 1947–50 by GAZ. No 125 at Irkutsk. These had upgraded electronics, and were powered by 2,100hp ASh-82IR engines driving AV-9VF-21K

propellers. In the 1950s the NATO ASCC (Allied Standards Co-ordinating Committee) allocated the codename 'Bat' to all versions. Postwar these outstanding aircraft were widely used for test programmes, as outlined later. Exports, mainly secondhand, were supplied to Bulgaria, China, Hungary, North Korea, Poland and Romania. Several are preserved in museums.

Dimensions as VS.

Weight empty (Block 10) 7,474kg (16,477lb), (Block 60) 8,404kg (18,527lb); fuel/oil (typical) 2,016kg + 300kg (4,445lb + 661lb); loaded weight (Block 10) 10,360kg (max 11,360kg, 25,044lb), (Block 60) 11,450kg (25,243lb).

Tu-2T, first aircraft, with two torpedoes, 11 February 1945. (V Nemecek/ RART)

The Tu-2 Paravan. (via Philip Jarrett)

GAZ No. 125 in 1947, serving with all three main AV-MF fleets until the mid-1950s.

Dimensions and engines, as Tu-2S.
Weight loaded 11,423kg (25,183lb) normal, 12,389kg (27,313lb) maximum.
Maximum speed 493km/h (306mph), or 505km/h (314mph) with only one torpedo; service ceiling 7,500m (24,600ft); range 2,075km (1,289 miles); take-off run 580m (1,902ft), landing speed 159km/h (98mph) landing run 480m (1,574ft).

Tu-2S modifications The Tu-2/ASh-83 (not, as in some accounts, called the Tu-2M) was the first series aircraft, 100716, retro-fitted in 1945 with 1,900hp ASh-83 engines, driving four-blade AV-5V propellers. Empty/maximum weights were 8,870/12,735kg (19,555/28,075lb), and maximum speed was 635km/h (395mph)

Air test of RD-20 turbojet. (M Passingham/RART)

The Tu-2Sh prototype. (Jean Alexander/RART)

An early air-refuelling test, showing a Yak-15 receiver. (Yefim Gordon archive)

in which the main landing legs were pivoted past the vertical to move the wheels forward 125mm (4.9in). NII-VVS testing was completed on 28 February 1947, the Il-10 continuing as the aircraft for this role.

Two aircraft, said unofficially to be designated Tu-2K (*Katapult*), were used to test ejection seats; the 1944 aircraft had the seat in the navigator cockpit and the 1945 example had an open cockpit at the radio operator station. Another, unofficially said to be called Tu-2N (*Nene*) was used in 1947 to test Rolls-Royce Nene and RD-45 turbojets slung under the fuselage; other Tu-2Ss tested five types of turbojet. Two Tu-2 *Paravan* were built to test barrage-balloon cable deflectors. The steel deflector cable was attached to the tip of a 6m (20ft) monocoque cone cantilevered ahead of the nose, leading back to a cable cutter on each reinforced pointed wingtip. The first, tested in September 1944 with 150kg (331lb) ballast in the tail, still reached 537km/h (334mph) at 5,450m (17,880ft).

In late 1943 System 104 was planned to install air interception radar into a large fighter, and the result was the 104 interceptor, first flown by Perelyot and engineer L L Kerber on

at 7,100m, but range was reduced to 1,660km (1,030 miles).

The Tu-2Sh was a *Shturmovik* (armoured ground attack) version proposed by A D Nadashkevich (Tupolev armament brigade leader) and tested in three forms: (1) 1944, with the bomb bay filled by eighty-eight modified PPSh-41 sub-machine-guns all firing ahead at 30 deg depression; (2) 1944, with a 75mm gun under the centreline reloaded by the navigator;

(3) 1946, with the devastating frontal armament of two ShVAK, two NS-37 and two NS-45; this third version reached 575km/h (357mph) at 5,800m (19,028ft). A related two-seat 1946 aircraft was the RShR or Tu-2RShR, in which the entire lower part of the forward and middle fuselage was occupied by a 57mm RShR automatic cannon, with the muzzle brake projecting 0.5m (20in) ahead of the nose. This was the first of several prototypes

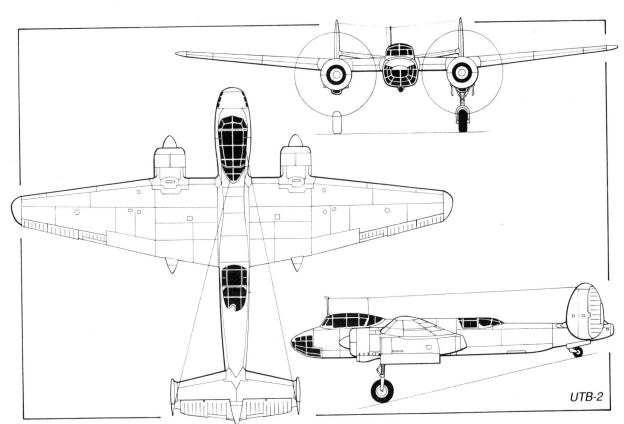
UTB-2

18 July 1944, with the radar simulated by ballast. The pilot had a modified sight, and two VYa-23 (23mm) guns were positioned under the forward fuselage. The rear fuselage was empty and faired over. The radar was fitted in 1945, the first in the USSR.

Several Tu-2Ss were gutted and equipped as freighters (not specially designated Tu-2G) for use by Aeroflot. They had an internal payload of 2,000kg (4,410lb). One Block-60 aircraft was used in 1948–49 for test dropping bulky loads carried externally, one such load being a GAZ-67b scout car, with which the aircraft reached 378km/h (235mph) and a height of 6,000m (19,685ft).

UTB P O Sukhoi used the Tu-2 as the basis for this series-built crew trainer, so it requires a brief mention. Powered by 700hp ASh-21 engines (each effectively half an ASh-82) driving two-blade VISh-111V propellers, it was stressed for lower weights and reduced design factors, with main and tail tyres 900mm by 300mm and 440mm by 210mm respectively. Ailerons, flaps, fuel and accessory systems were totally redesigned.

The large cockpit seated pilot and instructor side-by-side, with one or two navigator seats at the rear. A crawl tunnel through what had been the bomb bay led to the rear fuselage where the radio operator had a VUB-68 mount for a UBT (normally with only sixty rounds). There was no lower rear position. External fuselage

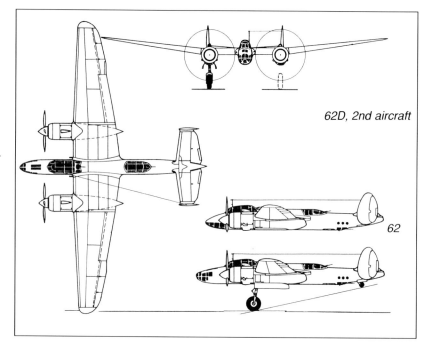

62D, 2nd aircraft

62

racks could drop four FAB-50s or FAB-100s aimed by the navigator through the glazed nose. First flown in summer 1946, the UTB (also called UTB-2) was also supplied to Poland.

62, Tu-2D

After A N Tupolev's release in July 1941 he was able to re-establish his OKB, with engineering teams both at Omsk and in the restored KOSOS

building. Designs continued the original numbering sequence, with the prefix Tu– gradually becoming standard. Another progressive change was elimination of designations based on function, the last being SDB.

From 1941 Tupolev had planned a long-range bomber with extended outer wing panels housing two additional tanks on each side. There appears to be no evidence for the designation 103D, which appears in unofficial accounts. Design effort for this project could not be spared until 1944, when two prototypes were quickly completed. The first, No.

The second 62, Tu-2D, No. 714. (via Philip Jarrett)

718 (originally the third production Tu-2S) was flown by Perelyot on 17 June (one document says July) 1944. The second, No. 714, had a redesigned forward fuselage. The nose was extended by 0.62m (23.6in) and widened to form a full-time compartment for the navigator/bomb-aimer. His original station was occupied by a second pilot, behind and to the right of the first, increasing the crew to five. Tail span was increased from 5.4m to 5.7m (17ft 9in to 18ft 8in), the vertical tails were enlarged, the aileron chord increased, the normal maximum bomb load increased to 4,000kg (8,818lb) and the main landing gears strengthened. Engines remained ASh-82FNs.

Both were put through NII-VVS testing between 20 October 1944 and 31 October 1945, being judged 'much better than the Il-4 or Yer-2', but there was little chance of production. In 1947 the second aircraft was modified as the 62T torpedo carrier, with Tu-2T torpedo installations, auxiliary tanks in the bomb bay, AV-9VF four-blade propellers and 1,260mm-diameter main tyres requiring bulged bay doors. Maximum speed was 501km/h (311mph) and range 3,800km (2,360 miles). Data are for No. 714 as tested in 1944.

Span 22.06m (72ft 4½in); length 14.42m (47ft 3¼in); wing area 59.05 sq m (636sq ft).

Weight empty 8,316kg (18,333lb); fuel/oil 2,820kg (6,217lb), weight loaded 12,290kg (27,094lb) normal, 13,340kg (29,409lb) maximum.

Maximum speed 465lm/h (289mph) at sea level, 531km/h (330mph) at 5,600m; climb to 5,000m in 11.8min; service ceiling 9,900m (32,480ft); range 2,790km (1,734 miles); take-off run 480m (1,574ft); landing speed 149km/h (92.6mph), landing run 610m (2,000ft).

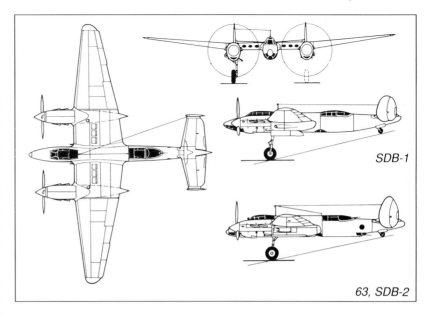

SDB-1

63, SDB-2

63, SDB

Towards the end of the war Tupolev produced two prototype SDBs (fast day bombers) with the *Izdelye* (product) number 63. The first, later designated SDB-1, was a modification of the original Aeroplane 103. It was arranged for a crew of two, the navigator/bomb-aimer being seated behind the wing and sighting through ventral windows. It retained the bulged rear canopy, single landing light and three rows of four windows under the nose. Differences included two AM-39 engines each rated at 1,870hp, driving 3.6m (142in) three-blade AV-5LV-22A propellers, modified centre-section

cooling ducts, and absence of rear guns or dive-brakes. The first flight, by Perelyot, was on 21 May 1944, and joint GAZ/NII-VVS tests took place from 5 June to 6 July 1944.

Span 18.86m (61ft 10½in); length 13.2m (43ft 3½in); wing area 48.52sq m (522 sq ft).

Weight empty 7,787kg (17,167lb); fuel/oil 1,767kg (3,896lb); weight loaded 9,800kg (21,605lb) normal, 11,800kg (26,014lb) maximum.

Maximum speed 527km/h (327mph) at sea level, 645km/h (401mph) at 6,600m; climb to 5,000m in 7.45min; service ceiling 10,000m (32,810ft); range 1,830km (1,137 miles); take-off run 470m (1,547ft); landing run 550m. (1,804ft).

The second Type 63, SDB-2, was distinguished by having redesigned main landing gears with a straight oleo leg connected to the inboard end

The AM-37-engined 63, SDB-1 (RART)

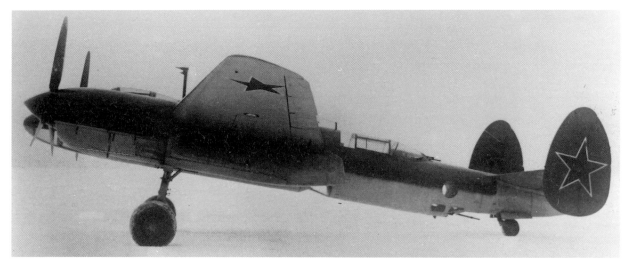

The second 63, SDB-2. (via PhilipJarrett)

of the axle. The wheel assemblies were unchanged, though the tailwheel carried a larger tyre, 480mm by 200mm. The engines were AM-39Fs with the same rating (but at higher altitude), in longer cowlings with improved oil coolers but driving unchanged propellers. Other differences included single UB guns in both upper and (now added) lower rear positions, an increase in reserve fuel, a bomb load from 1 to 4 tonnes (8,818lb), nose glazing with the rows of four windows replaced by rows of two double-size ones, and hydraulic tubing changed from steel to Dural. One document mentions the 5.7m (18ft 8½in) tail span of the Type 62, but this is not shown on three-view drawings. First flown on 14 October 1944, the SDB-2 completed a successful combined test programme by M A Nyukhtikov and engineer V A Shubralov on 30 June 1945. The later Type 68 was preferred.

Span 18.8m (61ft 8¼in); length 13.6m (44ft 7½in); wing area 48.8sq m (525sq ft).

Weight empty 8,280kg (18,254lb); fuel/oil 1,750kg (3,858lb); loaded weight 10,925kg (24,085lb) normal, 13,650kg (30,093lb) maximum.

Maximum speed 547km/h (340mph) at sea level, 640km/h (398mph) at 6,600m; climb to 5,000m in 8.7min; service ceiling 10,100m (33,140ft); range 1,530km (950 miles); take-off run 535m (1,755ft); landing speed 156km/h (97mph); landing run 650m (2,132ft).

63P, Tu-1 Apart from retaining Tu-2S landing gear, this later aircraft was based on the SDB-2, but was completed as a three-seat long-range radar-equipped interceptor and escort fighter. The engines were 1,950hp AM-43VS, driving four-blade AV-9K-22A propellers. The crew comprised a pilot on the centreline and upper and lower rear gunners, the upper also being the radio/radar operator. Armament comprised two NS-45s in the lower part of the nose, two NS-23s in

the wing roots and a single UBT in the upper and lower rear positions. An internal bomb load of 1,000kg (2,205lb) could be carried, and maximum fuel capacity was increased to approximately 4,000 litres (880gal). The nose, rear fuselage and tail housed components of the PNB-1 *Gneiss*-7 radar, based on the German FuG 220, with tail-warning. First flown on 22 March 1947, the 63P's test programme was completed on 3 November of that year. Series production as the Tu-1 was thwarted by unavailability of the engines, and the 63P was used for radar interception research. It is believed to have been the first aircraft in the USSR with operative radar.

Span 18.86m (61ft 10½in); length 13.6m (44ft 7½in) [including guns, 14.1m, 46ft 3in]; wing area 48.8sq m (525sq ft).

Weight empty 9,460kg (20,855lb); fuel/oil 3,000kg (6,614lb); loaded weight 12,755kg (28,119lb) normal, 14,460kg (31,878lb) maximum.

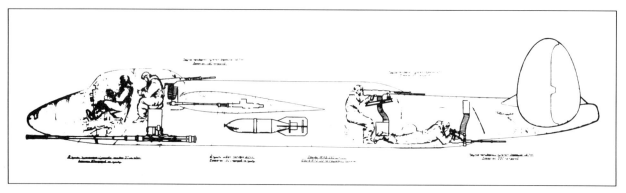

Armament of the first variant Tu-1 was two NS-45 under the nose, with 50 rounds per gun, each larger than a litre bottle, two NS-23 in the wing roots with 130 rounds per gun, a UBT 12.7mm with 200 rounds at upper rear and another UBT with 350 rounds at lower rear. This drawing shows the second variant, with a fourth crew member behind the pilot with a UBT (190 rounds), two ShVAK 20mm with 150 rounds per gun in the wing roots and two FAB-250 in the bomb bay. (G F Petrov/RART)

The 398mph 63P, Tu-1. (via Philip Jarrett)

Maximum speed 479km/h (298mph) at sea level, 641km/h (398mph) at 8,600m (28,220ft); climb to 5,000m in 11.6min; service ceiling 1,000m (36,090ft); range 2,250km (1,400 miles); take-off run 605m (1,984ft); landing run 560m (1,837ft).

64

In 1943 the Kremlin appeared to become aware once more of the importance of strategic bombers, and began a campaign to obtain Boeing B-29s under Lend-Lease. Stalin also ordered two Soviet aircraft in the same class. Myasishchev was ordered to build a close copy of the American aircraft as the DVB-202, while Tupolev was given a free hand to produce an aircraft in the same class. This was quickly launched as Type 64.

Tupolev assigned the most senior designers and brigade leaders to this enormous project. The wing was to have NACA-2330 profile, with aspect ratio 11.9, the centre section ending at the inner engines being rectangular and horizontal, and the outer panels having dihedral and taper on the leading edge only. The circular-section fuselage, with the same 2.9m (114in) diameter as the B-29, incorporated three pressure cabins, the forward section for a crew of four having a B-29 type nose. Two gunners occupied the rear fuselage, and a third gunner

was in the extreme tail. Defensive armament comprised nine ShVAK 20mm cannon in four twin turrets plus a single tail gun, and normal bomb load was to be 5,000kg (11,020lb), including if necessary a single FAB-5000.

Points of divergence from the US design were the liquid-cooled engines, 2,200hp AM-44TKs with twin turbos and underslung radiators, driving 4.5m (14ft 9in) four-blade propellers, large single-wheel main landing gears

retracting inwards into the wing, and a twin-fin tail resembling an enlarged Tu-2 design. The twin-wheel nose gear was double-jointed to fold straight up between the chin turret and the forward bomb bay. Auxiliary power services were hydraulic, except for the electrically powered turrets. The initial *Eskiznyi* (sketch) three-view showed a span of 39m (127ft 11½in), length 26.575m (87ft 2¼in), tail span 10m (32ft 10½in), track 7.6m (24ft 11¼in) and distance between

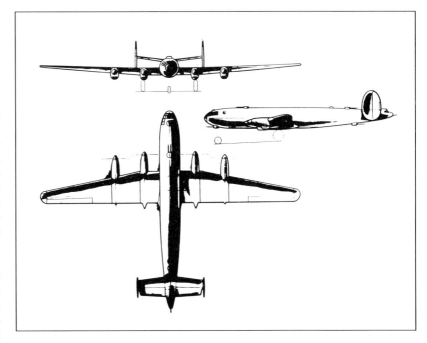

The original OKB drawing of the 64. (via Yefim Gordon)

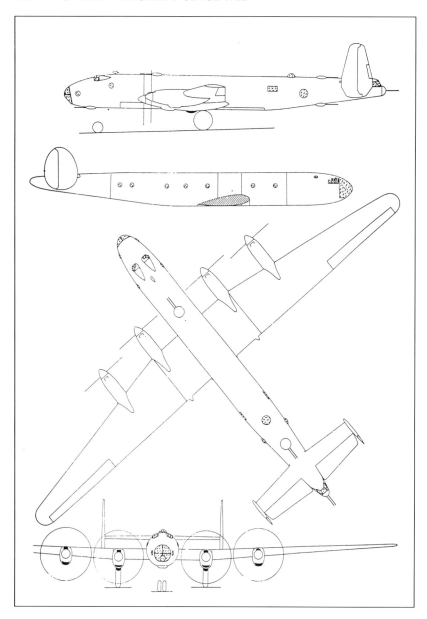

Weight empty 22,000kg (48,500lb); loaded (5-tonne bomb load) 36,000kg (79,365lb).

Maximum speed 600km/h (373mph) at 6,000m; range 3,000km (1,864 miles).

65, Tu-2DB

Type 65 was the penultimate DB (long-range bomber) derived from the Tu-2S. It used the long-span five-seat airframe of the 62, matched with the bomb load and defensive guns of the Tu-2S, but was powered by AM-44TK liquid-cooled engines. Each was rated at 2,200hp with the exhaust from the right bank of six cylinders driving a TK-1B (TK-300) turbo-supercharger. The propellers were 3.8m (12ft 5½in) AV-5LV-166B three-blade, with emergency mechanical control directly by the pilot. The main landing gears were set 125mm (4.9in) further forward, as on the RShR. The prototype, which was 100714 again modified, was flown by F F Opadchii on 1 July 1946. Testing was curtailed by engine problems, and the OKB turned its attention to the B-29.

Dimensions as 62, except wing area 59.12sq m (636.4sq ft).

Weight empty 9,696kg (21,376lb); fuel/oil 2,820kg (6,217lb); loaded weight (6,727lb) 13,205kg (29,112lb) normal, 15,962kg (35,190lb) maximum.

Maximum speed 579km/h (360mph) at 9,300m; climb to 5,000m in estimated 9min; service ceiling estimated at 11,000m (36,090ft); range estimated at 2,570km (1,597 miles) [also recorded as 2,670km]; take-off run 480m (1,574ft); landing run 490m (1,607ft).

A later drawing showing how the 64 would have been built, with an extra side elevation of the 66 transport.

inner/outer thrust lines 4.4m (14ft 5¼in).

Following the mock-up review, in 1945 the 64 was considerably modified, even as drawings were being issued to the shops. The fuselage was lengthened, and the nose was re-designed with only a glazed front cap for bomb-aiming, the pilots being given individual 'bug-eye' canopies of the type at that time coming into fashion in the USA (and first projected by Tupolev in 1935 for the ANT-26). Each pilot had flat windscreen panels giving a view ahead and to his own side only. The outer nacelles were made much shorter, stopping ahead of the flaps instead of forming part of them and projecting behind the trailing edge. The vertical tails were enlarged and given straight leading and trailing edges. The armament was changed to ten B-20 20mm cannon in five twin turrets, and the rear fuselage gunners were given additional windows. Provision was made for radar between the front and rear bomb bays. In January 1946 the whole project was cancelled, so that Tupolev could concentrate on producing a clone of the B-29. He was distressed, thinking this a retrograde decision, though he recognised that the American aircraft had greater fuel capacity and range. Data are for the 64 as it would initially have flown.

Span 42.8m (140ft 5in); length (excluding guns) 29m (95ft 1¼in); wing area 152sq m (1,636sq ft).

66

One drawing showing the definitive version of 64 also shows a side elevation of this derived transport version. It would have had a fully pressurised fuselage, with B-29 type nose, circular cabin windows and an empty tailcone carrying a tail with oval fins and rudders. The wing would have been lowered to give an unobstructed level cabin floor. No drawings were issued, but the fuselage greatly speeded design of the Tu-70.

The 65, Tu-2DB prototype. (via Philip Jarrett)

67, Tu-2D

A single example was built of this long-range bomber with the wing of the 65, fuselage of the 62, main tyres of 1,260mm (49.6in) diameter requiring bulged bay doors, and two 1,900hp Charomskii ACh–30BF liquid-cooled diesel engines. Flight testing by Perelyot began on 12 February 1946. He thought the 67 promising, but it was abandoned because of engine problems and the need to concentrate on the B–29.

Dimensions as 62 except wing area as 65.
Weight empty 8,323kg (18,349lb); fuel/oil 2,995kg; weight loaded 13,626kg (30,040lb) normal, 15,215kg (33,543lb) maximum.

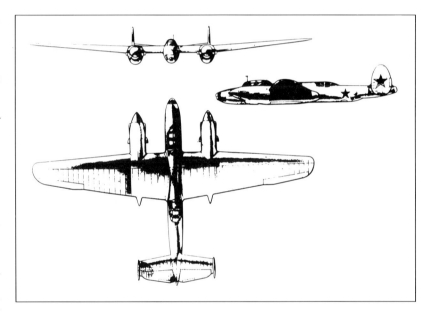

OKB drawing of Aeroplane 67. (via Yefim Gordon)

The diesel-engined 67. (RART)

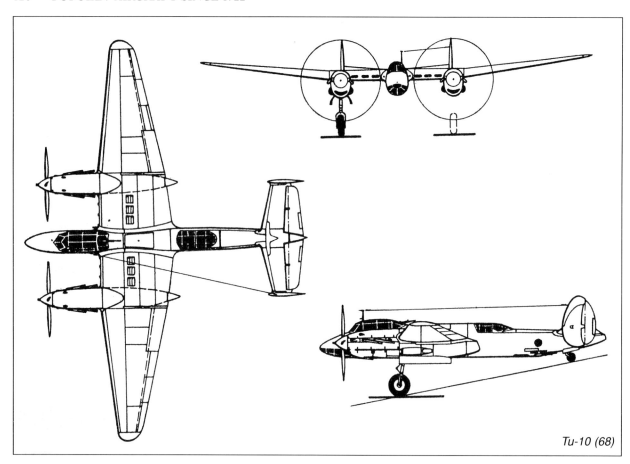

Tu-10 (68)

The 68 prototype. (via Philip Jarrett)

Maximum speed 509km/h (316mph) at 6,200m; climb to 5,000m in 13.6min; service ceiling 8,850m (29,000ft); range estimated at 5,000km (3,110 miles); take-off run 530m (1,738ft); landing run 700m (2,296ft).

68, Tu-10

This frontal (front-line tactical) bomber was essentially a Tu-2S re-engined with 1,850hp AM-39FNV engines driving 3.8m (150in) three-blade AV-5LV-22A propellers. Flown by Pere-lyot on 19 May 1945, it impressed him favourably, and it was subjected to progressive refinement. The propellers were changed to the AV-5LV-166B and

then to the 3.6m (142in) four-blade AV-9K-22A driven by 1,850/1,870hp AM-39FN-2 engines. Outer-panel dihedral was reduced to 1.5 deg, and the fins and rudders were enlarged. NII testing was completed on 20 November 1946, and to keep the Omsk line going ten series aircraft were built with Service designation Tu-10.

Dimensions as Tu-2S.
Weight empty 8,870kg (19,555lb); fuel/oil 1,630kg (3,594lb), weight loaded 11,650kg (25,683lb) normal, 12,735kg (28,076lb) maximum.
Maximum speed 520km/h (323mph) at sea level, 635km/h (395mph) at 7,100m; climb to 5,000m in 11.1min; service ceiling 9,800m (32,150ft); range 1,660km (1,030 miles); take-off run 525m (9,042ft); landing speed 190km/h (118mph), landing run 680m (2,230ft).

69, Tu-8

This long-range bomber was the ultimate development of the Tu-2 family, and also the heaviest. The wing was based on that of the 62 but with chord increased to give significantly greater area. The main landing gears were moved forward as on the RShR, the tail span was 5.637m (18ft 6in) carrying further-enlarged fins and rudders, and the landing gears were again strengthened and fitted with 1,170mm by 435mm main tyres and 580mm by 240mm on the tailwheel. The engines remained 1,850hp ASh-

The 68 after modification in 1946 as the prototype of Tu-10. (G F Petrov/ RART)

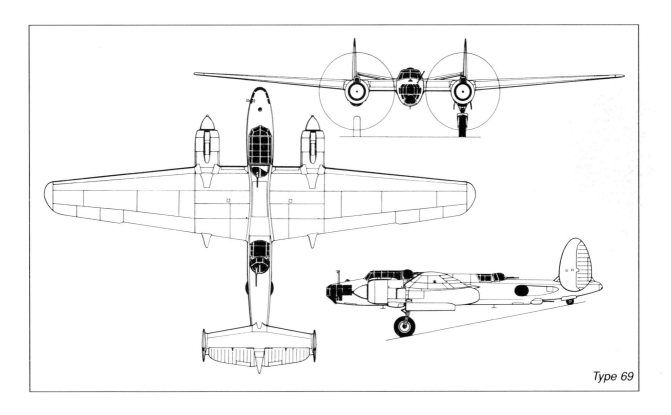

Type 69

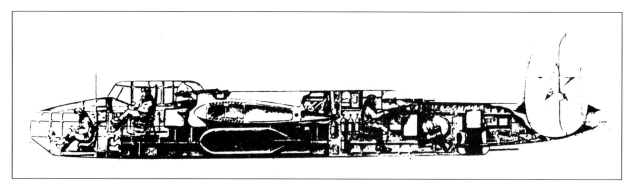

OKB drawing of 69, with the new nose and ventral power turret. (via Yefim Gordon)

82FNs, but driving four-blade (not square tip) AV-9K-22A propellers.

The defensive armament was completely revised. The pilot fired a 20mm B-20 on the right side of the fuselage. The second pilot could swivel round to fire a B-20 with 190 rounds. The radio operator had a B-20 with 250 rounds in an MV-11 electrically powered turret, and the lower rear gunner aimed his B-20 electric turret under remote control looking through diagonal portholes recessed into the sides of the rear fuselage (a scheme to be seen later on jet bombers). The bomb load was increased to 4,500kg (9,921lb). A single 69 was flown in May 1947, by which time such aircraft, however good, were of little interest. Projected versions with AM-42 engines (said to be designated Tu-8B) and ACh-30B diesels (said to be Tu-8S) were never built.

Span 22.06m (72ft 4½in); length 14.61m (47ft 11⅛in); wing area 61.26sq m (659sq ft).

Weight empty not found, but about 10,100kg (22,270lb); weight loaded (most documents) 16,750kg (36,927lb).

Maximum speed 507km/h (315mph) at 5,700m; climb to 5,000m in 17min; service ceiling 7,650m (25,100ft); range 4,100km (2,548 miles); take-off run 860m (2,825ft); landing run 632m (2,073ft).

Tu-4

The story of the Tu-4 is without parallel in technological history. It is the story of how what has been correctly described as 'the most complicated moving machine created by man up to that time' was appropriated without permission of the owners, dismantled and dissected in the most minute detail, to the extent of analysing all the materials used, and then copied in a form which was put into series production in the Soviet Union. It was the first and greatest example of what today is called 'reverse engineering'.

The resulting Tu-4 strategic bomber multiplied the radius of action of the DA (long-range aviation) so that in a war situation it could even have bombed the whole USA or Western Europe. This caused a wholly disproportionate expenditure on air defences by what in 1949 became the NATO powers. To the Tupolev OKB, at the cost of emotional problems and a gigantic discontinuity in technology, it provided the foundation for a totally new family of aircraft of increasing size and power which, by progressing in logical steps, has continued up to the present time.

A minor detail is that Western defence analysts at first were unanimous in dismissing the whole idea of a Soviet copy of the Boeing B-29 as a pipe-dream. They stuck to this belief even after there was irrefutable evidence that Soviet commercial agents were trying to buy B-29 spares, such as brakes and tyres. Thus, when three 'B-29s' flew over Moscow in the Aviation Day parade on 3 July 1947 there was utter amazement. Also in the parade was a transport version, the Tu-70. It was taken for granted this was an afterthought, and further astonishment ensued many years later when it was discovered that Tupolev had flown this *before* the first Tu-4.

By 1944 Stalin had little doubt the Soviet Union would soon get its hands on an actual B-29, even if only a crashed example. In the event, the windfall happened sooner than expected, and it brought not a crashed aircraft but three examples which had made normal landings in the far east of the USSR after suffering flak damage or merely running short of fuel. The aircraft were, on 29 July 1944, B-29-5-BW 42-6256; on 20 August 1944, B-29A-1-BN 42-93829; and on 21 November 1944, B-29-5-BW 42-6358 *Ding How*.

Though the property of an ally, these aircraft were immediately appropriated. All were made airworthy and

Type 69 Tu-8.

B-29 42-6358 operating as a trainer at Zhukovskii. (G F Petrov)

flown to Moscow by NII test pilots Reydel and Marunov. Aircraft 6256 was used as a carry-trials aircraft for missiles and the OKB-1 346 supersonic aircraft, 93829 was dissected at the Tupolev OKB and used as a model, its wing and engines being used in the Tu-70, and 6358 was retained for evaluation and crew training at UVVS-NII.

The task of dismantling and analysis was without precedent. A N Tupolev said that over 105,000 items had to be checked for material specification, function, manufacturing processes, tolerances and fits, and translated into Soviet drawings. Many parts were new to Soviet industry or alien to established Soviet practice. At the end of 1944 a full-scale programme was authorised to create a 'clone' of the B-29 for the VVS-DA. Tupolev told Stalin 'It will mean a three-year effort'; Stalin replied 'You have two years'. Tupolev told the author the meeting went precisely as he had inwardly predicted!

The original designation B-4 (*Bombardirovshchik*) was later changed to Tu-4 to reflect the immense 'copying' effort by the Tupolev OKB. In mid-

1945 a pre-production batch of twenty aircraft was ordered, with tooling at four GAZ and the main assembly line to be at GAZ No. 124 at Kazan. Few parts emerged identical to those of the B-29. Metric material gauges were generally overthick (the first Tu-4 airframe weighed 15,196kg, almost 500kg heavier than the target) and compromises had to be made in piping and cables. After tunnel tests of specimens the Boeing 117 wing profile was replaced by the RAF-34 with which Tupolev was familiar, with thickness/chord ratio of 20 per cent at the root tapering to 10 per cent at the tip. Whereas the B-29 had constant dihedral of 4 deg the Tu-4 was given dihedral of 4 deg 30 min to the outer engines and then 3 deg to the tip.

Difficulty was experienced with 75 micron bolt interference fits, and in the stress-relief and precision-machining of the new D16-ABTN light alloy, the nose and tail being D16-AT. The mainwheel tyres were metricised as 1,450mm × 520mm, inflated to 5–5.8at (73.5–85.3lb/sq in) pressure; the nose tyres were 950mm × 350mm, inflated to 3.6–4at (53–59lb/sq in). The front and rear

bomb bays were each redesigned to carry a maximum of four FAB-1000 bombs, a total nominal weight of 8 tonnes (17,637lb), with the prospect of nuclear weapons in due course. Normal bomb load was 6 tonnes (13,228lb).

It was commonly thought that the Tu-4 was powered by a 'crash programme' copy of the B-29's Wright R-3350–23 or -57. In fact, a suitable engine already existed. This had begun life as the ASh-71, which A D Shvetsov's KB had qualified at 1,700hp in 1941 using eighteen cylinders similar to those of the nine-cylinder M-63. By 1944 this had been developed into the ASh-73TK, with two TK-19 turbosuperchargers, with a continuous rating of 2,000hp and with 2,400hp (more than the B-29) available for take-off at 2,600rpm. The installation differed considerably from that of the B-29, though not much showed externally. Propellers were four-blade 5,056mm (16ft 7in, the same as the B-29) VZ-A3, VZ-A5 or VZV-A5.

Integral tankage was abandoned, and in the first three Tu-4 the fuel capacity was only 3,480kg (7,672lb).

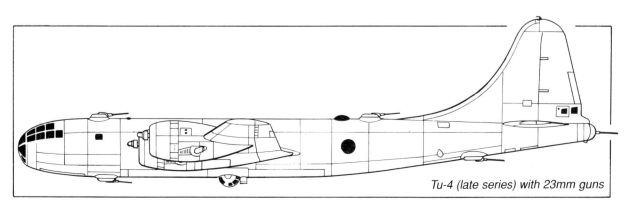

Tu-4 (late series) with 23mm guns

In the next seventeen it was 8,150kg (17,967lb), and in series aircraft it reached the design value of 14.9t (32,848lb) in twenty-two protected flexible cells in the wing, plus the option of 5,330kg (11,750lb) in three oval-section AMTs tanks in either the front or rear bomb bay. Another major task was the complete redesign of the defensive gun system (referred to later), fitting vhf radio and Soviet IFF (identification friend or foe).

The first three aircraft were assigned to OKB project pilots N S Rybko, M L Gallai and A G Vasilchyenko, respectively. The first was flown on 19 May 1947, six months after the Tu-70. Predictably, there were severe problems with many parts and systems, especially electrical items, but success was never in doubt. All three aircraft took part in the flypast over Red Square on 3 August 1947 (flying in the wrong direction and thus having to dive to 65m (200ft) to avoid the main 'parade' in which was the Tu-70)!

A demilitarised Tu-4 on air refuelling trials. (G F Petrov/RART)

Tu-4 1st series

Deliveries began in 1948 of the first series batch, capable of flying operational missions. The long-range training and development missions were very much a rerun of those flown by USAAF crews four years earlier. Defensive armament comprised ten UBT 12.7mm machine-guns in turrets not dissimilar to those of the B-29. During deliveries the remotely-sighted fire control was made operative, together with the pressure cabins (cabin height 2.5km [8,200ft] at an aircraft height of 7,000m [22,970ft]) with a 12mm (0.5in) layer of ATIM thermal insulation.

Tu-4 8th series

The 1949 production had almost all the necessary operating equipment. The defensive turrets were of VDB-2 type, each armed with two of the new B-20E guns of 20mm calibre, but with a reduced number of rounds per gun. This was the first batch to be equipped for nuclear bombs, and to have the *Kobalt* blind-bombing radar on a retractable mounting between the bomb bays, with the antenna group of the *Barii* M (Barium) IFF system immediately behind it. The *Kobalt* operator was seated right at the back of the rear-fuselage pressure cabin.

Tu-4 15th series

The final batch was cleared to an overload weight of 65,000kg (143,298lb). The defensive system was again redesigned, each turret being a PS-23 armed with two of the very powerful NS-23 cannon of

A Tu-4K with prototype MiG-designed KS-1 cruise missiles. (via Yefim Gordon)

23mm calibre, again with a reduction in number of rounds per gun. Another change was that the ailerons were enlarged, tapering from the root where their chord was the same as that of the outer end of the adjacent flap. Thus, in side elevation, the leading edge of the aileron sloped in the opposite direction to what it had before. The total number built of all version was about 1,000, ending in late 1952. All versions were given the ASCC reporting named of 'Bull'.

Tu-4R Under this designation were built strategic-reconnaissance aircraft with the forward bomb bay permanently housing the three auxiliary tanks and the aft bay carrying various groups of AFA/NAFA cameras and photoflash or flare magazines. (Tu-4 bombers had a standard camera installation immediately ahead of the aft ventral turret.)

Tu-4T A single prototype was flown in 1953 of an assault-transport conversion. All defensive armament and bombing equipment was removed, together with pressurisation, and simple inward-facing seats added for twenty-eight paratroops who dropped through a floor hatch. Later this aircraft flew with fifty-two seated passengers. In the same year a small series, possibly fewer than five, were built for the newly formed V-TA (military transport aviation).

Tu-4 tanker Several air-refuelling tankers, testing all known techniques, were converted from early Tu-4 bombers from 1952. Work centred on the tip-to-tip looped hose methods, followed by the probe/drogue methods perfected by Alekseyev and Yakovlev. There is no evidence for designation Tu-4N.

Tu-4K With designation K from *Kompleks* (electronic system), several aircraft assisted development of the Mikoyan KS-1 cruise missile, and then carried out Service trials.

Tu-4LL General designation of 'flying laboratory' for aircraft converted as carry-trials aircraft and for experimental or research payloads. The three most remarkable conversions had No. 3 engine replaced by a TV-12 (later called NK-12) turboprop more than five times as powerful as the other engines. With the first of these aircraft chief engineer D I Kantor and pilot M A Nykhtikov cleared the engine to power the prototype 95/II. Other Tu-4LL aircraft tested the AI-20 (both above and below the wing as in the Il-18 and An-10), AL-5, AL-7, AL-7F, AL-7P, AM-3, AM-5, AM-5F, NK-4, VD-5, VD-7, VK-2, VK-7, VK-11 and other engines. Three were used for cruise-missile trials, initially with captured Fi 103 ('V 1') flying bombs and later with the La-17 and (see above) KS-1.

Span 43.05m (141ft 2⅞in); length (1947) 30.18m (99ft), (1950) 30.8m (101ft 0⅝in); wing area 161.7sq m (1,740sq ft).
Weight empty (1948) 35,270kg (77,756lb); fuel/oil 11,200kg (24,912lb) max + 800kg; loaded (1947) 47,600kg, (1948 max) 54,500kg (120,150lb), (1951 max) 66,000kg (145,500lb).
Maximum speed 420km/h (261mph) at sea level, 558km/h (347mph) at 10,000m; climb to 5,000m 18.2min; service ceiling 11,200m (36,750ft); range (1-serie) 3,000km (1,864 miles) with 1½ tonne bomb load, (5-serie) 5100km (3,170 miles) with 2,000kg bomb load; take-off run 960–2,210m (3,149–7,250ft) depending on weight; landing speed 172km/h (107mph), landing run 1,070–1,750m (3,510–7,250ft).

A Tu-4 with civil registration (SSSR-92648) at Severnyi Polus 3, an Arctic station used as a secret airbase during the Korean War. (G F Petrov/RART)

Tu-4LL testing a prototype TV-12 engine. This became the NK-12, powering the Tu-95. (G F Petrov/RART)

70, Tu-12

Having dismantled B–29A No. 93829, Tupolev OKB rebuilt the complete wing, 2,200hp R–3350–57 engines, Curtiss Electric propellers, landing gear and tail into a passenger transport. Based on the Type 66, the completely new fuselage had a diameter of 3.5m (138in), and was intended to be pressurised from the B–29 type glazed nose back to a pressure bulkhead little more than 1m (39in) in front of the tailplane, but pressurisation was not at first installed. The nose housed the navigator, with a B–29 glazed end and dorsal astrodome. Next came the

Rollout of the Tu-12. (G F Petrov/ RART)

The Tu-12 on an early test flight. (G F Petrov/RART)

cockpit, again of B-29 type but moved up and to the rear and with airline-type windscreens to match the increased fuselage diameter. In the next fuselage section stations were provided for the radio operator, engineer and a second navigator. Next came a cabin with eight seats and a toilet, with two wide rectangular windows on each side and an underfloor hold. Then came a second eight-seat cabin, followed by steps up to a small galley above the wing centre section. Then back at the original level came a third eight-seat cabin, followed by a long cabin seating twenty-four, with seven small circular windows on each side and a second underfloor hold. Other changes affected the nose landing gear and the span of the centre section (which as on the B-29A, but not the B-29, extended only across the width of the fuselage).

The single Type 70 was intended to be the prototype of a series version with Service designation Tu-12 and this, with the 'Tu' logo, was painted

on the nose. Opadchii made the first flight on 27 November 1946, six months ahead of the first Tu-4. At this time the Goodrich pulsating-rubber de-icers on the leading edges had not been copied, though later full pressurisation and cabin air-conditioning was installed, and the forty-eight seat *lyuks* (luxury) interior was replaced by seventy-two seats. Later the prototype Tu-12 was given VVS markings to serve as a Kremlin VIP and staff transport. There was never any expectation of finding a civil application. The ASCC allocated the name 'Cart'.

Span 44.25m (145ft 2⅛in); length 35.4m (116ft 1¾in); wing area 166.1sq m (1,788 sq ft).
Weight empty 38,290kg (84,414lb); fuel/oil, 7,430kg or 16,030kg (16,380lb or 35,340lb) weight loaded 51,400kg (113,316lb) normal, 60,000kg (132,275lb) maximum.
Maximum speed 424km/h (263mph) at sea level, 568km/h (353mph) at 9,000m; climb to 5,000m in 21.2min; service ceiling 11,000m (36,090ft); range (maximum load, normal fuel) 4,900km (3,045 miles); take-off run 670m (2,200ft); landing run 600m (1,968ft).

71

This 1946 project was to have been a Tu-2S derivative with a new nose and either ASh-82M or M-93 engines in the 2,500hp class.

72

In 1946 this project was launched as a medium-range bomber, with power-driven gun turrets, powered by two ASh-2TK. These were essentially a pair of ASh-82 engines on a common crankcase, to be rated initially at 3,300hp. The project had not gone far when, in January 1947, it was cancelled and replaced by a far more effective twin-jet.

In one document this twin-jet 72 was called the Tu-8. This was not an

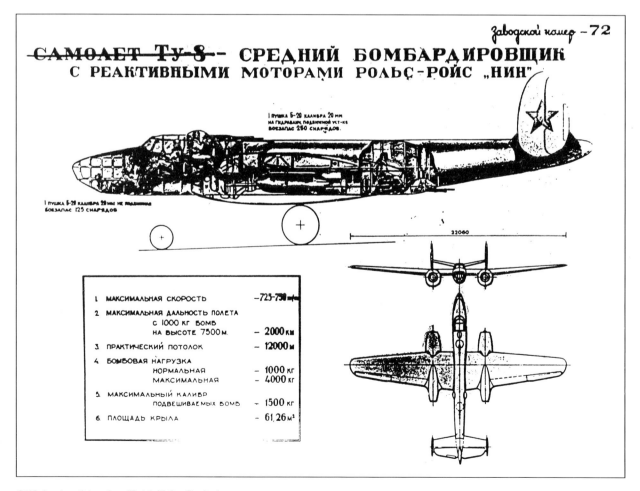

OKB drawing of Aeroplane 72. (via Yefim Gordon)

error, because the designation referred specifically to the *fuselage* of Type 69, and the same fuselage was the basis of the 72. Externally the wing and tail were unchanged, apart from modifications to suit the two 5,000lb Rolls-Royce Nene I turbojets in long nacelles at the ends of the centre section. They were sufficiently underslung for the spars to run unbroken above the jet-pipes, the engines being completely ahead of the wing. The jetpipe fairings broke the upper surface just behind the main spar and were centred on the trailing edge. The split-flaps were outboard of the nacelles only. The large mainwheels were to retract forwards to lie either beside or below the jetpipes immediately behind the engine. The nose landing gear was to retract to the rear. The bomb bay was to house bombs up to FAB-1500 size, normal and maximum loads being 1,000kg and 4,000kg, respectively.

Despite using a stretched Type 69 fuselage, the crew was to number only four. The nose, more streamlined than before, was to house the navigator/bomb-aimer, seated facing forwards. The single pilot left of centre was to have a sight for aiming the single B-20 forward-firing cannon, with 125 rounds. Behind the pilot was the radio operator, seated in a UST-K2 powered turret with a single B-20 with 250 rounds. Behind the wing the ventral gunner sighted through large bulged beam windows, with sightline recesses downstream as in the 69, to aim his inverted UST-K2, again with 250 rounds. There were to be four wing tanks and a large tank above the wing between the front and rear cabins, which despite the altitude capability appear to have been unpressurised. The 72 was never completed.

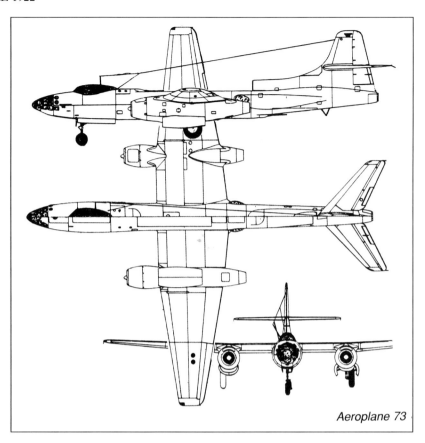

Aeroplane 73

Span 22.06m (72ft 4½in); length 16.11m (52ft 10¼in); wing area 61.26sq m (659sq ft).

Weight loaded, normal, 16,100kg (35,494lb).

Maximum speed (est) 725–750km/h (451–466mph); service ceiling 12,000m (39,370ft); range with 1,000kg (2,205lb) bomb load at 7,500m (24,600ft) 2,000km (1,242 miles).

73

In January 1947, with work on the 77 well advanced, Tupolev launched a second and larger jet bomber, intended to meet a VVS requirement issued the previous autumn for a bomber with a speed exceeding 800km/h (500mph) and a bomb load of 3,000kg (6,614lb). Using the 69/72 airframe as a basis, Arkhangel'skii's brigade scaled it up, and ended up with a completely new aircraft. The wing profile selected was SR-5S, with a thickness/chord ratio over the horizontal centre section of 12 per cent. B-29 techniques were used in the plate spars and machined wing torsion box skins forming integral tankage. The tapered outer panels had 5 deg dihedral. Inboard and

Internal arrangement of Type 73.

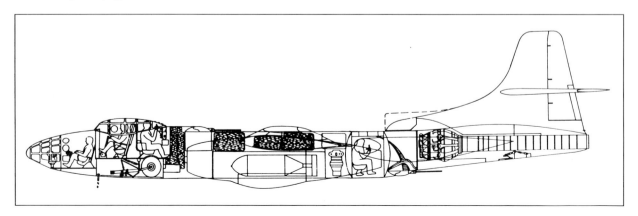

Prototype 73, bearing 'Tu-14' logos on nose and fin. (via Philip Jarrett)

outboard of the nacelles were large split flaps, driven to 45 deg, with narrow-chord ailerons extending to the tips. Twin landing lights were normally retracted into the outer starboard wing.

The fuselage was of circular section because front and rear compartments were pressurised. The forward cabin had a glazed nose, behind which was seated the navigator/bomb-aimer. A large glazed canopy covered the cockpits for the pilot and radio operator/gunner, seated back-to-back. Behind the pressure bulkhead was a full-section tank, followed by a bay devoted to the dorsal PS-23 turret, with twin NR-23 23mm guns and their ammunition, this being under the remote control of the radio operator. The centre fuselage housed two tanks in the upper half, with the wing box integral tank passing between them, and the bomb bay in the lower half, able to carry an FAB-3000 (6,614lb) bomb and with twin doors. Next came a camera bay, typically occupied by an AFA-33/75. Then came the rear pressure cabin, occupied by the ventral gunner managing a second twin NR-23 turret. He sighted through bulged beam windows downstream of which were sightline recesses. In the top of the compartment was the 23mm magazine, the belts from which passed through the aft pressure bulkhead to the turret in the bottom of the rear fuselage. Immediately behind a fireproof bulkhead was installed the centre engine, a 3,500lb thrust Rolls-Royce Derwent 5. This was fed via a dorsal inlet at the front of the single tail fin, fitted with a retractable fairing which could close off the duct in cruising flight, using just the two

wing engines. Its jetpipe went direct to the end of the fuselage. The fixed horizontal tail was mounted half-way up the fin. The elevators and rudder all had inset hinges and large tabs. Hot-air de-icing of all leading edges was to be installed later.

The main engines were again 5,000lb thrust Nenes, installed almost exactly as in the 72. On the right of each cowling was a ground electric power socket. Under the centre section were attachments for four PSR-1500-15 (*porokhovaya startovaya raketa*, 1,500kg for 15sec) auxiliary take-off rockets. The main landing gears were again pivoted to steel trusses triangulated down from the spars to retract forwards, the wheel turning to lie horizontal above the single leg. The long-stroke castoring nose landing gear retracted to the rear into a box with twin doors under the cockpit floor. Ahead of this was a hinged panel with integral steps for crew access. The rear gunner had his own floor hatch. Under the rear fuselage was a large removable panel for access to (or changing) the centre engine, followed by a sprung retractable tail bumper. The HF radio antenna was strung from the fin to a mast inclined to the right of the canopy, while a VHF antenna was built into the leading edge of the fin.

The first 73 was completed without its gun turrets. It was first flown by Opadchii on 20 December 1947. The second (*dubler*) 73 was equipped with turrets, and also had a plain inlet for the tail engine which increased dorsal fin area. It had an extra nose window on each side, and at the top of the fin was painted the Tupolev logo and the

intended Service designation Tu-14 NII-VVS State testing of both prototypes continued until 31 May 1949.

Span 21.71m (71ft 2¾in); length 20.32m (66ft 8in); wing area 67.38sq m (725sq ft).

Weight empty 14,340kg (31,614lb); fuel/oil (normal) 5,850kg (12,897lb); weight loaded 21,100kg (46,517lb) normal, 24,200kg (53,351lb) maximum.

Maximum speed 840km/h (522mph) at sea level, 872km/h (542mph) at 5,000m; climb to 5,000m in 9½min; service ceiling 11,500m (37,730ft); range 2,810km (1,746 miles); take-off run 740m (2,427ft); landing speed 173km/h (107.5mph), landing run 1,170m (3,838ft).

75

No information on Type 74 has been found. The 75 was a second transport derivative of the B-29. It was wholly Soviet-made, and differed mainly in being cleared to higher weights resulting from increased fuel capacity, in being powered by 2,400hp ASh-73TKNV engines, driving VZV-A5 propellers, and in having an unpressurised fuselage with a large rear ramp door for loading vehicles and other bulky loads. It was a military project, designed to carry 10 tonnes (22,046lb) of assault stores internally, on a floor with strong steel panels, or 100 (one document, 120) parachutists with their kit and weapons, on four rows of side-facing canvas seats. A drawing shows the intended series version (never built) with Tu-4 type sighting stations for dorsal, ventral and tail turrets each with twin B-20 guns. The aircraft actually built had provision for these turrets, but they were

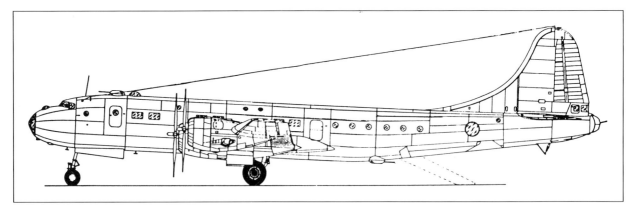

The intended armed production version of Aeroplane 75.

Unarmed 75 prototype as built. (Philip Jarrett)

not installed, so the crew numbered six (the same as the Tu-70) instead of eight. The rear fuselage had an ungainly bulged underside. This prototype was assigned to OKB pilot V P Marunov, but the first flight was in the hands of A D Perelyot on 22 January 1950. NII-VVS testing was by I Kabanov and M Melnikov. In 1951–53 dropping trials of heavy items took place through four hatches added under the fuselage. According to one document the 75 was given extended wingtips, increasing wing area from 166sq m (1,787sq ft) to the figure given in the data, but all drawings show span as either the same as or slightly less than the 70. No Service designation was allocated. The ASCC again assigned the name 'Cart'.

Span 44.25m (145ft 2⅛in); length 35.61m (116ft 10in); wing area 167.2sq m (1,800sq ft).

Weight empty 37,810kg (83,355lb); fuel/oil 10,540kg (23,237lb) normal, 24,900kg (54,895lb) maximum; weight loaded 56,660kg (124,912lb) normal, 65,400kg (144,180lb) maximum.

Maximum speed 545km/h (339mph) at 9,000m; service ceiling 9,500m (31,170ft); range 4,140km (2,573 miles) with normal fuel, 6,290km (3,910 miles) maximum.

77, Tu-12

Tupolev based this jet bomber on the standard Tu-2S, to get jet experience quickly. The airframe was stressed for considerably higher weights and greater indicated airspeeds. A major difference was the long underslung nacelles for the 5,000lb thrust Rolls-Royce Nene turbojets, though the nacelle centrelines remained 5.3m (208.7in) apart as on the Tu-2. These nacelles, and the strengthened forward-retracting main landing gears, were similar to those of the 72. Another difference was the long navigator compartment, making the nose resemble those of the 69 and 72. The pilot was seated above the aft-

retracting nose landing gear under a short hinged canopy whose rear view was hampered by a large tank immediately to the rear. The wing and stretched fuselage housed additional tanks totalling three times the capacity of the standard Tu-2S. The fuselage was both longer and deeper, raising the strengthened tail above the level of the wing. The bomb load of up to 3,000kg (6,614lb) was unchanged, as was the rear defensive armament of an upper VUB-68 (one 12.7mm UBT with 250 rounds) and lower Lu-68 (one UBT, but with magazine increased to 350 rounds). The pilot's forward-firing gun was a single NS-23 on the lower right side of the nose, replacing the ShVAK-20s in the wing roots.

The first 77 was flown by Perelyot on 27 July 1947. As it was the first Soviet jet bomber it was of interest to Stalin, who in 1946 ordered five to be built immediately. These were given Service designation Tu-12, duplicating that of the Type 70 which at the time was an active programme, though it

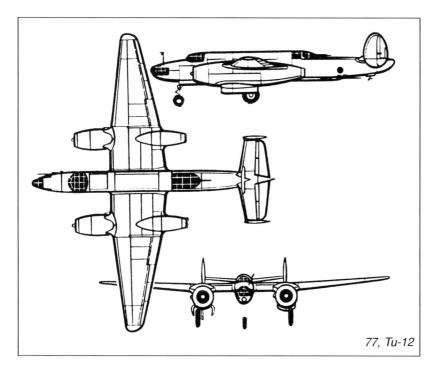

77, Tu-12

was not expected they would enter the VVS inventory. The first two took part in the Tushino parade on 3 August, and all had flown by late September 1947. They completed NII-VVS testing in October, and were of value in supporting development of later Tupolev bombers, leading to the Tu-14. One was used as the flying testbed for the RD-550 subsonic ramjet, which was mounted on a pylon above the fuselage.

Span 18.86m (61ft 10½in); length 15.75m (51ft 8in); wing area 48.8sq m (525sq ft)

Weight empty 8,993kg (19,826lb); fuel/oil 4,080kg (8,995lb) normal, 6,727kg (14,830lb) maximum; weight loaded 14,700kg (32,407lb) normal, 15,720kg (34,656lb) maximum.

Maximum speed 778km/h (483mph) at sea level, 783km/h (487mph) at 4,000m; climb to 5,000m in 8min; service ceiling 11,360m (37,270ft); range (normal fuel) 2,200km (1,367 miles); take-off run 1,030m (3,379ft) ; landing speed 163km/h (101mph), landing run 885m. (2,963ft).

Type 77, the fully-armed prototype Tu-12. (G F Petrov/RART)

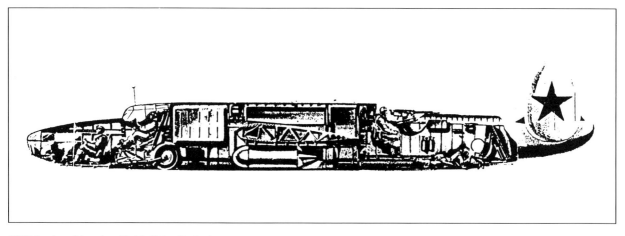

OKB drawing of Aeroplane 77. (via Yefim Gordon)

The fully-armed prototype 78. (via Philip Jarrett)

78

This tri-jet bomber was the next stage beyond the 73. Its airframe differed mainly in having the mid-fuselage stretched 0.3m (12in) and the chord of the vertical tail increased. The bottom of the rudder was raised, and the navigator had rectangular instead of circular side windows. Contrary to historian Shavrov's account, the principal reason for the 78 was that it was powered by Soviet-built engines, RD-45s replacing the Nenes and an RD-500 replacing the Derwent, with the retractable inlet fairing restored. The Soviet copies of the Rolls-Royce engines differed mainly in their accessories and fuel control, though the aircraft's auxiliary power systems were almost unaffected. The single 78 was first flown on 7 May 1948, and later underwent NII-VVS testing with the Service designation Tu-16 (later used for the totally different 88).

Dimensions as 73 except length 20.62m (67ft 7¼in).
Weight empty 14,290kg (31,505lb); loaded weight 23,790kg (52,447lb) maximum.
Performance similar to 73 except climb to 5,000m in 7.7min.

79

Never built, this aircraft would essentially have been a reconnaissance version of the 78. Among the external differences would have been a major extension of the front of each nacelle, so that in side view the fairing over the nose 'bullet' could not be seen. Behind the rear cabin would have been RSBN radar, with the rotating antenna in a ventral blister just ahead of what in earlier versions was the bomb bay, in the 79 to be filled with two auxiliary tanks. Between these tanks and the rear cabin would have been the cameras, a diagram showing two AFA-33/20, an AFA-33/75–50 and

an AFA-33/100. Armament would have comprised a fixed NS-23 with eighty-five rounds and two turrets each with twin NR-23, each with 200 rounds. The tail engine would not have had an inlet shutter.

80

In parallel with the task of converting the B-29 into the Tu-4 Tupolev schemed various improvements, as did Boeing. While the latter settled mainly for much more powerful engines (R-4360s), Tupolev initially concentrated on the rest of the aircraft, though completely new engines were in prospect (*see* 85).

An important advance was use of new aluminium alloys, which among other things enabled a wing of almost unchanged size to lift much greater weights and house 15 per cent more fuel. Seen from head-on, the kink at

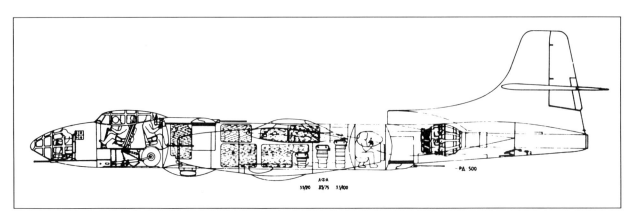

OKB drawing of Aeroplane 78, indicating AFA cameras and RD-500 tail engine.

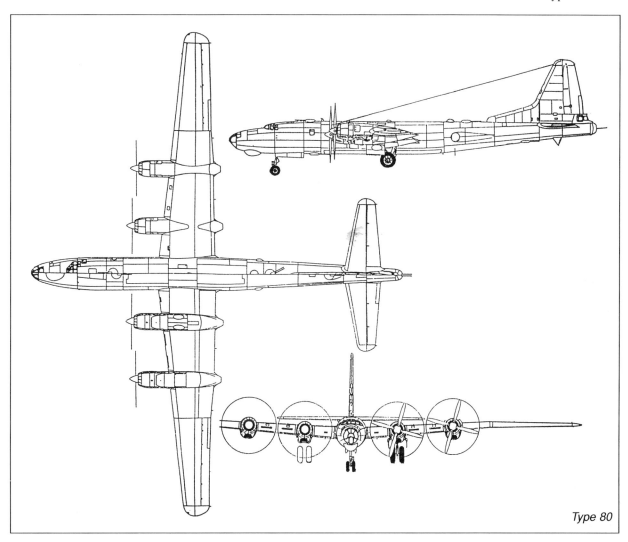

Type 80

The sole Type 80 prototype. (via Philip Jarrett)

each end of the centre section was made more pronounced by reducing dihedral of the outer panels to zero. The fuselage was considerably length-ened, and the nose was redesigned. The navigator occupied the large front compartment capped by the glazed end-cap. The two pilots were moved back and up as in the 70 and 75 with conventional flight-deck windows, to establish a design of nose that was to be used on Tupolev strategic air-craft for the next 25 years. At the back of this pressure cabin were the forward dorsal and ventral remotely controlled turrets, with emergency-escape hatches on each side. Then came the mid-fuselage with modified front and rear bomb bays, for a maximum bomb load of 12,000kg (26,455lb), and a communications tunnel joining the pressure cabins above the wing.

The longer rear fuselage had beam sighting stations with downstream recesses of the kind used on the 69/72/73. The next section to the rear had a door on the right, escape hatch in the roof and AFA-33/75 (or other) camera in front of the ventral turret. The guns were not 23mm but of the B–20E type, the total of eleven including a fixed gun aimed by the pilot. The vertical tail was completely redesigned, with different structure and greater rudder power. The engines, 2,400hp ASh-73FN, (sometimes called 73TKFN), were in completely different installations, externally distinguished by slimmer cowlings, ventral oil coolers and longer nacelles all projecting behind the trailing edge. Thin slit inlets in the leading edge between the engines led through heat exchangers to flush exits in the upper surface ahead of the front spar.

A single aircraft was built, lacking its gun turrets, making its first flight on 1 December 1949. Development was discontinued because it was self-evident that even improved versions would not have sufficient range to operate against the USA.

Span 43.45m (142ft 6⅝in); length 34.32m (112ft 7¼in); wing area 167sq m (1,798sq ft).

Weight empty 37,850kg (83,444lb), fuel/oil (maximum) 18,600kg (37,038lb), weight loaded 51,500kg (113,536lb) normal, about 60,600kg (133,600lb) maximum.

Maximum speed 428km/h (266mph) at sea level, 600km/h (373mph) at 9,000m; range with 3,000kg bomb load, 3,000km (1,864 miles) at normal weight, 7,000km (4,350 miles) at maximum weight; take-off run 1,200m (3,936ft.) (normal weight); landing run 505m (1,656ft).

81, 89, Tu-14

Rapid development of the imported Nene engine by Maj-Gen V Ya Klimov

led to clearance for production in January 1949 of the VK-1 version, rated at 2,700kg (5,952lb) thrust. This made it possible to eliminate the never-popular third engine. This was done with the 81, shown in an OKB drawing in its original form with the original defensive armament of two remote-control turrets each with twin NR-23s with 200 rounds. To try to bring the centre of gravity aft, the rear pressure cabin housed two crew, a gunner sighting through the beam windows and an operator to manage the RSBN and RiM-S radars, the former having its antenna in a blister ahead of the bomb bay. Problems with this layout were that the centre of gravity was still too far forward after removing the tail engine, and that a crew of five was needed.

The 81 was intended to be a bomber and (as the 81T) torpedo aircraft for the Aviatsiya-VMF (aviation of the Navy). With the customer's agreement, the balance problem was solved

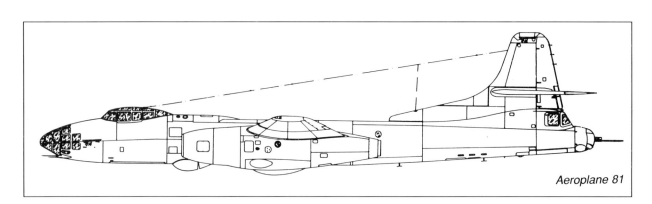

Aeroplane 81

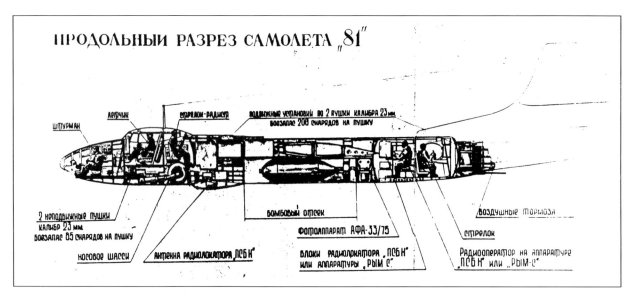

OKB drawing of Aeroplane 81, the first with no tail engine. (via Yefim Gordon)

by fitting a tail turret, which in turn resulted in a major redesign of the fuselage. The Ilyushin turret was originally related to that fitted to the Tu-80 and 85. It had electro-hydraulic drive, and armament of two NR-23s with 350 rounds, but the angular limits of traverse and elevation were rather limited. Despite this, both the previous turrets were omitted, the upper turret being replaced by a fairing as on the original 73. The rear pressure cabin was also eliminated, together with the beam sight stations. Instead, the rear fuselage swelled in cross-section under the tailplane to accommodate the much smaller pressure cabin for the tail gunner under the rudder. The forward pressure cabin continued to accommodate navigator, pilot and radio/radar operator. A single NR-23 was fitted firing ahead from the lower left side of the nose. As air-brakes during torpedo drops it was proposed to use the powerful split flaps depressed to 45 deg. The HF wire antenna ran from the fin to a small stub on the right of the canopy, while the UHF/VHF aerials were built into the vertical tail. The 81 first flew at the end of 1950.

Span 21.71m (71ft 2¾in); length (excluding guns) 21.4m (70ft 2½in); wing area 67.38sq m (725sq ft).

Weight empty 14,430kg (31,812lb); fuel/oil 4,000kg (8,818lb) normal, 8,250kg (18,118lb) maximum; weight loaded 21,000kg (46,296lb) normal, 24,600kg (54,233lb) maximum.

Maximum speed 800km/h (497mph) at sea level, 861km/h (535mph) at 5,000m; climb to 5,000m in 8.3min; service ceiling 11,500m (37,730ft); range (maximum) 3,150km (1,960 miles); take-off run 1,250m (4,101ft); landing speed 175km/h (109mph), landing run 1,120m (3,674ft).

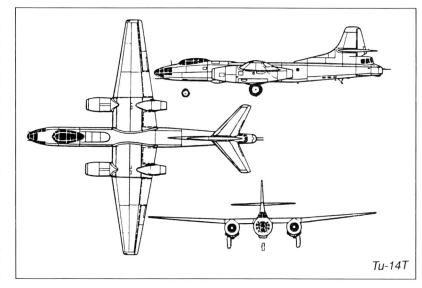

Tu-14T

89R, 89T Continued refinement led to the 89R naval reconnaissance aircraft and the 89T anti-submarine and anti-ship aircraft (still designated Tu-14, not Tu-14T as has often appeared). Externally the only obvious difference between the 89 and the 81 was that the bottom of the rudder was cut off diagonally and the tail turret was so large as to appear ungainly, but there were many other modifications. Rudder travel was increased, and the horizontal tails were appreciably enlarged by adding area at the root, sweepback remaining 43 deg. The main landing gear, with 1,150mm by 335mm tyres, retracted forwards, the wheel lying horizontally. The nose unit, with a 720mm by 310mm tyre, retracted to the rear.

The Ilyushin OKB had assisted the installation of the original design of turret, but in May 1950 the limited traverse of ±50 deg was one of the criticisms voiced at the MA evaluation. Accordingly, the fuselage was again rearranged internally to make better use of an increase in length, and a bulky new turret was fitted offering a traverse of ±70 deg, with the twin NR-23 guns replaced by B-20E type with greater ammunition capacity. The gunner's windows were made deeper, and the rear window was reduced in width. The pilot's forward-firing armament comprised two NR-23s each with eighty-five rounds. The 89T weapon bay was equipped to carry a wide range of ASW/ASSV torpedoes, bombs or mines (later including nuclear weapons of all three types) up to a maximum of 3,000kg

The 81 prototype. (Jean Alexander/RART)

Tu-14 on test with an unknown type of naval glider bomb. (G F Petrov/RART)

(6,614lb), loading being facilitated by modified doors able to open out to the horizontal position.

The crew was reduced to three: the navigator in the nose, with emergency escape straight down through the step-equipped entrance hatch, the pilot on an ejection seat under a hinged and jettisonable canopy with a circular clear-view window each side, and the gunner able to drop out by jettisoning his stepped entrance hatch. On the right side of the rear fuselage beside the dorsal fin was a compartment for an LAS-3 life raft. Fuel capacity was increased, partly by the extra fuselage bay and the elimination of the upper and lower turrets. An important late modification was addition of a hydraulically driven speed-brake door under the rear fuselage. Behind this was the camera bay, with ventral doors, for an AFA-33 of any focal length. Antennas including the fin/canopy wire for the RSB-5 guidance radio and ARK-5 radio compass, the flush fin antenna group for the *Materik* (Mainland) nav/com complex and the ventral blister ahead of the weapon bay for the RSBN-M surveillance and navigation radar. Along the flanks of the rear fuselage were three triple launchers for signalling rockets.

Production of 100 (later reduced to eighty-seven) aircraft for the Aviatsiya VMF inventory was assigned to GAZ No. 125 at Irkutsk, where Sukhoi was sent as programme director. The factory delivered forty-two in the first year, and the Tu-14 was popular in service, over the period 1952–62. Updates included an improved autopilot, engine-monitoring system, an upward-ejection seat for the navigator,

electronic countermeasures (including chaff/flare dispensers) and a braking parachute immediately ahead of the rear gunner's hatch. The planned more fuel-economical replacement, the Tu-91, was cancelled. In 1958–59 a used batch of fifty was supplied to China. A single prototype was tested of the Tu-14R, with auxiliary fuel, AFA-33-20, 30–50/75 and 33–100 cameras, a drift sight, flares and provision for later side-looking radar. To all aircraft of this family the USAF allocated the invented reporting designation 'Type 35', superseded by the ASCC name 'Bosun'.

Span 21.69m (71ft 2in); length (excluding guns) 21.95m (72ft 1¼in); wing area 67.36sq m (725sq ft).
Weight empty (T) 14,930kg (32,914lb), (R) 14,490kg (31,944lb); fuel/oil (T) 4,300kg (9,480lb) normal, 8,445kg (18,618lb) maximum, (R) 10,020kg (22,090lb) maximum; weight loaded (both) 21,000kg (46,297lb) normal, (T) 25,350kg (55,886lb) maximum, (R) 25,604kg (56,446lb) maximum.
Maximum speed (both) 800km/h (497mph) at sea level, (T) 845km/h (525mph) at 5,000m, (R) 859km/h (534mph) at 5,000m; climb to 5,000m in (T) 9½min, (R) 8.3min; service ceiling 11,200m (36,745ft); range (T, normal fuel) 2,870km (1,783 miles), (R, normal fuel) 3240km (2,013 miles); take-off run (T) 1,200m (3,936ft), (R) 1,655m (5,429ft); landing speed (both) 187km/h (116mph), landing run 1,100m (3,608ft).

509

So far as archivist V Rigmant can discover, no *Izdelye* (product) number was ever assigned to the ninth project of 1950 (hence designation 509). As explained in the story of the Tu-91,

this was one of two totally dissimilar attempts to design a multirole attack aircraft to operate from large aircraft carriers planned for the Aviatsiya VMF (Aviation of the Soviet Navy). Project leader for both was B M Kondorskii.

The 509 was to be based as far as possible on the Tu-14 naval land-based torpedo bomber. P O Sukhoi, who had managed Tu-14 production at GAZ No. 125 at Irkutsk, was recalled to Moscow to put his detailed knowledge of the twin-jet to use. The accompanying simple three-view shows that the Tu-14's pressurised crew compartment for the pilot and navigator/weapon-aimer was retained, as were the VK-1 engine nacelles and main landing gear. Outboard of the nacelles the wings were much smaller, and arranged to powerfold upwards. The remainder of the fuselage was as small as could possibly be managed to accommodate an internal 45–36 torpedo, with the radar moved to the rear. It was hoped no tail turret would be needed, but the very short moment arm meant that the fin and rudder had to be enlarged. Take-off weight would have been 15,000kg (33,070lb), maximum speed 900km/h (559mph), weapon load 1,500kg (3,307lb) and design range 1,500km (932 miles). The project was abandoned partly because the aircraft was considered marginally too big, but mainly because internal fuel capacity was inadequate even for the modest range demanded, especially as this thirsty jet was expected to operate mainly at low altitudes.

For drawing of 509 see page 158.

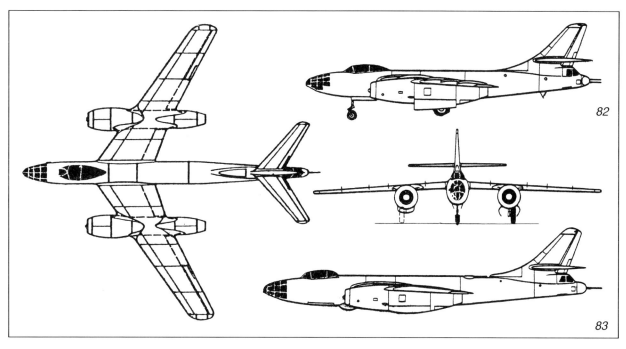

82, Tu-22

In parallel with the search for the definitive Tu-14, Arkhangel'skii led the design of this smaller aircraft, intended as a research tool to explore swept-back wings, using as many parts as possible from existing projects. Early in the programme the VVS took a more immediate interest and asked that it should be a frontal (tactical) bomber, with the designation Tu-22. Two prototypes were funded in summer 1948.

As far as possible Type 82 used fuselage, nacelle and tail designs from the previous jet-bomber programmes. This particularly applied to the engine installations. The two 2,270kg (5,000lb) thrust RD-45F (Soviet Nene) turbojets were installed ahead of the wing in underslung nacelles faired by large fillets into the leading and trailing edges. The inlets had gauze screens, the landing lights were in the 'bullet' fairings on the front of each engine wheelcase, an oil filler panel was at the front of the cowl, followed by four large engine-access panels, with a removable starting panel and cable connection under the leading edge. The main landing gears were similar to those of the 81, but shorter.

The wing profile was PR-10s, with a ruling thickness/chord ratio of 15 per cent, the sweepback along the aerodynamic centre (about 25 per cent chord) being 34 deg 5 min; leading-edge angle was about 36 deg. Aspect ratio was quite high at 6.9. The leading edge was fixed, but made so that droops or slats could be fittted. The trailing edge carried split flaps and inset outboard ailerons with power assistance. Fear of excessive spanwise flow led to the fitting of no fewer than four fences above each wing, each beginning just behind the leading edge and extending right across the chord. In the first studies the horizontal tail was only just above the fuselage, but eventually it migrated up to the same level as in the other jet bombers. Tailplane leading-edge sweepback was 40 deg, and for the first time in the Tupolev OKB the vertical tail was also swept. The rudder and elevators remained manual, with deeply inset hinges and large tabs. At the tip of the

Aeroplane 82, Tupolev's first swept-wing aircraft. (G F Petrov/RART)

fin on both sides was an exit for hot air from fin de-icing, and there is evidence similar exits (in these instances, slits round the tip) were provided on the wing and tailplane.

The crew of three occupied two pressure cabins as before. At the project stage the nose was made more pointed, though retaining an optically flat bomb-aiming panel, and the aft-retracting nose landing gear was moved to the rear to make room for an undernose boarding hatch. The pilot's canopy, offset to the left, was enlarged, and provided with optically flat circular windows in the windscreen and on the left side of the main upward-hinged portion. A single 23mm NR-23 gun was fixed in the lower right side of the nose. The rear gunner, seated above a jettisonable entrance hatch, was to have managed a new design of turret with two superimposed NS-23 guns, but in the event this was never produced, ballast being installed instead. On each side of the rear fuselage was a rectangular hydraulically driven speed-brake, and a retractable sprung tailskid was fitted underneath. The bomb bay had twin doors and carried 1 tonne (normal) or 3 tonnes (6,614lb, maximum). A maximum of 7,000 litres (1,540gal) of fuel could be carried in the wing integral boxes, four fuselage cells and an auxiliary bomb-bay tank. Apart from an RUSP gun-ranging antenna beneath the rudder, no radar was provided for, but the antennas were otherwise as in the Tu-14.

The first Type 82, bearing the OKB logo and 'Tu-22' on both nose and fin, made its first flight on 24 March 1949, the pilot being Perelyot. The second aircraft also flew, but OKB effort was quickly transferred to the Aircraft N, Type 88. No Western code name was allocated.

Span 17.81m (58ft 5⅛in); length 17.57m (57ft 7¼in); wing area 46.24sq m (497.7sq ft).

Weight empty 11,226kg (34,749lb); fuel/oil 2,250kg (4,960lb) normal, 5,670kg (12,500lb) maximum; weight loaded 14,919kg (32,890lb) normal, 18,339kg (40,430lb) maximum.

Maximum speed 870km/h (541mph) at sea level, 931km/h (579mph) at 4,000m; climb to 5,000m in 5.8min, service ceiling 11,400m (37,400ft); range (normal fuel) 2,395km (1,488 miles); take-off run 1,100m (3,608ft); landing speed 177km/h (110mph); landing run 550m (1,804ft)

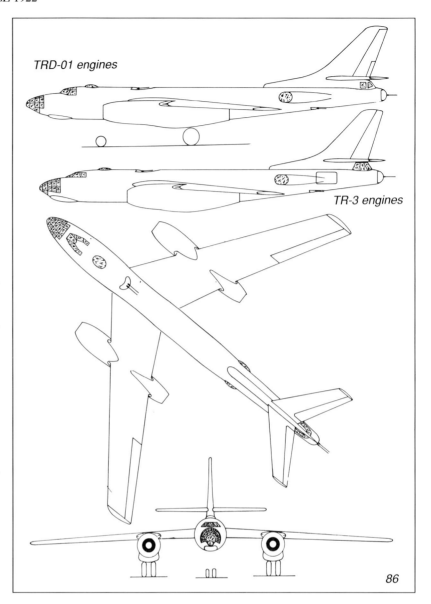

TRD-01 engines

TR-3 engines

86

83, 84, 86, 87

These unbuilt bomber projects were intermediate stages between the 82 and 88, and all dated from 1948–49. The 83 was merely an 82 stretched to 19.925m (65ft 4½in) in order to accommodate extra fuel, radar ahead of the bomb bay and a remotely controlled twin 23mm dorsal turret immediately ahead of the fin. One document suggests the 83 could have become a heavy all-weather interceptor. Type 84 would have been similar in size but powered by two 3,000kg (6,614lb) VK-5s and a tail-mounted 2,700kg (5,952lb) VK-1. Aircraft 86 would have been considerably larger (span 25.5m, length 24.15m), with a nose similar to the eventual 88 but outboard nacelles as on the 82. It would have had six NR-23s in remotely sighted forward dorsal and aft ventral turrets and a tail turret, with beam sighting stations as on the 69 and 72, and would have carried from 1 tonne to 6 tonnes (13,228lb) of bombs up to 4,000km (2,485 miles) at 1,000km/h (621mph), using two 4,600kg (10,140lb) Mikulin TRD-02 engines. The 87 would have been fractionally longer, with two Lyul'ka TR-3 turbojets of the same thrust.

85

This final extrapolation of the B-29 was in most respects the ultimate

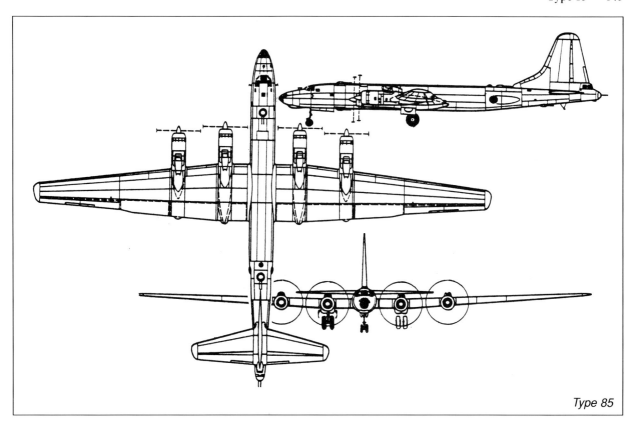

Type 85

Aeroplane 85 No. 1, probably before first flight. (G F Petrov/RART)

piston-engined bomber, more formid-able even than the B-36. Nevertheless, it was abandoned despite its excellence because of the dramatic potential of the turboprop-powered Tu-95. The 85 was made possible by the existence of new piston engines of unprecedented power: the ASh-2TK, VD-4K, M-51, M-501 and M-35. Inevitably, all were massive, complicated and thus poten-tially unreliable, but they enabled Tupolev's 'main team' to design a totally new bomber weighing twice as much as the Tu-4.

This demanded a completely new wing. The first drawings of the 85 showed a wing of 202.5sq m (2,180sq ft) area, representing a modest stretch of the Type 80. Later the outer panels

were redesigned, giving an area of 221sq m (2,379sq ft) and the final wing had the outstanding aspect ratio of 11.4, and gross area of 273.59sq m (2,945sq ft), compared with 161.7sq m (1,741sq ft) for the Tu-4, RAF-34 profile being retained. The wing was made in seven parts: a centre bridge the width of the fuselage, horizontal untapered centre sections carrying the inner engines, tapered mid-sections with 3 deg dihedral containing the flaps, sharply tapered at the outer trail-ing edge to meet the outer panels containing the ailerons, with the same mean taper and dihedral. There were fewer ribs than in the Tu-4 despite the greater span, skins up to 10mm thick at the root being secured by precision

bolts. Six sections of track-mounted slotted (virtually Fowler) flap were fitted, driven by electric screwjacks. The slender ailerons were each made in two tabbed sections.

The 2.9m (9ft 6in) diameter fuse-lage was almost that of the 80 but extended by fourteen extra frames, six ahead of the wing and eight behind, giving six main compartments. Of these, three were pressurised for the crew of eleven or twelve, comprising navigator, bomb-aimer, two pilots, engineer, radio, radar and technician in the main cabin, three gunners in the rear, plus the tail gunner. Three light-weight bunks were installed in the communication tunnel for off-duty members of a crew numbering up to

Aeroplane 85 No. 2. (Jean Alexander/RART)

sixteen on the longest missions. Front and rear bomb bays were improved and lengthened to carry a normal 5 tonnes, maximum 20 tonnes (one document, 18 tonnes) including the enormous FAB-9000 or nuclear weapons. The defensive sight/control system was as before, but the guns were ten NR-23s. The twin-wheel main landing gears had 1,580mm by 520mm tyres and retracted forwards pneumatically. The nose gear had hydraulic steering and retracted to the rear under the pressure cabin. The tail was based on that of the 80, with tailplane span 16.850m (55ft 3⅜in).

The list of available engines was whittled down to the Shvetsov ASh-2, which as noted under Aircraft 72 was a 28-cylinder radial of 3,300hp, and V A Dobrynin's VD-4. This was a basically later engine with four banks of six liquid-cooled cylinders with direct injection. When fitted with a giant turbocharger and gear-driven second supercharger the VD-4K was rated at 4,300hp for take-off and 3,800hp maximum continuous, and its performance at altitude surpassed that of every other piston engine to fly.

In late 1949 a static-test 85 was ordered, plus two flight articles, No. 1 with the VD-4K and No. 2 with the ASh-2. In the event, both were fitted with the VD-4K. The distance between the inner-engine centrelines was 9.1m (29ft 10¼in), and between the outers 20.38m (66ft 10in). Cooling air entered at the front, with forward gills around an oil-cooler ring, the two-stage superchargers being fed by a large dorsal inlet. The 4.5m (14ft 9¼in) reversing propellers had four black-

painted solid-dural blades. Fuel was housed in forty-eight flexible tanks, the total of 63,600 litres (13,990gal) being exceeded at that time only by the B-36. Later both aircraft were fitted with an air-refuelling probe on the starboard wingtip.

The No. 1 aircraft made its first flight on 9 January 1951, the crew comprising Perelyot, chief engineer N A Genov, navigator S S Kirichyenko and engineer A F Chernov. Tested to October 1951, it proved generally an excellent aircraft. On 12 September 1951 the No. 2 85 flew 9,020km (5,605 miles) in 20½hr with sufficient fuel remaining to have covered 12,018km (one document, 13,018km). Despite this, the decision was taken to rely on jet and turboprop bombers. The ASCC assigned the reporting name 'Barge'.

Span 55.939m (183ft 6⅜in); length (excluding guns) 39.905m (130ft 11in); wing area 273.59sq m (2,945sq ft).

Weight empty (No. 1) 54,711kg (120,615lb), No. 2, 55,400kg (122,134lb); fuel/oil 20,129kg (44,377lb), normal, 48,600kg (107,144lb) maximum; weight loaded 76,000kg (167,549lb) normal, 107,292kg (236,534lb) maximum.

Maximum speed 459km/h (285mph) at sea level, 638km/h (396mph) at 10,000m; climb to 5,000m in 8.9min; service ceiling 11,700m (38,390ft); range (5-tonne bomb load) 12,000km (7,457 miles); take-off run 1,640m (5,380ft); landing speed 185km/h (115mph), landing run 1,500m (4,921ft).

88, Tu-16

Partly because of the typically Russian large size and simplicity of its engine, and partly because of the aerodynamic

and structural excellence of its airframe, this jet bomber sustained the biggest production programme in financial terms in the entire history of the Tupolev bureau. Though originally based largely on the systems technology of the B-29, the swept-back aerodynamics, propulsion system, landing gears and many other features were pure Tupolev.

They proved amenable to progressive modification for fresh roles – even to the extent of serving as the starting point for families of passenger airliners – and repeated updating over a long period. Together with the excellent inspectability and anti-fatigue properties of the structure, they enabled some 2,000 of these substantial aircraft to serve in harsh environments for some 40 years, whereas equivalent Western aircraft, such as the B-47 and Valiant, were consigned to the scrapyard after ten years or less.

Tupolev naturally sought to combine the range and bomb load of the Tu-4 with the increased flight performance possible with jet engines, and in mid-1948, before the Tu-4 was in service, Stalin ordered him to do just this. He promptly transferred project engineers from the smaller Types 83 and 86/87, and set them to study possible aircraft using engines available or expected in the near future. One of the few answers appeared to be an aircraft powered by six VK-1 engines. This was unattractive and clumsy, and Tupolev became convinced that the required combat range could never be met except with high-compression axial engines, with better specific fuel consumption.

Accordingly, attention switched to Project 90/88, with four Lyul'ka TR-3F turbojets each rated at 4,600kg (10,140lb) thrust, two buried in the wing roots and two installed in under-wing pods. This nearly went ahead with the first *Rubin* radar and the OPB-11r sight (in fact Tupolev was to use a later version of the same engine in the four-engined Tu-110 derived from the Tu-16 bomber). The break-through was the availability of a large new turbojet developed principally for the projected Tu aircraft by the engine KB of A A Mikulin, who had provided piston engines for most of Tupolev's early large aircraft.

This engine began life as the M-209, but received the later factory designation of AM-3 and the VVS Service designation of RD-3. It was designed for high power simply by being big. It had an eight-stage compressor handling an airflow of 135kg/s, rising in production versions to 150kg/s (331lb/s), with a pressure ratio of 6.4 at 6,500rpm. (For comparison, the most powerful production engine then available in Moscow, the Nene-derived RD-45, had an airflow of 36.3kg/s.) Prototypes of the AM-3 were run in 1950 at 6,750kg (14,880lb) thrust, and air-tested in a pod carried under a Tu-4LL, but chief designer P F Zubets was confident the engine could be developed to much greater power. This turbojet enabled the new bomber to be twin-engined.

In late 1948 the Tupolev OKB received an order for two 88 prototypes plus a static-test specimen. Internally the project was called Aircraft N, the design team being led by Dmitri S Markov, who remained to become a Chief Designer of ANTK Tupolev in the 1990s. Tupolev himself decided the location of the engines. It was, of course, a requirement to be able to maintain height at maximum weight with one engine inoperative. It was a remarkable achievement to do this with a twin-engined medium bomber of this era. In contrast, a B-47 with three engines out on one side would mean eventual ejection. A A Yudin then led the design of the novel main landing gears, setting a fashion which for over 30 years became an OKB 'trademark'.

The basic requirement of the VVS-DA (air force long-range aviation) was to carry 5 tonnes (11,020lb) of bombs over a range not less than 5,000km

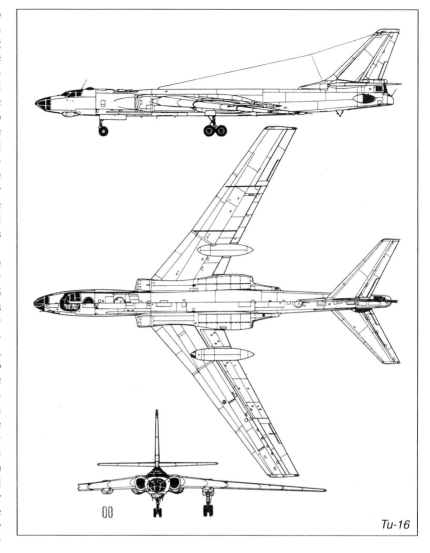

Tu-16

(interpreted as a combat radius of 2,500km, 1,553 miles), and also to carry the largest bomb, the enormous FAB-9000, weighing 9,000kg (19,840lb). This dictated the width of the bomb bay. Crew requirements made it logical to select a ruling fuselage diameter of 2.5m (98½in), with a forward pressure cabin not very different from that of the 85, but there were significant differences in that the sweep of the wing spars made it possible to place the bomb bay immediately behind them. Despite this, the wing was mounted relatively higher than in the piston-engined bombers, so that the top of the wing could be level with the top of the outer fairing of the inlet ducts with the engines at mid-height on the fuselage.

Tupolev found that drag could be reduced by discarding the circular section in the centre fuselage and moving the engines inwards close beside the bomb bay, only 1.9m (75in) apart. This

involved curving in the inlet ducts and curving out the jetpipes. This gave the desired inward (Küchemann) curvature of the streamlines around the wing roots, and also brought the jetpipes well away from the sides of the rear fuselage, which avoided scrubbing by the jets on the skin.

The inlets and ducts were not circular but resembled a D with the flat surface horizontal at the top. The ducts were split so that part of the airflow passed under the spars and part passed through. Apart from (in series aircraft) the gas-turbine starter in the nose bullet, the engines were entirely aft of the strong spars and spectacle frames needed to transmit the wing loads. They were attached by upper and side anti-vibration mounts to fuselage frames, with a row of five double doors underneath providing servicing access and, when necessary, for dropping the engine out straight downwards.

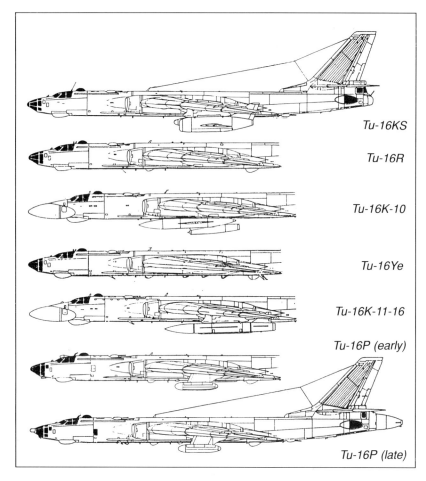

Tu-16KS

Tu-16R

Tu-16K-10

Tu-16Ye

Tu-16K-11-16

Tu-16P (early)

Tu-16P (late)

and static anhedral –3 deg. The main structural box comprised four spars inboard, the front pair meeting at the outer-wing joint to become three spars running to the tip, with thick machined rectangular upper and lower skin panels, forming integral tanks throughout. Hot-air anti-icing was provided for the fixed leading edge, with exhaust from multiple slits round the tips.

On the trailing edge were mounted large tracked slotted flaps inboard and outboard of the main landing gears, the total area on each wing being 25.17sq m (271sq ft), driven by electric ball-screws to a maximum setting of 35 deg for landing. From the flaps to the tips were balanced and tabbed Frise ailerons each of 14.77sq m (159sq ft). Fences 200mm (8in) deep (extended to the leading edge during production) were added at the structural break and at the flap/aileron junction.

The tail was scaled-up from the type 82 to give a fin area of 23.3sq m (251sq ft), the leading-edge sweep being 40 deg; rudder area was 5.21sq m (56.1sq ft). The area of the fixed tailplane was 34.45sq m (371sq ft), leading-edge sweep being 42 deg; each elevator measured 8.65sq m (93.1sq ft). All control surfaces were tabbed and balanced for manual control; production aircraft added hydraulic boost on all axes, retaining the tabs for trim, and also gave the tailplane variable incidence as the primary longitudinal trimming surface.

The fuselage was of almost circular cross-section, with a diameter of 2.5m (98.4in), made in five sections. The first section contained the double-glazed nose, with the main radar under the cockpit and, immediately

To a considerable degree the wings and tail were based aerodynamically and structurally on the 82, though they were much larger and refined in detail. As finally developed, the wing consisted of an inboard portion made integral with the fuselage, with incidence of +1 deg, and two outboard panels. The latter could be disconnected at a double rib perpendicular to the rear spar immediately outboard of the attachments for the main landing gears, 6.5m (256in) from the centre-line. The aerofoil profile of the inboard wing was PR-1-10S-9, giving a theoretical (theoretical because much of the airflow passed through the engine installations) thickness/chord ratio on the aircraft centreline of 15.7 per cent. The profile on the outer panels was the more common SR-11–12, giving a t/c ratio of 12 per cent.

Leading-edge sweep was 41 deg from the root to the structural joint, and then 37 deg to the tip. Mean aerodynamic chord was 5,021mm (198in),

A Tu-16KS with KS-1 cruise missiles in place. (RART)

to the rear, a ventral crew door with an integral telescopic ladder under the avionics/systems compartment, bounded by the convex rear pressure bulkhead. This pressurised section was constructed mainly of 3mm (0.12in) magnesium alloy, with thick ATIM insulation. The next section housed the nose landing gear, forward fuel tanks, dorsal turret, cameras and retractable landing lights. This section tapered in width towards the rear, in addition having dished sides to improve airflow into the engines. Next came the long centre section with protected tanks and the 6.5m (256in) bomb bay with two hydraulically-driven outward-hinged doors, followed by a bay for target indicators (or other loads in later versions). Next came Nos. 5 and 6 tanks and the ventral turret, followed by the rear pressurised compartment with a ventral door, side blister windows, retractable tail bumper, PTK-16 braking parachute and tail turret. There were a total of thirty-six main frames, those in the centre section giving way to spar bridges across the fuselage made of KhGSA high-strength steel to minimise the volume unavailable for fuel.

In the nose was the station for the navigator/bomb-aimer on an armoured seat, with a work table and OPB-11 optical sight. Next came the side-by-side pilots, then the *radist* to operate the RBP *Rubin* navigation and bombing radar, and the dorsal gunner

A Tu-16P with its high-power antenna pallet, taken from a Tu-16KS without missiles. (G F Petrov/RART)

with a sight under a transparent dome who also managed the electrical system and signals. In the pressurised tail compartment was a ventral gunner with optical sights at lateral blisters and a tail gunner manning the turret, the defensive system being described later. The pilots could blow off roof hatches and eject upwards, other crew-members ejecting downwards.

Fuel was housed in twenty-seven tanks with inert-gas protection, the fuselage tanks being self-sealing flexible cells and the wing tanks being integral. Total capacity was initially

43,800 litres (9,635gal), corresponding masses for two common fuels being given in the data. All fuel was fed through heated filters and flow-proportioners, and could be jettisoned from the wingtips. Later aircraft were equipped for in-flight refuelling.

This aircraft had the first Soviet bogie main landing gears. Each leg had four wheels with multi-disc anti-skid brakes. Tyre size was 1,100mm by 330mm (43.3in by 13in), the pressure being 6.5 bars (94.3lb/sq in), compatible with unpaved runways. Each main gear was rotated backwards by a hydraulic jack, the truck somersaulting to lie inverted in a box fairing with twin doors behind the wing. This (then strikingly unusual) solution avoided cutting into the highly stressed wing skins, and caused remarkably little drag. The nose gear had twin hydraulically steered wheels with tyres 900mm by 275mm (35.4in by 10.8in), hydraulically retracting to the rear into a bay with twin doors behind the pressure cabin. The wheelbase was 10.91m (35ft 10in) and track 9.775m (32ft). The large nose-up attitude required on take-off made it necessary for the landing gear to support the aircraft much higher off the ground than its piston-engined predecessors, and fortuitously this later made it possible to hang large missiles externally.

Each engine drove a pump in an independent hydraulic system with an operating pressure of 210kg/sq cm

Most of this modification line are Tu-16PPs (1880904 nearest), with a Tu-22M3 in the background. (G F Petrov/RART)

(2,987lb/sq in). Each served as the main or standby system for operating the landing gear, bomb doors, brakes and other services, and in series aircraft for flight-control boost. Electric power was generated as DC at 28V for the flaps and turrets, and for starting the main engines and (in later versions) the S-300M gas-turbine starters. Raw AC was supplied for de-icing the tail and windscreens. The air-cycle cabin environmental system energised by engine bleed operated with a maximum dP (differential pressure) of 0.5kg/sq cm (7.11lb/sq in). A KP-23 gaseous-oxygen system was fitted, with banks of spherical bottles, and two LAS-5M five-place liferafts were carried.

Tupolev was distressed by the fact that the normal loaded weight grew to over 80,000kg (176,367lb). At that time there was no expectancy that bombers would have to penetrate hostile territory at low level. He therefore contested the requirement for Mach 0.9 to be demonstrated at low level, and got this downgraded to 700km/h (435mph). This enabled the weight of the bare airframe structure to be reduced by no less than 5,500kg (12,125lb), but only in the No. 2 prototype. The heavy first prototype was flown by a test crew headed by N S Rybko on 27 April 1952. This aircraft reached 1,015km/h in a shallow dive, but level maximum was only 945km/h (587mph) and the heavy structure prevented attainment of the design range. Despite this, and the arguably better handling of the rival Il-46, the potential of the larger and more powerful Tupolev was much

An M-16 pilotless target conversion (8204203). (Yefim Gordon)

greater, and production was ordered in December 1952.

The OKB was permitted to delay the programme while the No. 2 aircraft was built. This was not only 5,500kg (12,125lb) lighter but it also had AM-3M engines rated at 8,200kg (18,078lb). This aircraft was flown in early 1953. Its maximum take-off weight was reduced to 71,560kg (157,760lb) and it reached 992km/h (616mph) and exceeded the specified range. Despite this, Markov was reprimanded by the Ministry for not getting it right first time.

With Service designation Tu-16, production with the RD-3M engine began at GAZ No. 124 at Kazan in late 1953, followed nearly a year later by GAZ No. 18 at Kuibyshyev. Nine series aircraft flew over Red Square on May Day 1954, causing a great stir in Washington. A small number of Tu-16s were supplied to the DA and also to the A-VMF (naval aviation), but most were retained for test and training. The USAF alotted the designation Type 39, replaced in

A Tu-16K-26, with missiles loaded. (Yefim Gordon)

1954 by the ASCC reporting name 'Badger'.

Tu-16A With designation from *Atomnykh*, this replaced the Tu-16 in production at the end of 1954. The Tu-16A was the first operational version, powered by RD-3M-200 engines rated at 8,700kg (19,180lb), started by a 100hp S-300M gas-turbine in an enlarged nose bullet. The weapon bay was configured for five types of nuclear bomb or a single FAB-9000, or other loads to a normal limit 9,000kg (19,840lb).

For the first time the operative PV-23 fire control was installed, representing a major advance over the Tu-4 fire control derived from that of the B-29. It governed all the defensive guns, and was linked to the PRS-1 *Argon* tail radar located immediately beneath the rudder. At the rear of the main crew compartment was installed the DT-V7 dorsal turret. This was armed with twin short-barrel AM-23 cannon, each with 500 rounds, with 360 deg traverse and elevation limits of +90 deg/–3 deg, an electric safety cutout preventing the turret from firing on the aircraft's own tail. Under the rear fuselage (Section 4) was installed the DT-N7S ventral turret, with similar guns each fed by a magazine of 700 rounds, slewing ±95 deg, and with elevation limits of +2 deg/–90 deg. The DK-7 tail turret was armed with twin long-barrel AM-23 guns, each with 1,000 rounds, slewing ±70 deg, and with elevation limits of +60 deg/–40 deg.

On the right side of the nose was a PU-88 installation of a single long-barrel AM-23, fed from a 100-round box, aimed by the pilot with a PKI reflector sight. The PV-23 system incorporated S-13 (forward) and PAU-457-1 and -2 (turret) ciné cameras.

Standard avionics, retained as the baseline in subsequent versions, comprised SPU-10 intercom, R-807 and (pilot) -808, RSIU-3 UHF, RSIU-4 VHF, an RBP-4 *Rubin* main radar, RV-17 (high-attitude) and RV-2 (low-altitude) radar altimeters, an AP-28 autopilot linked to the NAS-1 navigation system including a DISS-*Trassa* doppler, MRP-48P ADF, KRP-F VOR, RSBN Tacan and SD-1M DME linked via a GRP-2 glide-slope receiver into the SP-50 ILS, the SRO-2 IFF subsystem with the usual triple-rod antennas, plus the Sirena-2 RWR (radar warning receiver). Standard cameras were the AFA-33/50M, 75M and 100M and AFA-42/50 (day) and NAFA-8S (night). During production, upgrades included the AP-6E autopilot, RV-5 and -18 radar altimeters, SOD series ATC/SIF, ARK-15 ADF, ARK-5 tanker homing, the SRO-2M *Khrom-Nikel* IFF, and various installations of ASO-series chaff/flare cartridge dispensers. In 1965 the ILS was upgraded to SP-50M standard for landing in Cat III (almost blind) conditions. A proportion of aircraft were fitted with RSDN *Chaika* Loran, with a 'towel rail' antenna, and, from 1983, a *Glonass* navigation-satellite receiver.

Production of the Tu-16A was approximately 700, out of a total for all versions of some 1,515 built at Kuibyshev, Kazan and Voronezh. A total of fifty-four, including some Tu-16s, flew over Moscow on Aviation Day 1955, causing alarm in NATO circles. Many Tu-16As were re-engined with the RD-3M-500, rated at 20,950lb, and after 1960 most were rebuilt for other purposes. Egypt received twenty, many of which were destroyed on the ground in the 1967 Six-Day War. The ASCC name was 'Badger-A'.

Tu-16M Officially called a sea bomber, this variant was based on the Tu-16A with minor modifications for A-VMF service.

Tu-16KS From the outset the Aviatsiya-VMF (naval air force) played an increasingly important role in the programme. From 1948 it funded the development by the Mikoyan/Guryevich OKB of the KS-1 *Komet* swept-wing cruise missile for stand-off attacks on ships. To carry and launch this 3,000kg (6,614lb) weapon the Tu-16KS prototype was flown in August 1954. It carried one KS-1 under each outer wing on a pylon 7.75m (25ft 5in) from the centreline, plus a *Kobalt-N* guidance transmitter from the Tu-4K. When attached to the bomber at cruising speed most of the missiles' weight was supported by their own wings, so that the Tu-16 pilot would have little difficulty holding the aircraft level in the asymmetric

Testing the engine installation of the M-17 (6401410). (Yefim Gordon)

state, with one missile launched. Like later missile-carrying versions, the flaps of the KS incorporated a slot cut out of the trailing edge to clear the missile's vertical tail. Operating radius was 1,800km. Over 100 of this version were built in 1954–57 for the A-VMF, plus twenty-five of an export model for the AURI (Indonesian Air Force) Nos. 41 and 42 Sqns, delivered in mid-1961. Those in A-VMF service were converted into later versions. The ASCC name was 'Badger-B'.

Tu-16R, Type 92 This major reconnaissance variant was produced from 1955 in day and night photographic versions. Most were later converted with different sensors and equipment. The pilot's gun was removed, and pallets were tailored to the bomb bay carrying from five to nine AFA-series cameras, including two NAFA-MK-75, from 1961 adding other sensors including two steerable Comint/Elint (communications or electronic intelligence) antennas in ventral blisters. The ASCC name was 'Badger-E'.

Tu-16RM This was the corresponding A-VMF maritime version, the camera fit always including AFA-33/20M and AFA-42/75.

Tu-16Z In 1955 this tanker version was evaluated, using a looped hose from the right wingtip of the tanker to the left wingtip of the receiver, as developed with the Tu-2 and Tu-4.

Additional transfer fuel was carried in AMTs auxiliary tanks in the bomb bay, which could be removed in order to restore bombing capability. Total transfer fuel 19t. No ASCC name is known.

Tu-16T This A-VMF torpedo bomber prototype was tested in early 1956. It carried four 533mm TAN-53 or RAT-52 torpedoes, and/or bombs or sea mines. Several more were built new, all later being converted to Tu-16S standard. No ASCC name is known.

Tu-16K-10 This major A-VMF weapon system carried a single K-10S turbojet supersonic cruise missile, weighing 4,400kg (9,700lb), on the centreline. The weapon required a multipin guidance link and fuel connection. The K-10S *Kompleks* included an A-329Z guidance system replacing the nose navigator station by a YeN search/tracking radar with a 2m (80in) antenna scanning through a limited forward arc to provide offset guidance to the missile autopilot out to a theoretical distance of 250km (155 miles). The guidance radar itself was below the cockpit. The bomb bay housed the upper part of the missile fuselage and its vertical tail, its top-up fuel tank and a pressure cabin for the YeN operator. From this variant onwards the pilot's gun was usually omitted. The Tu-16K-10S prototype was flown in 1958. Tests took place to see if there was any advantage in using

the missile engine to assist take-off. About 220 were built from mid-1959, but Fleet trials of the K-10S were not completed until October 1961 when production of the missile began. Most of these aircraft were rebuilt as EW (electronic warfare) platforms. The ASCC name was 'Badger-C'.

Tu-16Ye The first of several A-VMF electronic-warfare variants, the Ye (written E in Cyrillic characters) dated from 1961. The designation comes from the *Yolka* (fir) ECM installation. They were rebuilt A, KS or K-10 aircraft, with the bomb/missile capability deleted but retaining the original types of radar and adding a row of three steerable receiver antennas under the bomb bay, which was occupied by multi-waveband (A through I) analysis/classification/threat libraries. At the rear of the bay, in what had been the target indicator and photoflash bay, was usually installed a bulk chaff cutter/dispenser. The chaff was automatically cut to match the wavelengths of the perceived hostile threats, and fed out at rates giving rapid-bloom clouds for maximum protection. The ASCC name was possibly 'Badger-D'.

Tu-16P The second type of A-VMF conversion for electronic warfare, these also began life as A or KS aircraft, rebuilt for the new role from 1961. The sensing and jamming equipment, mainly in the bomb bay, was served by a total of twelve receiver/jammer antennas on

Testing the bicycle landing gear of the Myasishchev M-4 at reduced scale. (G F Petrov/RART)

the nose, on small pods under the centre fuselage and (usually) in larger pods on underwing pylons. From 1965 the tail turret was often replaced by a fairing for an SPS tail-on reception/DF/jamming installation (*see* Tu-95KM). Some of this type were equipped for Elint (electronic intelligence) missions only. About sixty served to at least 1990. Combined production of the Ye and P was 135. The ASCC name was probably 'Badger-F'.

Tu-16K-11-16 From 1959 some Tu-16A and KS aircraft were refitted with *Rubin-1* radar. Later in that year most unconverted Z and KS aircraft were rebuilt for A-VMF service with the K-11-16 *Kompleks* to interface with two new types of rocket-propelled cruise missile. These were the KSR-11 (K-11 system), each weighing 5,950kg (13,117lb) or KSR-2 (K-16), each weighing 3,800kg (8,377lb), two of either type being carried on underwing pylons. In theory these aircraft retained free-fall bombing capability, though for weight reasons no bombs could be carried if the missiles were installed. A simplified K-16 version was used by Egypt in the Yom Kippur war in October 1973. A total of twenty-five of the supersonic missiles were said to have been launched, of which the Israeli Air Force claimed twenty shot down (which is hard to believe), and two of the Tu-16s were also claimed. The A-VMF aircraft were later rebuilt for further roles. The ASCC name was 'Badger-G'.

Tu-16K-10-26 This second type of A-VMF missile carrier was produced in small numbers from 1962. They were Tu-16K-10 aircraft modified to launch a K-10S carried on the centreline or two KSR-5S (option of KSR-2) on wing pylons. They were replaced by the K-26. The ASCC name was probably 'Badger-C Mod'.

Tu-16N These aircraft were the first dedicated air-refuelling tankers of the DA and A-VMF, dating from 1963. Almost all were rebuilt Tu-16A bombers, and they were initially required to support Tu-22 regiments. The bomb bay was occupied by a hose-drum unit for the Yakovlev-developed probe/drogue system, together with special lighting and an ARK-5 receiver-aircraft homing beacon. The usual fuel capacity was 54,000 litres

A sensor testbed with the nose of the Myasishchev M-55. (RART)

(11,880gal), the normal single-hose transfer time for 20 tonnes (about 25,000 litres) being 15min. Over seventy were still airworthy with the CIS Navy in 1994. No ASCC name is known.

Tu-16Sh These aircraft were Tu-16A rebuilds fitted with the radar, navigation and bombing/missile *Kompleks* of various versions of the Tu-22, for use as crew trainers. No ASCC name is known.

Tu-16P Under this unchanged designation various further rebuilds of the Tu-16P were delivered as Elint platforms with a self-protection active jammer antenna in a nose 'thimble' and additional receiver antennas, mainly for VHF. All had the tail turret replaced by the ECM tailcone. The ASCC name was 'Badger-L'.

Tu-16S An important A-VMF SAR (search and rescue) version, with an operating radius of 2,000km (4,409 miles). All are rebuilds from about 1965 – believed to include the remaining Tu-16T aircraft – with additional radio, auxiliary fuel in the bomb bay and a *Fregat* radio-controlled rescue boat para-dropped from under the fuselage and then guided to the survivors by a *Reya*' transmitter. This type is still in service. The designation comes from *Spasitelnyi* (help). There is no apparent evidence for the Western name 'Korvet'. No ASCC name is known.

Tu-16K-26 This remains the principal A-VMF missile-launch version with an operating radius of 2,100km (4,630 miles), dating from 1965. These aircraft, a few of which are still airworthy, have the K-26 *Kompleks* and are able to launch either the KSR-2, KSR-5N, KSR-5P or KSR-11 rocket missile, each weighing about 5,950kg (13,117lb) from pylons under either the right wing or both wings. A large steerable antenna for the guidance radar is installed ahead of the bomb bay. The ASCC name is 'Badger-G Mod'.

Tu-16PP These are further rebuilds of A or KS versions, with bombing or missile equipment removed, to fly stand-off jamming missions in support of attacks by other aircraft. Tandem Elint receivers are installed in front of and behind the bomb bay, which is rebuilt to dispense up to 8,000kg (17,640lb) of bulk chaff cut to length to match the received threat-wavelengths. The ASCC name is 'Badger-H'.

Tu-16P mod Yet another A-VMF EW platform, these are second rebuilds of Tu-16K-10 aircraft from 1969. Their primary mission is active jamming with interlinked groups of palletised receivers and transmitters. These are mainly in the bomb bay, though some are served by large horizontal plate antennas projecting beyond modified wingtips, with a radiation pattern similar to those mounted between the jetpipes of Avro Vulcans. The installation usually covers Bands A or B to J. Most of this variant have a *Chaika-M* Loran rail antenna. The ASCC name is probably 'Badger-J'.

Tu-16KRM Basically, these are modified Tu-16RM aircraft transferred to the PVO and modified yet again to launch and guide rocket target drones. A small number remained in use until the 1990s.

M-16 A number of 'tired' aircraft, not worth further conversion, were gutted of reusable equipment and fitted with a remote-pilot receiver station for use as high-altitude missile targets. Features include a nose 'pimple' and rod antenna, an antenna group in a new tailcone and three antennas in a group projecting vertically under each wing.

Tu-16 Tsyklon These are weather-reconnaissance conversions, dating from about 1977. They are equipped for a range of meteorological research and measurement tasks and also for cloud-seeding by various chemical dispensers.

Tu-104G Under this designation a number of demilitarised aircraft, believed all to be Tu-16s though possibly including some 16As, were used by Aeroflot to carry urgent cargo, in particular matrices for printing *Pravda* and *Izvestiya* in distant parts of the USSR. Others, dubbed 'Red Riding Hood', were used as trainers.

Tu-16LL
As soon as Tu-16s became available they were used as flying testbeds (LL, flying laboratory), and LII at Zhukovskii was eventually to use nearly thirty, for a total of thirty years. As early as 1954 one aircraft was fitted with an experimental air-refuelling hose and drogue extended from a tube at the extreme end of the tail for tests with probe-equipped fighters. Though many became the standard aircraft for particular types of radar and systems test, the most important single use was to fly new types of turbojet and turbofan engine. The engine under test was installed in one of two sizes of standard nacelle, mounted on a pair of hydraulic jacks in the manner pioneered with the Tu-4LL so that it could be retracted into the bomb bay or, in the air, fully extended away from

airflow disturbance upstream of the inlet. The chief engineer could observe the nacelle by various means, including a periscope from the rear gunner's position. The engines first tested on the Tu-16 were: the AI-25, AL-71F1/F2/F4, AL-31F, D-30, D-30K and KP, D-30F6 (Mig-31 installation), D-36, NK-8-2, R-11AF-300 (Yak-28 nacelle), R-15-300 (Ye-150 and MiG-25), R-21-300, and R-21F (Ye-8), RD-3M, RD-15-16, RD-16-17, RD-36-41, RD-36-51, VD-7M and VD-19 (intended Tu-128 installation). A Tu-16LL with outer wings removed tested the complete powerplant of the V/STOL Yak-38. One TU-16LL flew with a zero-track 'bicycle' landing gear.

H-6 In September 1957 China was granted a licence to manufacture the Tu-16A. In 1959 two pattern aircraft were delivered, and one of these was assembled at Harbin and flown on 27 September 1959. In 1961 the Chinese decided to build the Tu-16 as the H-6 (*Hongzhaji*-6, bomber 6) at what became XAC, the Xian Aircraft Mfg Co. The engine was produced as the WP-8. The first H-6A flew on 24 December 1968, and by 1987 a total of 120 had been delivered to the PLA air force and navy, plus four to Iraq and spare parts to Egypt. The current version is the H-6D (B-6D in Westernised notation). This has an increased wing area, and most have a large Chinese navigation and bombing radar, with a rectangular antenna rotating in a drum-shaped radome, and a computer providing targeting information for two C-601 anti-ship cruise missiles carried on wing pylons. Unlike the 6A, the 6D has no fixed nose gun. There are several other Chinese variants either studied or flying, including a tanker version.

Span (88) 35.5m (109ft 11in), (production aircraft, excluding any wingtip antennas) 33.4m or (most) 32.989m (108ft 2¾in); length (basic) 34.8m (114ft 2in), (modified versions) up to 37.902m (124ft 4in); height (usual) 10.36m (34ft 0in); wing area (standard) 164.65sq m (1,772sq ft), (H-6D) 167.55sq m (1,804sq ft).

Empty weight (88 No.1) 40,200kg (88,624lb), (Tu-16A) 36,600kg (80,688lb), (maximum for EW conversions) 38,000kg (83,774lb), (H-6D) 38,530kg (84,944lb); fuel (Tu-16A) 34,360kg (75,750lb), equivalent to 41,400 litres, 9,107gal of T-1 fuel or 43,750 litres, 9,624gal of TS-1, (H-6D) 33,000kg (72,751lb); normal loaded weight (88 No.1) 80,000kg (176,367lb), (production, all versions) 72,000kg (158,730lb); maximum loaded 75,800kg (167,108lb), maximum

landing weight (paved runway) 50,000kg (110,229lb), (unpaved) 48,000kg (105,820lb), (H-6D) 55,000kg (121,252lb).

Maximum indicated airspeed (except 88 No.1) 700km/h (435mph); maximum speed at 6,000m (88 no.1) 945 km/h (587mph), (Tu-16A) 1,050km/h (652mph); maximum speed at 10,000m (Tu-16A) 1,010km/h (628mph, Mach 0.95); maximum cruising speed of H-6D with C-601 missiles, 786km/h (488mph); service ceiling (88 No.1) 11,000m (36,090ft), (Tu-16A) 15,200m (49,870ft), (Tu-16KS) 12,800m (41,995ft), (H-6D) 12,000m (39,370ft); range with 3,000kg bomb load (Tu-16A) 5,760km (3,580miles), with 9,000kg bomb load 4,400km (2,735 miles); combat radius (Tu-16KS, H-6D, with missiles) 1,800km (1,118 miles); take-off run (typical) 1,250m (4,100ft); landing speed 223km/h (139mph), landing run 1,100m (3,600ft).

90

This project of 1951 was essentially a version of the 88 (Tu-16) with four engines. Drawings show an installation of AL-7 engines similar to Type 110.

91

In 1945 Stalin ordered that the VMF, the Soviet Navy, should be 'still stronger and more powerful', and – though hampered by the need to rebuild the construction yards – huge shipbuilding programmes were authorised. These concentrated on submarines and destroyer/frigate types, but in 1947 it was decided also to build several giant 80,000 tonne aircraft carriers 'to project Soviet power to all corners of the globe'.

The Tupolev OKB was involved from the start. One possibility was a ship-based version of the Tu-14. Though this was achievable, there was an obvious need for a smaller aircraft able to fly ship-based attack, bomber, torpedo, minelaying and reconnaissance missions at less cost, with much lower fuel consumption, especially at sea level. It was expected that the same aircraft might be capable of serving as an air-defence fighter, though there was an obvious need for a smaller swept-wing jet purely for the fighter role. As noted previously, the carrier-based Tu-14 was launched with OKB designation 509, signifying 1950 project No. 9, but it was abandoned.

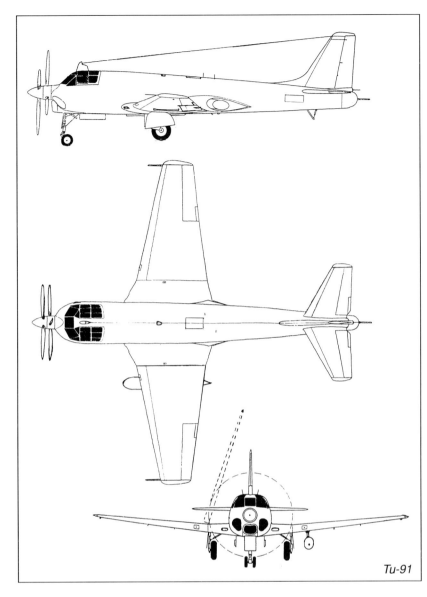

Tu-91

influenced by the Douglas XA2D Skyshark, designed under Ed Heinemann and first flown in May 1950. The missions, propulsion, power, dimensions, plan-view shape, weights and performance were remarkably similar, the only significant difference being that the US Navy aircraft was a single-seater while the Soviet navy demanded a pilot and navigator. The point can also be made that this aircraft was unlike anything previously attempted by the Tupolev OKB, and far removed from the large multi-engined aircraft for which the OKB had become famous.

When Stalin died in February 1953 the outline design of Project 507 had been settled. The only way to meet the conflicting demands of high take-off thrust, high speed at sea level and long endurance and mission radius was to use a single TV-2 turboprop. Its design team, under Kuznetsov, was led by Austrian Ferdinand Brandner. The prototype engine ran in 1948, and in 1950 flew in the No. 3 position on a Tu-4LL testbed. By 1952 the uprated TV-2F was on test, with a take-off power of 6,000hp.

The Advanced Projects team submitted their proposal to the Aviatsiya-VMF (naval aviation) and to A N Tupolev in late 1952. Features included a low-mounted unswept wing with outer panels housing no fuel and arranged to power-fold upwards, a single engine mounted amidships near the centre of gravity driving contra-rotating propellers via a long shaft, a crew of two on side-by-side ejection seats in the nose, tricycle landing gear, a remotely aimed tail gun turret, and, after careful analysis of tradeoffs, external carriage of weapons. The design was accepted with the Service designation of Tu-91. Originally odd numbers were reserved for fighters; like the later Su-25 for similar roles, this aircraft was loosely judged to be 'fighter size' rather than a bomber.

Design went ahead in the main OKB under Chief Designer V A Chizhevskii (once a pioneer of pressurised stratospheric balloon gondolas). Section heads were V I Bogdanov (airframe), A M Shumov (propulsion) and M G Pynegin (systems/equipment). M M Yegorov was Chief Engineer charged with solving hardware problems and getting flight tests started. The difficult engine and propeller drive was flight-tested from

The smaller multirole aircraft became Project 507 and from the start it was a major challenge to the Advanced Project Section directed by V M Kondorskii. The first drawings were produced by A A Yudin.

There appeared to be no way the missions could be flown by a jet. To achieve the necessary endurance the engine had to be turboprop (even a sea-level version of the 4,500hp VD-4 piston engine was briefly considered). The choice appeared to lie between the same engines as were later to compete to power the Tu-95 strategic bomber, the TV-2 (or the twinned 2TV-2F) and TV-12. Both were single-shaft engines based on German wartime technology. They were the assigned task of some 800 German engineers, almost all ex-Junkers or BMW designers, sent under armed guard in 1946 to form two special bureaux, OKB-1 and

-2, at Kuibyshev (unrelated to the German aircraft-design bureaux with the same designations, at Podberezye). General Constructor in charge of both bureaux was Nikolai D Kuznetsov.

The TV-2 was started at once, the design being based on that of the Junkers 109-022, and aimed at an initial rating of 5,000shp. The even larger TV-12, a completely fresh design, was designed under intense pressure from 1951 for the Tu-95 strategic bomber. Its design power of 12,000shp was so impressive that in 1952, when the Project 507 had reached an advanced stage, a sudden decision was taken to examine an enlarged version with the TV-12 and a wing having the same 35 deg sweepback as the Tu-95, though of lower aspect ratio. This, was finally judged overweight, and reluctantly dropped.

The Advanced Projects team were

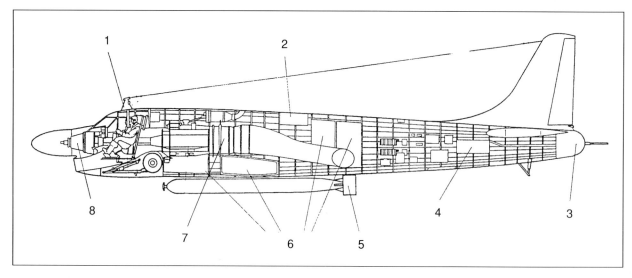

Internal arrangement of 91: 1, rear-view periscope; 2, LAS-5M liferaft; 3, tail turret; 4, gun magazine; 5, Type 45-36AN torpedo; 6, tanks; 7, engine; 8, propeller reduction gear.

1952 on a Tu-4LL. On 24 April 1953 the Tupolev OKB received an order for a single flight-test article plus a static-test airframe.

Within a month of Stalin's death the Kremlin cut the naval building programme by roughly half, a major casualty being cancellation of the carriers. Tupolev himself led a delegation to plead the cause of the 91, and it was eventually agreed that this should go ahead with some changes as a land-based aircraft. It was to equip coastal A-VMF regiments tasked with level and dive bombing attacks on hostile ships, torpedo attack, minelaying, strafing with cannon and rockets and coastal reconnaissance. An anti-submarine version, with radar and sonobuoys as well as weapons, was to be studied.

Structurally the Tu-91 was conventional, the ruling material being D16-T Duralumin. The low wing had three spars, a sharply tapered centre section housing four fuel tanks and the retracted main landing gears, outer panels with a fixed leading edge, track-mounted slotted flaps and manual tabbed ailerons. The tailplane, with a span of 6m (19ft 8⅛in) and an area of 10.2sq m (110sq ft) was fixed horizontally on the rear fuselage, carrying manual elevators. The single fin carried a manual balanced rudder, the combined area being 5.05sq m (54.4sq ft).

The nose landing gear had twin wheels, with tyres 570mm by 140mm, hydraulically steerable and fitted with disc brakes. This unit retracted hydraulically to the rear into an unpressurised bay under the cockpit. Consideration was given to stressing this unit for nose-tow catapult launching, but after cancellation of the carriers short take-offs were to require the use of two SPRD solid-propellant a.t.o. rockets. The main landing gears had short legs with pivoted levered-suspension arms carrying single mainwheels, again with disc brakes. Track was 3.37m (133in) [an OKB plan gives 3.75m, 148in]. The high-pressure tyres, 1,050mm by 300mm, were intended to be replaced for land operation by larger tyres inflated to a lower pressure, but this was not done in the first prototype. The main gears retracted inwards, the levered legs shortening the unit to keep the wheel outboard of the root rib. The high design rate of descent for carrier operation fitted the landing gears for use from rough surfaces.

The arrangement of the cockpits in this first aircraft is not known for certain. The presence of the propeller drive–shaft would probably have made it difficult to accommodate tandem ejection seats on the centreline. All that is known is that the nose was redesigned in a second aircraft, and the cockpit of that aircraft is described later.

CAHI assisted in the prolonged investigation into how best to feed air to the engine. Remarkably, the Tu-4LL flying testbed was eventually required to fly with one of its engines replaced by a simulated 91 nose, with cockpits and operative propulsion system! The unique final choice was to use three chin inlets to ducts between and beside the crew to a plenum chamber ahead of the engine. The turboprop, initially a TV-2F rated at 6,000shp, was carried on anti-vibration mounts resting directly on the wing spars. The long drive-shaft passed above the nose-gear bay, between the seats, to the epicyclic nose gearbox. From here concentric shafts drove the 5.7m (18ft 8in) six-blade reverse-pitch contra-rotating propeller, which was designed to be used as a brake in dive attacks. The jetpipe was bifurcated to exit on both sides of the fuselage, the gas passing below the tailplane.

Fuel was housed in four centre-section cells plus two between the jet-pipes, all protected and inerted. A drop tank could be carried under each wing. Access doors were provided beside and above the engine. Engine-driven accessories and the gas-turbine APU/starter were mounted above the engine. An oil cooler was installed in each wing root, with its inlet in the leading edge and a louvred exit above the wing.

Two NR-23 cannon, each with 100 rounds, were installed in the roots of the outer wings. In the tail end of the fuselage was fitted a DK-15 powered turret with two NS-23 guns, each with a magazine of 150 rounds in the rear fuselage. This turret was aimed by the navigator, using a rear-view periscope and electro-hydraulic control.

There was no internal weapon bay, but external racks could be configured

91 No. 2 with auxiliary fuel tanks. (RART)

for a single TAN-53, jet-propelled RAT-52 or classic 45-36 torpedo under the centreline. Alternative loads could include a single FAB-1500 or twelve FAB-100s, or three 500kg (1,102lb) anti-ship bombs, or three mines or various unguided rockets (eight TRS-212, thirty-six TRS-132 or 120 TRS-85 or ARS-85) on two or four wing pylons outboard of the power fold. A sting-type hook was provided at the rear, a camera in the rear fuselage with ventral doors, an LAS-3M three-man dinghy in a dorsal box, retractable landing lights under the outer wings, a mast above the cockpit for an hf radio wire, and a vhf blade antenna above the rear fuselage.

The prototype 91 was rolled out in spring 1955. Thanks to a great deal of prior systems testing, the flight-test programme began on 17 May, the test crew being factory pilot D V Zyuzin and flight-test observer K I Malkhas-yan. They pronounced it highly manoeuvrable, though barrel rolls were forbidden. Factory testing was completed about September 1955. Results were generally encouraging, though Tupolev compiled a list of desirable modifications, and a second flight article was built to a tight schedule. This was regarded as a pre-series aircraft, incorporating the features needed for land-based operation as well as modifications shown to be needed by flight test.

Most important of the changes in the second aircraft was a completely redesigned nose and pressurised cockpit to improve crew comfort and field of view. Certainly in this second aircraft the crew sat side-by-side, pilot on the left. They boarded via ladders and entered by upward-hinged gull-wing canopies hinged on each side of the centreline. Each crew-member had an ejection seat mounted on slightly oblique rails so that, even if fired simultaneously, the seats and occupants would not interfere with each other. The entire cockpit was encased in a 'bath' of ANBA-1 light-alloy armour, from 8mm to 18mm thick, and except for those overhead all windows were bulletproof.

Other major modifications in the No. 2 aircraft included a non-folding wing, provision for a.t.o. rockets, wider tyres of unchanged diameter (fitting the original wing bays), deletion of the arrester hook, and a detailed rework throughout the aircraft to improve accessibility and reduce weight (empty weight was cut by 500kg). A major change was that the engine was the TV-2M, uprated to 7,650shp, driving an unchanged propeller.

This second aircraft had a big-headed appearance, tapering there-after to a slim tail, and this gave rise to its popular name *Golavl'* (chub). No evidence has been found for the

Western story that it was known as the *Bychok* (young bull).

The second aircraft was submitted to NII-VVS testing in January 1956. The assigned flight crew, Lt-Col Alekseyev and Maj Sizov, wrote a favourable assessment. Subject to certain modifications, such as improved engine-starting at −30°C and replacement of outdated navigation equipment, production was duly authorised, including versions for ASW, ECM jamming and dual training.

Whether or not the 91 would have been a good investment, it is on record that the whole programme collapsed because the officer detailed to explain the aircraft to the newly elected Party General-Secretary, Nikita S Khrushchyev, said 'This aircraft can replace a heavy cruiser' (because its firepower was judged similar to a salvo from eight 203mm guns). Khrushchyev said 'But nobody needs heavy cruisers any more', and re-marked what a ridiculous machine it was. His entourage had to smile and nod at the new leader's instant judgement, which ensured that the programme would go no further.

Time was to show that many modern wars need tough multi-role attack aircraft able to operate from rough airstrips. Few since the 91 have had the ability to destroy large surface vessels or lay mines, and the vulnerability of such machines to fighters would have been reduced in this air-craft by its hard-hitting turret. The 91 was the most powerful single-engined propeller aircraft ever built, and could have carried much heavier weapon loads than the stipulated 1½ tonnes. The second 91 was shown to a Western delegation at Kubinka in June 1956, though its appearance remained unknown until 1993. The ASCC assigned the reporting name 'Boot'.

Span 16.4m (53ft 9⅝in); length 17.7m (58ft 0⅞in) (another document 17.19m, 56ft 4¾in); height 5.06m (16ft 7¼in); wing area 47.48sq m (511sq ft).
Empty weight *c.*8,000kg (17,640lb); normal loaded weight 12,850kg (28,329lb); maximum take-off weight 14,400kg (31,746lb).
Maximum speed (clean) 800km/h (497mph); dive attack speed (with weapons) 750km/h (466mph); economical cruising speed 250–300km/h (155–186mph); service ceiling 11,000m (36,000ft); range (maximum) 2,350km (1,460 miles); range with weapons 1,600–1,900km (994–1,180 miles); take-off run (MTO weight, a.t.o. rockets) 518m (1,700ft), landing run 552m (1,811ft), (using reverse pitch) 438m (1,437ft).

92–94

The 92 number was assigned to the Tu-16R (variant of 88); 93 was a 1952 version of the Tu-14 intended to have Klimov VK-5 turbojets. Aeroplane 94

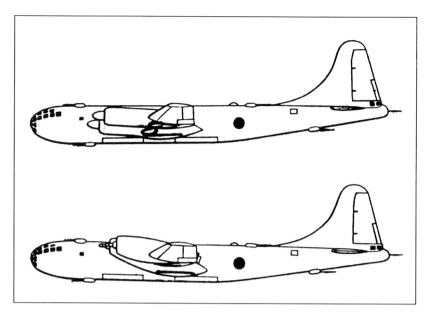

Tupolev drawings showing the unbuilt Tu-4/TV-02 (upper) and Tu-4/TV-4 (also designated VK-2); both engines were flown on Tu-4LL flying testbeds.

was the 1952 project for a turboprop Tu-4, with either TV-02 (later called VK-2) or TV-4 (NK-4) engines. Estimated figures included maximum speed 650–680km/h (404–423mph), range 5,500–6,300km (3,418–3,915 miles) at 9,500m (31,168ft), altitude cruising at 580–600km/h (360–373 mph), and normal and maximum bomb loads for this range of 1,500kg (3,307lb) and 6,000–12,000kg (13,228–26,455lb).

Tu-95 and Tu-142

The creation by the USAF of SAC (Strategic Air Command) in 1946, with nuclear weapons, naturally spurred the Kremlin to rethink their whole idea of strategic airpower. Within months the ADD had been reorganised as the DA (*Dalniya Aviatsiya*, long-range

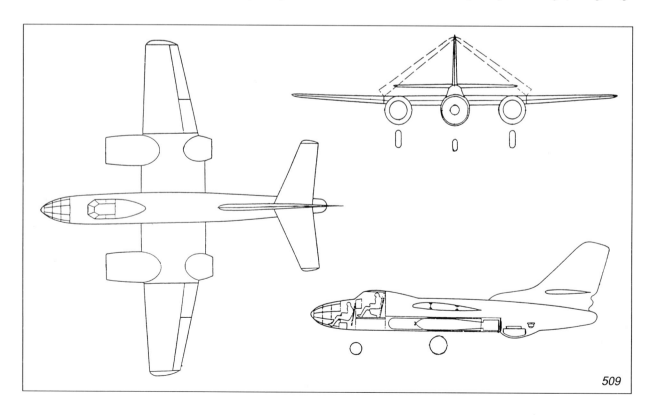

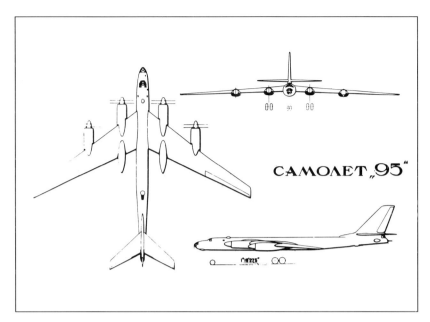

OKB drawing of 95/I.

aviation), with global aspirations almost rivalling those of SAC.

Despite the ability of the Type 85 to reach the USA and return, it was increasingly recognised from 1949 that future strategic bombers would have to be powered by gas-turbine engines. Even the newest and most powerful piston engines, notably the VD-4K, were massive, complicated and insufficiently powerful. The ideal was jet propulsion, but because of their high specific fuel consumption none of the jet studies then made in the USSR or elsewhere could find a way to achieve useful intercontinental range, even with a single air refuelling.

Pressure was put on A N Tupolev personally by Stalin's direct order in late 1949 that his OKB should produce 'a turbine-engined strategic bomber'. The outline requirement was a range of 14,000–17,000km (8,699–10,563 miles) at a height of 10,000–15,000m (32,808–49,212ft) and cruising speed of 750–800km/h (466–497mph), whilst carrying a bomb load of 12 tonnes (26,455lb). This was far beyond the capability of the first model of Boeing B-52.

There was no reason to change the fuselage, so it was based on the 85 but lengthened. The rest of the airframe appeared to be a choice between turbojets and swept wings and tail, or turboprops and a direct extrapolation of 85 surfaces matched to much greater fuel capacity and take-off weight. The Korean war began in June

1950, and the vulnerability of the B-29 quickly swung the scales against propellers and unswept wings.

The speed demanded sweptback surfaces, and CAHI helped to define a wing derived from that of the Type 88, swept at 35 deg with an aspect ratio which was initially 7 (already higher than for the 88) and was later increased. Studies with various engines finally led to two which were examined in depth: six AM-3 turbojets (superimposed pairs in the wing roots plus two under the rear fuselage) or turboprops totalling about 40,000hp. The jet study looked exotic, but could not meet the required range. For the turboprops there was a choice of two engines being developed by N D Kuznetsov's bureau (and discussed in the entry on the Tu-91): eight TV-2F in the 5,000–6,000hp class or four TV-12 of 12,000hp. The rival VK-2 by V Ya Klimov was too small, in the 4,800hp class, and in any case fell by the wayside.

The crucial breakthrough was the realisation that it would be possible to put a sea-level power of 12,000hp through front and rear contra-rotating co-axial propellers of a practical (such as 5.6m, 18ft 4in) diameter, and to fly over the target at full power with the blades at such a coarse setting that the true airspeed would be not greatly inferior to that of a jet. Thus, for the first time with a propeller aircraft (the Bell L-39 being merely a wing test-bed), swept wings and tail were the correct choice. Development of the

vital propellers, and of the reduction gearbox which was to drive them, was assigned to K Zhdanov. The airframe was a direct extrapolation of Type 85 and 88 practice, the aerodynamic design being closely similar to the latter aircraft. Take-off weight was calculated to be in the 150-tonne class, refined in January 1951 to 156 tonnes (343,915lb).

On 11 July 1951 the new strategic bomber was officially authorised, and Type 85 work stopped. The OKB designation number was 95, the project leader being N Bazyenkov, with structural design overseen by the veteran Arkhangel'skii. Prudently, the project went ahead with both types of turboprop, the 95/I being planned with eight TV-2s, and the 95/II to have the option of four TV-12s. In 1950 Tupolev and Kuznetsov studied long push/pull nacelles extending ahead of and behind the wing, as well as other arrangements of eight engines. In January 1951 these led to Kuznetsov's development of the 2TV-2F, built as a single engine unit with the left power section, with its inlet and jetpipe on the left, driving the front four-blade unit of 5.6m (220½in) AV-60 coaxial propellers via a shaft passing through the drive from the right power section, with its inlet and jetpipe on the right. Each power-plant had a broad triangular cross-section. Distance between the outer thrust lines was 24.4m (80ft 0in). This engine, without its two propellers, weighed about 3,674kg (8,100lb), and the shafts and gears appeared likely to cause trouble. Whichever engine was used, the propulsion system would be entirely new and posed high risk, whereas the airframe was based mainly on previous OKB practice.

The wing was designed with the same SR-5S profile as the Type 88, thickness/chord ratio being 12.5 per cent at the root and 10 per cent at the tip. The reduction of t/c ratio on the outer panels enabled their sweepback to be reduced to 33 deg 30 min. Aspect ratio finally settled at the very challenging value for a swept wing of 8.39, the wing being made as a *centroplane* and inner and outer panels on each side. The wing was naturally set at the mid level, with three immensely strong bridging spars across the width of the fuselage. From each heavy root rib three straight spars extended to the tip, plus a fourth ending at the struc-

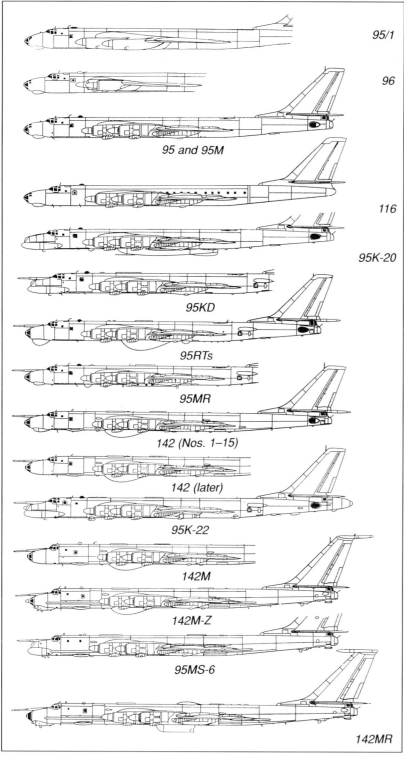

95/1

96

95 and 95M

116

95K-20

95KD

95RTs

95MR

142 (Nos. 1–15)

142 (later)

95K-22

142M

142M-Z

95MS-6

142MR

tions, the inner section having a trim tab, and the main surfaces having hydraulic boost. A single hydraulically actuated spoiler was added ahead of each inner aileron as an all-speed air-brake and roll assister. After flight testing, three 200mm fences were added above the wing from just behind the leading edge to the trailing edge, one between the propeller wakes, one in the flap/aileron gap, and the third at the joint between the mid and outer ailerons. Structurally, much attention had to be paid (led by Tupolev personally) to solving new problems in absorbing very powerful low-frequency engine/propeller vibrations into the airframe. Perhaps ahead of most Western designers, the OKB was concerned not only to improve the operating environment of the crew but also to minimise the rate of onset of fatigue damage.

The fuselage retained almost the same 2.9m-diameter circular section as on the Tu-85, but the greater length demanded thicker skins in the mid section to take the higher bending stresses. Close-pitched hemispherical-head rivets were used in the pressure cabins, with a thick (15mm) layer of ATIM thermal insulation. Planned defensive armament comprised two dorsal, two ventral and tail turrets, each with two NR-23 cannon. Only the tail turret was to be manned, the others being controlled as in the Tu-4 from optical-sight stations above the rear of the forward compartment and blisters on each side of the tail section. There was no requirement for a fixed gun.

The pressurised nose was arranged for six crew: two pilots, a navigator-operator (of the radar), second navigator, flight engineer and radio operator/gunner. They entered via a ladder in the nose-gear bay. The small unpressurised tail compartment, cut off from the front, housed a 'lower gunner' with lateral sights in blisters on each side and a stern gunner with bulletproof glazing controlling the tail barbette. Unlike the Type 88, no ejection seats were fitted. Emergency escape for the main crew involved extending the nose gear and triggering a hinged section of floor, pushed down by hydraulic rams. The tail gunners dropped through their floor hatches.

Conveniently, sweeping the wing put the transverse bridging spars ahead of the large and unobstructed

tural break at 90 deg to the rear spar outboard of the outer engines. The joint between the middle and outer panels, again at 90 deg to the rear spar and like the first capable of being disconnected, was inserted between the outer flap and the aileron. The slightly kinked leading edge had no movable surfaces. The trailing edge was made straight. Static dihedral was set at –1 deg (i.e., anhedral).

Large flaps of the tracked slotted type were provided inboard and outboard of the landing gear fairings. Their effectiveness was enhanced by the very powerful slipstream from the propellers. Each of the four sections was driven by duplicate ballscrew jacks to a maximum (landing) angle of 35 deg. To avoid jamming due to flexure of the wing the long all-metal ailerons were arranged in three sec-

This Tu-95 appears to have additional blister antennas under the fuselage. (G F Petrov/RART)

bomb bay evenly disposed ahead of and behind the CG (centre of gravity). The bay was sized to take the TN (thermonuclear) bomb, then in the design stage, or an FAB-9000, the largest conventional weapon weighing 19,840lb. The agreed maximum bomb load was 12,000kg (26,455lb). Immediately aft of the bay was a compartment for pyrotechnics, especially large target indicators. The centre fuselage also housed Nos. 2 and 5 groups of fuel cells and the two LAS-5M-2 dinghies.

The tail was scaled up from that of the 88, with control movements transmitted by rods and bellcranks. The tailplanes, with a span of 14.78m (48ft 6in), were joined on the centre-line and carried hydraulically boosted tabbed elevators. The fin, swept at 40 deg, carried an inset-hinge boosted tabbed rudder. The landing gears again followed 88 practice, but Hydromash upgraded the legs and bracing struts to match the greater weights. Each main unit was fitted with four tyres 1,500mm by 500mm (59in by 19.69in), each wheel having hydraulically-powered internal expanding brakes. The complete unit was arranged to retract to the rear, pushed by an electric screw-jack, the bogie somersaulting to lie inverted in the rear of the extended inboard nacelle (track 9.4m, 30ft 10in). The tall hydraulically steerable nose gear was inclined forwards, carrying twin tyres 1,100mm by 330mm (43.3in by 13.0in). From the front, the twin struts

formed a V-shape, that on the left having an arm projecting above the hinge pushed by a jack from the rear to fold the unit back into a bay with tandem left/right pairs of doors. The tailskid bay had twin doors, and further back was a compartment for twin braking parachutes.

As work progressed, the 2TV-2F appeared increasingly unattractive, even though it was hoped to be able to maintain flight with any four power sections shut down. The TV-12, though being developed to a much later schedule, appeared potentially superior, though no decision on the engines for the 95/II were yet taken. Construction of the 95/I began at GAZ No. 156 in September 1951. On 20 September 1952 it was conveyed in sections to LII Zhukovskii, where it made a successful 50-minute first flight on 12 November 1952, in the hands of A D Perelyot and test crew. This was also the first flight of the 2-TV-2F. Flying continued until 11 May 1953 when, on the 17th flight, the No. 3 engine intermediate gearbox failed, causing seizure of one power section which caught fire. All efforts to extinguish or isolate the fire failed, and, though the engine eventually fell off, the aircraft crashed with Perelyot and NISO engineer A Chyernov still trying to save it, six crew escaping. Coming at such an early stage in perhaps the most important aircraft programme in the Soviet Union, this was a major blow. The decision was

taken to abandon the eight-engine formula, even though this left the OKB with a hiatus of almost three years before test flying could be resumed, with what was hoped with increasing confidence would be a less-troublesome powerplant.

Span (project) 49.8m (163ft 4⅜in), (as built) 50.1m (164ft 4⅜in); length (fuselage) 44.35m (145ft 6in), (overall) 46.17m (151ft 5⅜in); wing area 284.9sq m (3,067sq ft).

Weight empty 75,950kg (167,438lb); fuel *c*70,000kg (154,323lb); maximum loaded weight 156,000kg (343,915lb).

Maximum speed 890km/h (553mph) at 9,500m; service ceiling 13,500m (44,290ft); range 14,200km (8,824 miles).

The second prototype, 95/II, was completed in July 1954, but painstaking development of the TV-12 engine, its tandem planetary reduction gear and the AV-60 propeller delayed the first flight to 16 February 1955. This time the test crew was headed by M Nyukhtikov and co-pilot I Sukhomlin. The engines were mounted on long steel trusses off the wing top skin and front spar to put the propellers in approximately the same position as on the 95/I, the inners being set at +3 deg incidence. Zhdanov succeeded in retaining the key feature of extremely coarse pitch, giving high thrust at 750rpm to Mach 0.835.

This time the engines were fed by circular annular inlets behind the large spinners. To start each engine, a 65hp

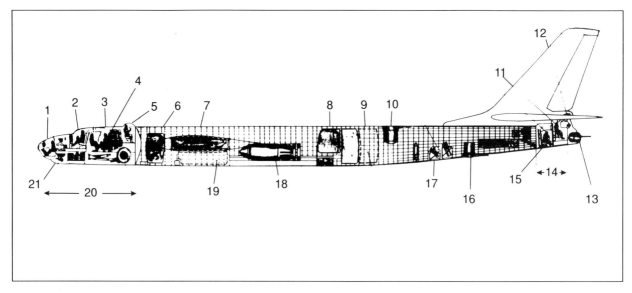

OKB drawing of Tu-95 crew stations: 1, navigator/bomb aimer; 2, two pilots; 3, engineer; 4, nav/operator; 5, radio/gunner; 6, forward tank; 7, wing tanks; 8, tank 4; 9, tank 5; 10, twin 23mm turret, 350 rounds per gun; 11, rear gunner; 12, Argon radar; 13, twin 23mm, 500 rounds per gun; 14, tail pressure cabin; 15, mid gunner; 16, twin 23mm, 400 rounds per gun; 17, day/night cameras; 18, nuclear weapon; 19, tank 2; 20, forward pressure cabin; 21, Rubidii MM-II radar.

TS-12 gas-turbine (started on battery power) was fitted above the compressor casing. The jetpipe was split to exhaust on both sides of the nacelle, the gas passing under the wing. Downstream each nacelle was skinned in nickel-alloy 'shingles'. Heat exchangers in the jetpipes could provide hot air for de-icing the wing. The tail was de-iced by a TA-12 APU in the dorsal fin.

Total fuel capacity of the 95/II was 72,980 litres (16,055gal) in sixty cells in the wing box and Nos. 2 and 5 groups in the mid-fuselage, with heated filters and flow proportioners. In cruising flight all engines were governed at a constant 8,300rpm, power being varied by adjusting the fuel flow, the maximum fuel giving a power of 12,000ps (11,833shp) at sea level. The engine had been developed to a tight schedule by German engineers at Kuibyshev. During initial flying of the 95/II and early production aircraft it was limited to a turbine gas temperature that gave a maximum power of only 9,750shp. This was inadequate to meet the VVS performance requirement, but Kuznetsov was confident that greater power would be forthcoming. Accordingly, from spring 1955 this very challenging programme never looked back.

The new long-range bomber was accepted for service, with a temporary certificate recognising the shortfall in power. The Tupolev *Izdelye* (product) number was adopted for Service use,

even though odd numbers were traditionally reserved for fighters. Production was organised at GAZ No. 18 at Kuibyshev, near the engine KB. The 95/II was included in the Aviation Day parade at Tushino in July 1955, causing intense interest to Western observers, who failed to understand how a swept-wing aircraft could have propellers. This big bomber could be regarded as the first aircraft to be powered by propfans.

Tu-95 The first two series Tu-95 bombers were flown in October 1955. NII State acceptance trials with the first two series aircraft plus 95/II were successfully completed on 20 December 1955, followed by the end of OKB testing in January 1956. A formation of five, including 95/II, participated in the 1956 Aviation Day parade at Tushino. By this time the original turbine temperature limitation had been removed, enabling full power to be produced for short periods. Production continued to mid-1958, but the Tu-95 did not enter DA service. Some were used for trials or rebuilt as Tu-95U trainers. Upon this aircraft's appearance at Tushino it was given the USAF designation Type 40. This was superseded by the ASCC reporting name 'Bear', which in fact became popularly used in the USSR.

For many years it was supposed in the West that the true designation was Tu-20.

Span 50m (164ft); length 47.3m (155ft 2in); height 12.5m (41ft); wing area 284sq m (3,057sq ft).

Empty weight 77,480kg (170,811lb); maximum take-off weight 172,000kg (379,189lb).

Maximum speed (95/2 with fully rated NK-12s) 882km/h (548mph) at 7,000m; service ceiling 11,300m (37,075ft); range with 5 tonnes bomb load 13,460km (8,364 miles).

Tu-95M In 1955 testing had begun of the TV-12M engine. Without increasing the frame size, so that it was installationally interchangeable with the TV-12, this had an improved compressor handling an increased airflow with higher pressure-ratio, giving more power with reduced specific fuel consumption. Its take-off power of 14,795shp enabled the take-off weight (and thus fuel capacity) of the Tu-95 to be increased, and it also raised flight performance until it fully met the official demands. Because of this, in 1957 the engine was redesignated NK-12M in honour of General Constructor Nikolai Kuznetsov.

Between August 1956 and February 1957 one of the series aircraft was largely rebuilt. It received the fully-rated NK-12M engines, driving AV-60 propellers. This enabled weight to increase greatly (*see* data), the fuel system holding from 88,500–100,000kg in seventy-four flexible cells. Nosewheel tyres were enlarged to 1,180mm by 350mm size. Each engine drove two GSP-18000M generators to charge the twelve SAM-55 batteries. Secondary supplies came from PO-

4500 and PT-1000 rotary converters. Each engine also drove a variable-frequency alternator type SGO-30U. All leading edges, and the propellers, were de-iced electrothermally, while the engine inlets used bleed air. One of the last major changes was to pivot the tailplane as a trimming surface, driven by an irreversible ball-screwjack over the small range +3 deg/−1 deg.

When fitted with the NK-12M engine, and improved AV-60N propellers with reverse-pitch braking capability, the Tu-95 was redesignated Tu-95M (Modernised). The braking parachutes could now be deleted. Adding the operative radar shifted the CG, making it possible to remove the forward dorsal and forward ventral turrets. This reduced the number of guns to six, but this was considered adequate, especially since they were henceforth of the newly introduced AM-23 type. The rear dorsal turret, with 350 rounds per gun, was retained and made retractable, normally controlled from an improved optical sight in the dome at the rear of the main cabin. The rear ventral turret had 400 rounds per gun, and the tail turret 500 rounds per gun. The aft PV-23 fire control was modified similar to that of the Tu-16, using a PRS-1 *Argon* radar at the base of the rudder. The weapon bay with two doors was equipped to carry the specified 12-tonne bomb

load including FAB-9000s; provision was made (but seldom used) for a further 6 tonnes to be carried externally. Tyre pressures were increased to match the higher weights, and the tail-skid was replaced by twin retractable tailwheels with 480mm by 200mm tyres.

Basic avionics comprised SPU-10 intercom, RSIU-3 uhf and RSIU-4 vhf short-range radio, R-803/807/808 communications radio, an RBP-4 *Rubidii*-MM-II main navigation and bombing radar, linked via the *Tsezii* connector to an optical bombsight of the OPB-5 (later OPB-11) family in the glazed nose, an integrated navigation system with an AP-15 autopilot linked to a DISS-1 doppler, ADF, VOR/DME (plus later RSBN Tacan), SP-50 ILS, high- and low-range radar altimeters (usdually RV-17 and RV-2, respectively), an SOD series ATC/SIF transponder and the standard SRO-2 IFF. Many aircraft received an RSDN *Chaika* Loran with a rail antenna. As well as the usual group of cameras – typically AFA-50M, 75M and 100M, AFA-34/OK-100 and AFA-42/50 – the unpressurised rear fuselage housed a dispenser of precut chaff, and various groups of ASO chaff/flare cartridge dispensers could be installed in the main-gear fairings. The normal crew remained 8/9 as before.

Loss of a Tu-95M in March 1957 led to the NK-12MV engine with rapid autofeathering capability backed up by positive manual feathering. This engine, with the AV-60N propeller, was used in almost all subsequent versions. NII testing of the Tu-95M was completed in October 1957. Manufacture of the 95 and 95M ceased at about the 55th aircraft in 1959. All were initially delivered as free-fall bombers to the VVS-DA. Survivors were recycled through repair centres over a period of nearly 40 years for rework and role conversion, some being transferred to the A-VMF. The ASCC name was 'Bear-A'.

Span 50.04m (164ft 2in); length 46.17m (151ft 5¼in); height 12.5m (41ft); wing area 283.7sq m (3,054sq ft).

Empty weight 79,600kg (175,485lb); maximum take-off weight 182,000kg (401,234lb).

Maximum speed 910km/h (565mph) at 7,000m; service ceiling 12,500m (41,000ft); range with 5 tonnes bomb load 14,960km (9,296 miles).

Tu-95K In late 1955 a development prototype was modified as the Tu-95SM-20 to carry a Mikoyan SM-20 (experimental MiG-19S) recessed into the belly, start its engines and launch it at altitude. This piloted aircraft tested the guidance for the Kh-20 cruise missile, and the carrier aircraft had instrumentation and communication systems to monitor its flight after release. The designation was changed to Tu-95K. This aircraft was flown from 1 January 1956, and led to the Tu-95K-20.

Tu-96 Discussion in 1952 regarding the tradeoff of reduced range in order to get still higher over-target height and speed resulted in the even more powerful NK-16 engine rated at 18,000hp. This was fitted to the single Tu-96 prototype, designed to carry a 5,000kg (11,020lb) bomb load 10,000km (6,210 miles) cruising at 850km/h (528mph) at heights up to 17,000m (55,775ft). Slightly larger than the Tu-95, the Tu-96 had a new centre wing, forward fuselage, engine installation and propellers. Factory flight testing began in 1956, but the project was soon abandoned because it was clear (six years before this was realised in Britain) that SAM defences made extra over-target height pointless.

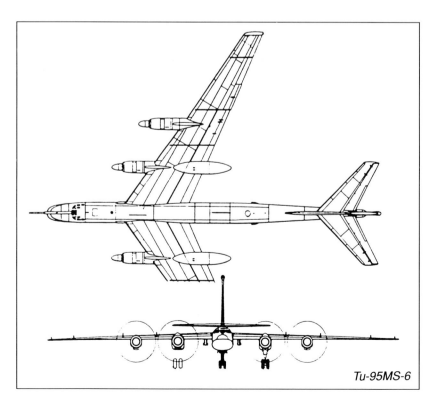

Tu-95MS-6

Tu-95RTs with SPS tailcone. (MoD)

Tu-114, Tu-116 Described separately.

Tu-95K-20 From 1956 increased altitude was replaced by supersonic cruise missiles as the best way to deliver nuclear weapons against heavily defended targets. This first missile-launching variant was based on the Tu-95K, with the K-20 electronic *Kompleks* to carry and launch a single Kh-20. This large (8,950kg, 19,730lb) turbojet-engined swept-wing cruise missile was designed by what was then the Mikoyan/Guryevich OKB to fly 350km (217 miles) at Mach 2. It was carried recessed into the bomb bay with its nose air inlet sealed against a quarter-spherical fairing which rotated up into the fuselage prior to starting the missile's engine at high altitude using the carrier aircraft's electrical power. The bomber's glazed nose was replaced by the required *Rubin-N* radar and *Kripton* guidance installation, with a very large 'duckbill' radar antenna scanning over a limited forward arc to illuminate the target. An Elint (electronic-intelligence) passive receiver antenna group, often accompanied by an AFA-33 camera, was installed in a shallow blister on the right side of the rear fuselage. Many innovations were required, and though the K-20 version flew in 1958 it was not certified operational until 1960. Few were built, production switching to the KD, and all were later modified for other roles. The ASCC name was 'Bear-B'.

Tu-95N A single Tu-95 was modified in 1959 as the carrier aircraft for the Tsybin NM-1/RSR supersonic-aircraft programme.

Tu-95KD The large protruding missile and unshapely guidance radar of the Tu-95K-20 significantly reduced the combat range. In May 1960 this led to air-refuelling tests, using the Yakovlev-developed probe/drogue system, resulting in a retrofit to aircraft in service. The resulting Tu-95KD featured a large and strong fixed probe at the top of the nose on the centreline, feeding via an external pipe on the right of the pressure cabin to No. 2 tank behind. An ARK-5 tanker homing receiver was also added. This became standard on most subsequent versions, together with a crew rest area for missions which could now last up to 36 hours. About fifty KDs were produced alongside other versions in 1961–65. The ASCC name remained 'Bear-B'.

Data, generally as for the Tu-95M except maximum take-off weight 184,000kg (405,644lb), with maximum fuel load of 89,000kg (192,211lb).

Tu-95M mod A single bomber was modified from 1959 to carry a giant hydrogen weapon weighing 24 tonnes (52,910lb). In 1961 the first bomb of this type was dropped over the Novaya Zemlya range. The actual aircraft designation remains classified.

Tu-95MR This strategic-reconnaissance version was the first supplied to the A-VMF, also now called the VVS-VMF. A small batch was quickly produced at Kuibyshev during the Cuban crisis in late 1962 by modifying Tu-95M bombers. Attack capability was replaced by sensors, including the main radar, tuned to US surface warships, especially carriers. The bomb bay was occupied by a large pallet loaded with up to seven optical cameras of three types (later infrared linescan was added), plus an air-refuelling probe, Elint/ECM blisters on both sides of the rear fuselage and, usually, a camera installation on both sides also. A long hf strake antenna was added above the fuselage, and other flush or blade antennas appeared over the years, to some degree depending on the Elint fit. The dorsal turret was usually removed because of the rear-fuselage electronics. Some MRs had different aft fire-control radar, and a few had the tail turret replaced by an extended tailcone housing further Elint and high-power defensive ECM jammers, as retrofitted to the KM and later versions. Usually the ventral turret and lateral sighting blisters were retained. Modifications after 1975 included addition of a chaff-dispensing pod under each outer wing. The ASCC name was 'Bear-C', later changed to 'Bear-E'.

Tu-95RTs This widely used VVS-VMF version was the second series type to have the bomb/missile capability deleted. The primary mission was reconnaissance and target indication for submarines and surface warships armed with P-6 and P-7 cruise missiles. Two large radars were fitted for this purpose, one under the nose (*c*14.5GHz, with four PRF modes, for missile guidance) and an even larger radar in the former weapon

A late-production Tu-142M-Z. (R van Woczik)

bay (*c*9GHz) for targeting ships on or over the horizon. Basic radio comprised R-837, 1RSB-70 and RSIU-5, with SPU-14 intercom. Radar altimeters were RV-UM and RV-25A, and primary navaids comprised the ADSNS-4 Loran and Tacan. Twin dorsal antennas were later added for radio links with Soviet communication and surveillance satellites. The upgraded Elint and reconnaissance fit included a receiver blister and camera on both sides of the rear fuselage and a 'farm' of up to seven blade antennas under the rear fuselage. A streamlined antenna pod for the *Lira* (Lyre), a passive sensor to give range and bearing of hostile emitters, was added on each tailplane tip, and the hf strake antenna was retained above the fuselage. The dorsal turret was deleted. Though the prototype was being built in 1959, it did not fly until 1962, and intensive testing took two further years before this version was cleared for oceanic service in 1964, about fifty being built. The Tu-95RTs operated from Cuba, Libya, Angola, Mozambique, Guinea, Ethiopia and Vietnam. Though this was the version most commonly encountered by Western fleets and aircraft, the thirty-seven on strength in the 1980s were naturally excluded from the SALT/START treaties. The RTs was the basis of the anti-submarine Tu-142. The ASCC name was 'Bear-D'.

Tu-126 Described separately.

Tu-95U From 1965 several Tu-95s and 95Ms were gutted and refurbished with the weapon bay sealed and the interior furnished for various forms of crew training (U = *Uchebno*, training). They were visibly distinguished by a red band around the rear fuselage.

Tu-95KM Modernised K-20/KD aircraft were reissued from 1962 with new multifunction main radar and greatly augmented defensive electronics. A visible modification was to fit an extended tailcone, replacing the tail turret. This SPS (*Stantsiya Pormekhovikh Signalov*, station for interference signals) was structurally an almost standard Tupolev design for

all subsequent Tu jet bombers, but the high-power ECM *Kompleks* differs for each aircraft. The KM installation has a group of ten aft-pointing antennas forming a drum. The KM also received upgraded radio and navigation systems for precision flying, particularly in Polar regions where magnetic compasses are less reliable. ASCC name 'Bear-C'.

Dimensions as 95M.
Weight empty 81,200kg (179,012lb); maximum loaded weight 182,000kg (401,235lb).
Maximum speed 870km/h (541mph); service ceiling 11,600m (38,060ft); range 12,500km (7,767 miles) cruising at 700km/h (435mph).

Tu-142 Under this designation the

Line of Tu-142s and 142Ms. (R van Woczik)

Tupolev OKB developed a considerably modified aircraft for the VVS-VMF. Since 1959 the Navy had searched for the best available long-range airborne ASW (anti-submarine warfare) platform, to meet the serious threat of US Navy Polaris/Poseidon submarines. Though the Tu-142 was funded as a new design, with a different *Izdelye* number, maximum use was made of the Tu-95RTs airframe and systems experience.

The biggest change was that the wing was redesigned with an improved (so-called supercritical) aerofoil profile, double-slotted flaps and long-life structure, increasing lift at all speeds, reducing cruise drag and with the structural box forming integral tanks, increasing fuel capacity. Flight controls were made fully powered, and rudder chord was increased progressively towards the top. The VVS-VMF wished to be able to operate from foreign bases with unpaved runways, so to reduce footprint pressure the main gears were redesigned with six twin-tyred wheels, requiring wider and longer fairings, while the nosewheel tyres were enlarged to 1,140mm by 350mm, requiring bulged doors. The tailwheels were omitted.

The fuselage was completely rearranged, with a glazed nose compartment, a J-band navigation radar with a slim radome under the nose and a large *Berkoot* search radar (*c*21GHz) ahead of the weapon bay, with several blisters and ram-air cooling inlets upstream near the leading edge of the wing. The weapon bay was modified to carry ASW torpedoes and nuclear

and conventional depth charges. The rear-fuselage Elint and camera installation of previous naval versions was retained, but it proved impossible to fit all the desired equipment and crew of ten into the aircraft. Accordingly, the forward pressure cabin was extended by 1.78m (70in), and both upper and lower turrets were deleted and replaced by a sonobuoy bay with twin doors, and new electronics with a row of ventral blisters. A ram-air inlet on the left of the rear fuselage served for electronics cooling. Twin dorsal hf strakes, satcom antennas and tailplane tip fairings (modified to house MAD [magnetic-anomaly detection] sensors) were similar to the RTs.

The prototype Tu-142 flew in July 1968, but the development task was so great that certification for service was not granted until 1972. Initial aircraft were built at Kuibyshev, but all subsequent Tu-142 versions were built at the Taganrog factory named for Dimitrov. The ASCC name was 'Bear-F'.

Span 50m (164ft 0½in); length 48.17m (158ft 0½in); height 12.59m (41ft 3½in); wing area 289.9sq m (3,121sq ft).
Empty weight 84,160kg (185,538lb); fuel 100,000kg (220,490lb); maximum take-off weight 182,000kg (401,235lb).
Maximum speed 850km/h (528mph); service ceiling 11,200m (36,745ft); range (3 per cent reserve) 12,000km (7,460 miles).

Tu-142 After producing about fifteen Tu-142s production was switched to a simpler standard version with no separate chin-mounted navigation radar, and a single cooling inlet on each side for the main radar. Another

change was that, reflecting upgrading of runways, the last Kuibyshev aircraft, and all those built at Taganrog, reverted to the four-wheel main landing gears fitting the smaller fairings. Tyres were reduced in size to 1,450mm by 450mm. The ASCC name was 'Bear-F Mod I'.

Tu-95M-5 In late 1970, to meet the threat of the USAF B-52 with cruise missiles, the decision was taken to integrate into the Tu-95 electronic *Kompleks* 95K-26, based on that already in production for the Tu-16K-26, including the wing pylons for two KSR-5 rockets. The OKB designation of Tu-95K-2 was changed to Tu-95M-5, but in 1971 it was decided not to produce this version in quantity, leaving the K-26 missile to the Tu-16 and Tu-22M.

Tu-95K-22 To replace the M-5, further major rebuilds of the Tu-95 KD were authorised to carry the K-22 weapon-system *Kompleks* already used by the Tu-16 and Tu-22/22M. The Tu-95KD nose was retained, but a completely different radar with improved performance and ECCM (electronic counter-countermeasures) qualities was installed in an even larger radome 3.5m (138in) wide, operating in various modes for navigation and, especially, for targeting the Kh-22 rocket-propelled missile. One missile, weighing 6,800kg (14,990lb), could be carried semi-internally, or one on a pylon under each wing root. An ECM thimble was added on the nose,

Tu-95MS-6 at Zhukovskii in 1993. (G F Petrov/RART)

left/right pairs of small pylon-mounted pod antennas under the forward and rear fuselage, and an Elint fairing and camera on each side of the rear fuselage. The ventral turret and sighting/observation blisters were retained, but the dorsal turret was removed, and the tail turret replaced by the SPS tailcone. An external fairing was added on the right side to carry cables and fuel pipes past the forward pressure cabin. The tailwheels were removed. The prototype Tu-95K-22 was flown in October 1975, and most K-20/KD/KM aircraft were rebuilt in the late 1970s to this standard. About forty-five remained operational with the Irkutsk Air Army until the early 1990s. The ASCC name was 'Bear-G'.

Data as Tu-95K except length overall 51.5m (168ft 11½in).

Tu-142M In this, the longest of all operational versions, the tailplane-tip MAD system was replaced by a single very sensitive *Ladoga* MAD sensor projecting aft from the tip of a slightly taller fin, which also carries a swept-back VHF antenna. The 142M entered production in 1975, with the rear-fuselage sighting blisters omitted. The aft (sonobuoy) bay is doubled in length, but its doors are slightly narrower, and the ram air inlet above this bay is removed. The ASCC name is 'Bear-F Mod III'.

Span 50.04m (164ft 2in); length (with probe) 53.07m (174ft 1⅜in); height 14.47m (47ft 5⅝in); wing area 289.9sq m (3,121sq ft).
Weight empty 83,050kg (183,091lb); fuel 77,000kg (169,756lb); maximum take-off weight 185,000kg (407,848lb).
Maximum speed 855km/h (531mph); maximum cruising speed 735km/h (457mph); range 11,900km (7,395 miles).

Tu-142M-Z In late 1974 the OKB, at this time led by Dr A A Tupolev, was called upon to meet the threat of new low-noise submarines. This required the otherwise excellent Tu-142 to be fitted with upgraded avionics and sensors (especially a new *Korshoon* sonar system), as well as precise triple inertial navigation and new part-automated secure communications systems. The forward fuselage was redesigned to improve crew efficiency still further. The nose back to Frame 3 was revised with the refuelling probe angled down 4 deg; between Frames 3 and 6 the cockpit

roof was raised 0.36m (14in), providing room for deeper windows and new cockpit displays; aft of Frame 6, the fuselage was structurally altered with a rearranged interior, the available length being increased by 0.23m (9in), and the previous satcom antennas were replaced by a steerable antenna inside a dome. Other features included the small main gears, and square access doors along the underside of the wing leading edge as far as the structural break beyond the outer engines. The Tu-142M-Z prototype was flown by I Vedernikov and test crew on 4 November 1975, but systems development was a giant task and this version was not declared operational until 1980. About thirty are airworthy. The ASCC name is 'Bear-F Mod II'.

Data as Tu-142 except maximum take-off weight 185,000kg (407,848lb) and service ceiling reduced to 10,700m (35,100ft).

Tu-142MK This is a variant of the Tu-142M. In 1983 eight of this type, with certain classified equipment items deleted, were exported with the designation Tu-142MK-E. They serve Indian Navy No. 312 Sqn, at Dabolim.

Tu-142M-Z In 1976 Bazyenkov died 'in harness' and programme leadership passed to N Kirsanov. At this time the chief demand was the VVS-VMF requirement for further upgrading of the ECM sensors and jammers, and without changing the designation the M-Z was re-equipped. The main updates involved adding an ECM

thimble radome on the nose, a large sensor group including a FLIR (forward-looking infra-red), communications antennas and an extra radar altimeter in a new chin compartment linked by an external cooling-air and multi-cable duct along the left side of the fuselage to a generating and air-conditioning group in the rear fuselage (with a cooling-air ram inlet on each side) and an ESM passive receiver installation under the tail with twin aft-facing planar-spiral antennas in small pylon-mounted pods under the tailcone. This was the final production anti-submarine version for the Navy. In 1994 about thirty-nine to this standard remained operational, the ASCC name being 'Bear-F Mod IV'.

Tu-142MP This prototype, incorporating several further equipment updates, was offered to the VVS-VMF in the early 1980s. The customer decided to stay with the M-Z.

Tu-142MR In 1982 Taganrog began production of this small batch to provide a secure communications link between the USSR national command centre and strategic naval units, especially missile submarines submerged on station around the world. The airframe is to Tu-142M-Z standard, with large fuselage areas occupied by a powerful VLF relay station transmitting via suspended wire antennas. The main wire, about 8,000m (26,250ft) long, is unreeled from an unstreamlined external box under the centre

Were he alive today, A N Tupolev would probably pick the Tu-95 as his greatest single design. This MS-6 was built 40 years after the 95/I. (Neville Becket)

fuselage. A weight on the end keeps it almost vertical while the aircraft flies small circular orbits. Dorsal antennas comprise the usual hf rails, a satcom dome and a large Glonass navsat (navigation-satellite) blister. The unglazed nose has a pimple radome, refuelling probe, chin sensors and cable duct. The rear fuselage and tail remain as in late Tu-142M versions, but the extended fin carries a forward-facing hf probe antenna, terminated at the rear by a circular spiral passive receiver antenna as on the Tu-95MS. About ten are divided between the Northern and Pacific Fleets. The ASCC name is 'Bear-J'.

Tu-95M-55 In 1976 the imminent threat of USAF BGM-109G land-based mobile-launch cruise missiles led to the urgent development of this new version with updated navaids, a group of extremely sensitive sensors and precision cruise missiles. The prototype, a rebuilt Tu-95M, completed factory testing in 1978, but in that year the Tu-95MS was adopted instead.

Tu-95MS The final new production version for the CIS Air Armies, this long-range missile carrier was first flown at Kuibyshev in September 1979. It is based on the airframe of the Navy Tu-142M-Z, but with the original shorter forward fuselage (though the new cockpit adds the 1.78m) with a diameter of exactly 2.9m (114in). Features include the refined long-life NK-12MA engine, still rated at 14,795hp, the taller fin, a new main radar in the chin position with a deeper antenna scanning 20 deg left/right, a reinforced horizontal refuelling probe, nose thimble, left-side duct fairing and some ECM/ESM installations as on the Tu-142M, two hf dorsal strakes, and a satcom dome on the rear fuselage. Meteor-NM comprehensive defensive avionics include high-power computer-controlled jammers and a new pattern of main-gear chaff/flare dispensers, various antennas under the nose and tail, SO-type hemispherical multi-facet IR (infra-red) warning receivers under the nose and above the rear fuselage, and an aft-facing spiral receiver antenna in the fin tip. The ventral turret was deleted, and from 1988, instead of fitting an SPS tailcone, the tail turret is retained in a new design fitted with two GSh-23 23mm twin-

barrel guns fed by enlarged magazines. The crew was reduced to seven. The initial series MS-16 had the weapon bay modified to contain a rotary launcher for six RKV-500A cruise missiles, and four pylons added under the wings for the same missile. Pylons under the wing root each carried two RKV-500As and one between each pair of engines each carried three, for a total of sixteen missiles. All have now been modified to MS-6 standard, to reduce the number of nuclear warheads in conformity with the SALT-2/START treaties. The ASCC name is 'Bear-H'.

Tu-95MS-6 The final missile-launch version is the MS with all wing pylons removed, leaving just the six-round launcher in the weapon bay. Production at Kuibyshev was completed in early 1992, after 38 years. About eighty MS versions are operational with ADD *Polks* (regiments) of the CIS Air Armies. Though a handful of Tu-160 aircraft are operational, the MS-6 remains the backbone of CIS Strategic Aviation. The ASCC name remains 'Bear-H'.

> Span 50.04m (164ft 2in); length 49.13m (161ft 2¼in); height 13.3m (43ft 7¼in); wing area 289.9sq m (3,121sq ft).
> Empty weight 87,700kg (193,342lb); fuel 77,000kg (161,756lb); maximum take-off weight 185,000kg (407,848lb); weight after air refuelling 190,000kg (418,871lb); maximum landing weight (common to all late series versions) 135,000kg (297,619lb).
> Maximum speed 845km/h (525mph); service ceiling 10,500m (34,450ft); range (16 missiles) 10,900km (6,775 miles); take-off speed (approximately common to all late series versions) 300km/h (186mph), take-off run 2,450m (8,040ft); landing speed (ditto) 270km/h (168mph); landing run 1,400m (4,600ft).

Tu-95LL, Tu-142MLL Apart from the Tu-95N, at least eleven aircraft were modified for special test purposes. A modified 142M air-tested pods containing the NK-25 and NK-32 engines for the Tu-22M3 and Tu-160. Many test projects never came to fruition.

Tu-119, 95LAL Described separately.

Tu-95I This high-performance version with NK-20 engines in the 20,000shp class was never built.

The 'Bear' began as a high-risk solution to a challenging problem, but

went on to become one of the A N Tupolev bureau's greatest and proudest successes. In 1956 the competition against the 3M strategic jet bomber designed by V M Myasishchev – once one of Tupolev's principal aides – was intense, and it was the ability of the Tu-95 to carry large missiles that opened up roles which its rival could not perform. This kept it in production 40 years after it first flew. The author's best estimte for total production is 505. In 1989 the Tu-95MS established about sixty class records, and in November 1990 the Tu-142M added ten more. It all rested on the unique propulsion system, whose distinctive low-frequency rumble can be heard from many miles away, and on the extraordinary longevity of the splendid airframe.

96, 97

96, ultra-high-flying Tu-95, described in that section; 97, no information.

98

In late 1952, shortly before Stalin's death, the Kremlin issued a VVS (air force) requirement for a bomber capable of supersonic speed at high altitude over the target area. Submissions were received quickly (having regard to the fact that such an aircraft was then a severe challenge) and two prototypes were ordered from the Ilyushin OKB and two from Tupolev. The rival Ilyushin Il-54 was rather smaller and lighter than the Tupolev 98.

Both aircraft were crucially dependent upon their engines. Though such engine KBs as Dobrynin, Kuznetsov and Mikulin were planning extremely powerful engines suitable for supersonic aircraft, the only engine actually available was the TR-7 by the small Leningrad-based KB of Arkhip M Lyul'ka. This recently-appointed General Constructor had worked on axial turbojets since 1936, but though several of his engines had powered prototypes, none had been chosen for production. He had already worked with Tupolev in the planning stage of

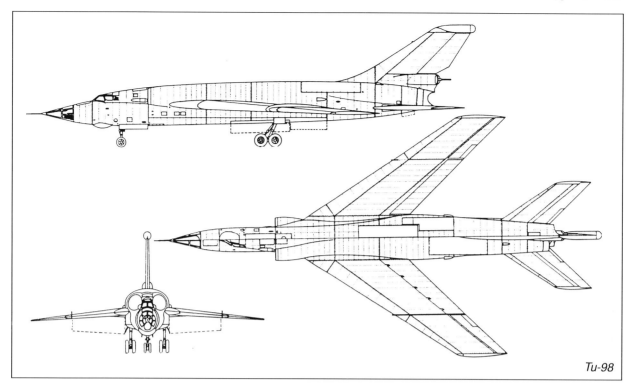

Tu-98

Aircraft N, the 90 version of which was to have had his TR-7 engines. The TR-7, soon redesignated AL-7 in his honour, was notable for having supersonic flow through the first two stages of the compressor. Two, fitted with afterburners, were chosen to power the Tu-98.

As the first in the OKB to need supersonic aerodynamics, throughout the preliminary design of the 98 the CAHI played a major role. At this time the OKB did not possess a supersonic tunnel, even for small models. Where possible the Advanced Projects staff adhered to features already established. The fuselage resembled a smaller, pointed-nose version of that of the 86/88, the swept wing passed through in the mid position, and the

tricycle landing gear had main units with four-wheel bogies. Almost everything else had to be different from prior Tu experience.

The wing had to be much thinner than that of any previous Tu project, the aerofoil profile being one of the SR-12s series but with a thickness/chord ratio of only 7 per cent at the side of the fuselage and 6 per cent at the tip. The sweep angle on the leading edge was no less than 58 deg. Partly because of the high design indicated airspeeds, and partly because of the concentration of mass in the fuselage, the stresses in the wing were much higher than in any previous Tupolev aircraft. Thus, the upper and lower skins of the main torsion box were machined from thick stretch-levelled

plate. A shallow full-chord fence was added across the wing between the fully powered tabbed aileron and the large one-piece track-mounted slotted flap. Static anhedral was –5 deg.

The fuselage was unusual in that, as in combat aircraft of 30 years later, the forward fuselage was relatively small in comparison with that downstream of the engine air inlets. The first section featured a small, pointed glazed nose for the navigator/bomb-aimer, forming a cone of only 15 deg semi-angle, with a narrow optically flat panel along the underside for the OPB-11 optical sight. This cone defined the compact dimensions of the pressurised crew compartment, which basically had a circular cross-section above which the

98 on roll-out. (RART)

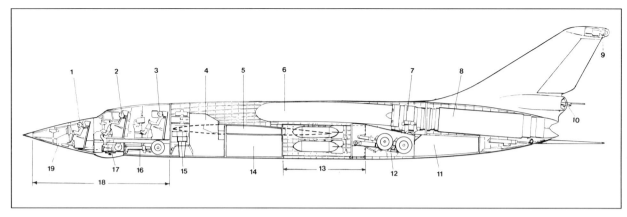

OKB drawing showing layout of 98. 1, navigator/bomb-aimer; 2, pilot; 3, radio operator/gunner; 4, No. 1 tank; 5, wing tanks; 6, air duct; 7, engine; 8, afterburner; 9, Argon radar; 10, twin 23mm guns; 11, No. 4 tank; 12, main landing gear; 13, bomb bay; 14, Nos. 2 & 3 tanks; 15, camera; 16, nose landing gear; 17, Initsiativa radar; 18, pressure cabin; 19, OPB-11 sight.

glazed canopy over the pilot projected, as in a fighter. The pilot was seated on the centreline behind a birdproof V-windscreen and hinged glazed roof. Immediately behind him was a mass of electronics racking managed by the radio-operator/gunner whose narrow canopy was all-metal except for a small window on each side. All three crew-members were seated on forward-facing KT-1 upward-ejection seats firing through the roof hatches which were used for boarding.

Under the floor was an unpressurised bay for the *Initsiativa* navigation/bombing radar (later used in the Yak-28I), a single fixed AM-23 gun and the nose landing gear. The rear pressure bulkhead was flat and vertical. Immediately behind it was a camera bay, with No. 1 protected fuel tank just behind, extending up between the engine ducts. Next came the front part of the centre fuselage, with the shallow wing torsion box passing through in the mid-position. The whole space under the wing box was occupied by two further protected tanks, Nos. 2 and 3.

In the next fuselage section the upper part was occupied by the engine air ducts, which swept inwards to lie close together on each side of the central keel member. The lower half of this section was the short (3.68m, 12ft 1in) bomb bay, which had a maximum capacity of one TN (thermonuclear) weapon or four FAB-1000 bombs each with a nominal mass of 2,205lb. The hydraulically powered doors folded up without projecting into the slipstream, and it was intended in production aircraft to expel the bombs by powered rams.

In the rear fuselage, built integrally with the tail, the upper part was occupied by the aft portion of the inlet ducts, curving down from their passage over the bomb bay, and the long afterburning engines, which were installed at a slight nose-up angle matching that of the ducts. At the extreme tail the afterburners and variable nozzles were close together, separated by a 'pen-nib' fairing. The nozzles were set at an angle to the afterburners to direct the propulsive jets horizontally. The lower half of the rear fuselage began with the bays for the retracted main landing gears. Next, under the engines, was the large No. 4 protected tank. Behind this, under the afterburners, was the horizontal tail. The whole fuselage conformed to the supersonic Area Rule, the inlets sweeping in over the wing and the rear fuselage being noticeably wider than the central portion.

While the Il-54 was initially flown with AL-7 turbojets, the heavier Tupolev bomber required the uprated AL-7F version, rated at 6,500kg (14,330lb) dry and 9,000kg (19,840lb) with maximum afterburner. Each engine was fed from a high oval inlet mounted at an oblique angle matching the angle of the adjacent upper part of the fuselage. The inlets had sharp lips but fixed geometry, and stood well away from the fuselage so that boundary-layer air could be diverted above and below. Immediately ahead of each inlet was a small part-cone blister to focus oblique shockwaves on the sharp lips and thus improve pressure-recovery during brief supersonic dashes near the target. Doors in the side of the fuselage offered access to the engine fuel system, starter and other accessories.

Total fuel capacity was about 23,500 litres (5,169gal), representing a mass of about 19,500kg (43,000lb). This was distributed between the four large fuselage tanks and two integral tanks filling the torsion box of the inboard wing. There was no provision for external fuel or for flight refuelling.

One of the major design problems was how to arrange the landing gear. The rival Il-54 had an extraordinary 'bicycle' gear which caused fuselage sagging problems. The final choice for the 98 was to use a conventional tricycle arrangement but with all three units retracting into the fuselage. The nose gear was fitted with a conventional twin-wheel truck with hydraulic steering, the tyre size being 660mm by 200mm. It retracted backwards hydraulically into a narrow unpressurised bay with twin doors. Originally it was hinged immediately behind the radar under the cockpit floor, but an extra bay was added to move the pressure cabin forward in relation to the leg. The main landing gears had vertical legs carrying four-wheel bogies. Each wheel had a hydraulic disc brake and a tyre 900mm by 275mm. Front and rear dampers and snubber struts controlled the truck beam attitude, the whole unit retracting to the rear hydraulically to lie inverted in the bay with side and bottom doors. The chief shortcoming of this otherwise neat arrangement was that, while the aircraft stood high off the ground, the track was only 2.5m (98in). In contrast, the wheelbase was 10.901m (35ft 9in); on take-off the maximum permissible ground angle was 11 deg 30 min.

The low-mounted tailplanes had a span of 7.7m (25ft 3in), with no dihedral and the same 58 deg leading-edge sweep as the wing. They were driven

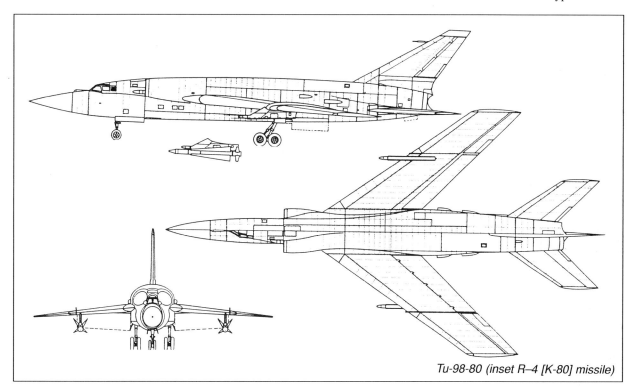

Tu–98-80 (inset R–4 [K-80] missile)

in unison by electro-hydraulic power units as primary control and trimming surfaces, carrying tabbed elevators. The large fin, again with a leading edge swept at 58 deg, carried a tabbed rudder and a tip fairing for an *Argon* radar. This aft-facing radar directed the fire of the twin remotely-sighted AM-23 guns in a power turret immediately below the rudder, with the magazines in the fairing further forward. Twin braking parachutes were normally stowed in a box under the rear fuselage. The aircraft was stressed for a.t.o. (assisted take-off) rockets.

The first of the two prototypes was first flown in summer 1956 (it is

believed, in June) by V Kovalyov, accompanied by navigator/engineer K I Malkhasyan. They completed OKB factory testing in the same year. The programme was one of several abruptly terminated by N S Khrushchyev before the start of NII testing. With other non-starters, it was shown to the US delegation at Kubinka in June 1956. It was promptly misidentified as the Il-140 or Yak-42. The ASCC reporting name was 'Backfin'.

Span 17.274m (56ft 8in); length 32.065m (105ft 2½in); height 8.063m (26ft 5½in); wing area 87.5sq m (942sq ft).
The only weight known with assurance is maximum take-off 49,500kg (109,127lb); empty weight was about 28,000kg (61,730lb).

Maximum speed 1,365km/h (848mph, Mach 1.29); service ceiling 12,750m (41,830ft); range with maximum bomb load 2,440km (1,516 miles).

Though some documents state that both Tu-98 prototypes were completed, V Rigmant insists that only one was flown. Accordingly, it was this aircraft which, in 1958, was largely rebuilt to serve as the testbed for *Kompleks 80*, for the Tu-128 interceptor. The nose was rebuilt to carry a radome and inert mass simulating the *Smerch* radar, with an optical/IR sight in front of the windscreen. The tandem cockpits were rebuilt to simulate the Tu-128 pilot and radar operator. The

The grossly-modified 98-80 after its landing accident at Zhukovskii. Note the P–18/PRW–13 surface/air missile radar in the background. (RART)

bomb bay was faired over, and the rear fuselage was shortened by one frame, moving the tail forwards and modifying the fairing between the nozzles. The inlets were modified and the engine bays given extra cooling inlets. The vertical tail was totally new, with no fillet, reduced chord and area and no gun turret or radar, but with the pitot head relocated near the top. The wing was given a new leading edge of reduced chord and with sweep reduced to the 56 deg 22 min 24 sec of the Tu-128. Inboard of the fences, pylons were added for dummy (later test) R-4 missiles. The rebuilt aircraft, probably designated Tu-98-80 was damaged in a landing accident. Data as before except length 31.85m (104ft 6 in).

99–103

Type 99 was a 1955 project for a high-altitude four-jet strategic bomber; 100, no information; 101, twin-turboprop assault transport; 102, pressurised passenger version of 101; 103, four-jet development of 88 (Tu-16) with transonic performance, the engines being superimposed at the wing roots.

Tu-104

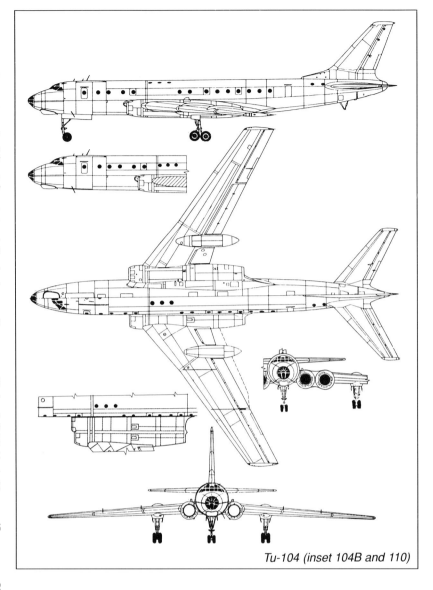

Tu-104 (inset 104B and 110)

No other large jet transport has been produced for such a modest cost in time and money as this derivative of the Type 88 (Tu-16). Tupolev showed the *Eskiznyi* (rough outline three-view) drawing to Stalin a few days before the latter died in February 1953, and gained his approval, though not long beforehand Stalin had recommended to Ilyushin that he should give up the idea of a transport able to carry perhaps 100 people. Tupolev assigned S M Yeger the task of turning the bomber into a passenger aircraft.

The main task was to design a new fuselage, with pressurisation and environmental control. This was facilitated by the prior work on the Tu-70, with the same 3.5m (11ft 6in) diameter and almost the same length, though the new jet, 104, demanded the higher pressure differential of 0.5kg/sq cm (7.1lb/sq in). This was

maintained from the nose to the convex bulkhead in front of the tail-plane, with an airtight bulkhead inserted at Frame 11 aft of the cockpit whose door was to be kept sealed in cruising flight. The main cabin was 16.11m (52ft 10¼in) long, 3.2m (126in) wide and 1.95m (76.8in) high, the volume being 142.3cu m.

The navigator station in the nose and the cockpit for the two pilots were based closely on the Type 70, though a new addition was the RBP-4 mapping radar with the rotating antenna under the floor. Behind the radar was the nose landing gear, similar to that of the 70 but angled slightly back instead of forwards. Behind the pilots were stations for the radio operator and engineer. Influenced by the Comet disasters, all windows in the passenger cabin were circular. The forward passenger cabin had four on each side

and could be equipped for sixteen passengers seated 2+2. Next came a step to a section where headroom was reduced by the presence of the wing centre section between the engines. This was used as the galley with three roof windows left of the centreline and two high on the right. Next came an eight-seat cabin with two windows each side, followed by the main cabin for twenty-eight in seven 2+2 rows each with a window. Next to the left-side entrance door were a wardrobe and baggage area, and further aft roof windows for two washrooms and two toilets. Three windows on the left and four on the right were mounted in pull-in emergency hatches, which were square with radiused corners. Except in the first prototype there were to be six small underfloor baggage/cargo holds with plug doors on the ventral centreline.

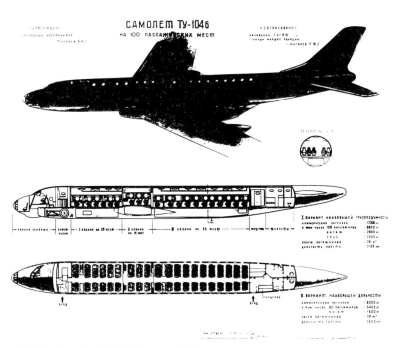

OKB drawings showing Tu-104B accommodation.

1955, and on 22 March 1956 L-5400 visited London Heathrow, only the second Soviet aircraft to visit Britain. Observers could not fail to notice this was the only jet airliner in service in the world, despite its furnishings of mahogany, brass and lace. Line service began to Irkutsk on 15 September 1956, followed by the first international route, to Prague, on 12 October. These aircraft had an impact on Aeroflot services, carrying three times (later six times) the typical Il-14 load and cutting scheduled time from Moscow-Irkutsk from 13hr 50min to 5hr 30min.

Span 34.54m (113ft 3¼in); length 38.85m (127ft 5½in); wing area 174.4 sq m (1,877sq ft).

Weight empty 39,500kg (87,081lb); fuel/oil 26,500kg + 470kg (59,524lb + 154lb); weight loaded 72,500kg (159,832lb).

Maximum speed 870km/h (541mph) at 10,000m; cruising speed 750km/h (466mph); range 2,800km (1,740 miles).

Tu-104A While engine TBO (time between overhauls) rose from 300hr to 2,000hr, the engine was fully rated as the RD-3M, at 8,700kg (19,180lb), which enabled take-off weight to grow considerably. The immediate result was that at the eleventh aircraft production switched to the Tu-104A, which though slightly reduced in empty weight was furnished, in a more modern style, for seventy passengers. The low-headroom centre fuselage remained a galley, but seats were provided for sixteen (2+2) in the front cabin and sixty-four (first row 2+2, others 3+2) in the rear. Class records set in September 1957 included: on the 6th, an altitude of 11,221m (36,814ft) with a payload of

Port and starboard wings were similar to the Tu-16 but joined to a wide centre section mounted low on the fuselage, with three heavy plate spars with KhGSA booms linking the banjos encircling the engines and their ducts. The engines were RD-3s rated at 6,750kg (14,881lb), with electric starting, and total fuel capacity in series aircraft was to be 33,500 litres (7,292gal). Unlike the bomber, each engine was fed by a straight duct from a de-iced circular inlet well outboard of the fuselage. The tailplane was lowered to the fuselage and mounted on fixed root sections, span being 13.4m (43ft 11½in). Area of the horizontal tail was 41.7sq m (449sq ft), and of the vertical tail 23.525sq m (253 sq ft). Main landing gears were unchanged, with 1,100 mm by 330 mm (KT-16/2U) tyres, but the nose gear was longer than in the bomber, with steerable 900mm by 275mm tyres, all units being retracted rearwards by the 17-MPa (2,466lb/sq in) hydraulic system. Track was 11.325m (37ft 1⅞in) and wheelbase 14.1m (46ft 3⅛in). On icy runways two 40sq m braking parachutes could be deployed from a compartment under the tail, behind the retractable sprung tailskid. As in the bomber, the flaps were driven electrically to 20 deg or 35 deg, electrical systems being the usual 28V dc and 115V ac. An AP-6E autopilot was

fitted, and among twenty-two electronic items were British VOR/ILS and SSR.

While production tooling was set in place at Kharkov, the prototype, L-5400, was built at Vnukovo and flown by Yu T Alasheyev on 17 June 1955. Few problems were encountered, though the wings were given shallow full-chord fences outboard of the landing gears and at the flap/aileron junction. An initial series of fifty-seat aircraft was put in hand, while Aeroflot crews were trained with the Tu-104G (demilitarised Tu-16). NII-GVF testing began in October

The Tu-104 prototype parked at London Heathrow on its first international flight. (Charles E Brown)

A Tu-104B landing on snow. (RART)

20 tonnes; on the 11th, 847.498 km/h (526.6mph) over a 2,000km circuit with 2,000kg load; on the 24th, 970.821km/h (603.3mph) over 1,000km with 10 tonnes. Despite its low initial cost, published as £425,000 in 1956 and at £715,000 for the stretched Tu-104B in 1960, no export sales were achieved except five (plus a later replacement) Tu-104As to CSA (Czech airlines). From 1960 a few 104As were converted to Tu-104D standard with a thirty-nine-seat rear cabin and luxurious forward cabins with side-facing divans.

Dimensions unchanged.
Weight empty 41,400kg (91,270lb); fuel unchanged; maximum loaded weight, initially 74,000kg (163,139lb), later 75,500kg (166, 446lb); maximum landing weight 53,850kg (118,717lb).
Maximum speed 1,000km/h (621mph) at 10,000m; cruising speed 775–850km/h (482–528mph); operational ceiling 12,000m (39,370ft); range 3,125km (1,942 miles); take-off run 1,950m (6,397ft); landing speed 260km/h (161.5mph), landing run 1,410–1,520m (4,625 – 4,986ft) (with parachute, 980–1,060m, 3,215–3,477ft).

Tu-104B In 1958 production was switched to the Tu-104B. Except for the first few, this was powered by the RD-3M-500, rated at 9,500kg (20,950lb). This made possible a further increase in weight, to make use of which Yeger's team stretched the fuselage ahead of the wing by 1.21m (3ft 11½in), at the same time increasing dP (cabin differential pressure) to 0.57kg/sq cm (8.11lb/sq in). By installing 3+2 seating over the whole length of the cabin, including the section over the wing, seating could be

provided for 100, which made a major difference to the previously uneconomic seat–km cost. The mid section had three windows on each side, higher than the rest, the roof windows being omitted. By slightly raising the level at which the main floor was installed the underfloor holds were enlarged, and they were now accessed by more useful doors on the right side. Among other modifications the most important was a considerable increase in chord of the inner wing and flaps, the outer flap tapering at its outer end to reach the original trailing edge at its junction with the aileron. The B version entered service on the Moscow-Leningrad route on 15 April 1959. Two class records set in August 1959 were: on the 1st, 1,015.86km/h (631.24mph) over a 2,000km circuit with 15 tonnes load; on the 4th, 12,799m (41,991ft) with a 25 tonnes load. A single aircraft designated Tu-104Ye (in Russian, 104E) was fitted with special RD-3 engines designated RD-16–15, using a B airframe. Production was completed at just over 200 of all versions in late 1960.

From 1962 many 104Bs were modified to Tu-104V (in Russian, 104B) standard with 3+3 seating for thirty-five in the front cabin, eighteen in the centre and sixty-four at the rear, a total of 117. Others were given VIP interiors as Tu-104D, repeating the short-body VIP designation. A few of several versions were delivered or transferred to the VVS and A-VMF for use as staff transports, and Tupolev built a military cargo version, *see* 107. A few of the air-force aircraft were

greatly modified as Tu-22 crew trainers, with designation Tu-104Sh. Others were used to give Cosmonauts experience in weightless (zero-g) flight. At least two were used by *Gidrometsovcentr* as meteorological research aircraft, with a pointed hail-resistant nose radome and underwing pylons for cloud-seeding rockets. The last was withdrawn from Aeroflot service in 1981, after carrying more than 90 million passengers. About six are preserved. The ASCC assigned the reporting name 'Camel'.

Span 34.54m (113ft 3¼in); length 40.05m (131ft 4¾in); wing area 183.5sq m (1,975sq ft).
Weight empty 44,020kg (97,046lb); weight loaded (B) 78,100kg (172,178lb), (V) 75,500kg (166,446lb).
Maximum speed 950km/h (590mph); economical cruising speed 770km/h (478mph, Mach 0.72); range with maximum payload 2,500km (1,553 miles); take-off run 2,500m (8,202ft); landing as 104A.

105, Tu-22

In the early 1950s the war in Korea and the intensification of the Cold War resulted in the Soviet Union exploring every possibility for increasing its striking power, and in this pre-ICBM era the only long-range answer was newer types of aircraft. In early 1954 *Izdelye* number 105 – in one record Tu-105 – was assigned to a supersonic bomber much larger and more powerful than Aeroplane 98. This was spurred by increasing OKB capability in supersonic design, by the knowledge that in the USA Convair were building the XB–58 and by the development by the Lyul'ka and Dobrynin KBs of much more powerful afterburning turbojet engines designed for supersonic propulsion.

In August 1954 Tupolev authorised an internal project study, called Yu, for a theatre-type bomber and missile carrier to fly Tu-16 missions but at supersonic speed over the target. It was assumed it would be required to operate from existing or planned VVS and A-VMF bases. It was accepted that, compared with the Tu-16, the combat radius would be shorter and the costs greater. Both aerodynamically and structurally the maximum use was made of solutions already found for the Aeroplane 98, and of course CAHI was involved from the outset.

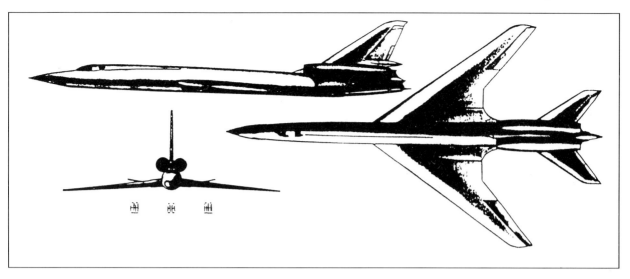

OKB drawing of Aeroplane 105.

Initial studies revolved around a fuselage resembling a more streamlined Tu-16 with a wing swept at 45 deg and powered by four Dobrynin VD-5 or VD-7 engines in vertical pairs at the wing roots. Continued work led to different answers, and by August 1955 Project Yu had assumed a radical form with either two or four engines mounted outside the fuselage beside the fin, and with wing sweep increased to 55 deg. The OKB's commitment and self-confidence played a part in securing a prototype contract in December 1955, the designation being Tu-22, which had previously been used for the Type 82. The following description refers to the prototype, the design of which was repeatedly modified even until the mock-up review in December 1957.

The wing was of modified SR-5s profile, with a thickness/chord ratio of 9 per cent at the root and 7 per cent at the tip, anhedral being –2.5 deg. Leading-edge sweep was 70 deg over a small fillet at the root, quickly changing to a constant 54 deg 30 min to the tip. Quarter-chord sweep was 52 deg 14 min 30 sec. The 105 was the OKB's first design with machined integrally stiffened wing skins, applied as usual in parallel rectangular strips. These enclosed integral tankage bounded by spars at 10 per cent and 60 per cent chord. Three intermediate spars extended to a structural joint which, following OKB policy, was inserted at about half the semi-span, with bolted double ribs arranged perpendicular to the rear spar. From the joint to the tip there were only two intermediate spars, the front one fading out into the front spar half-way to the tip.

Though various movable high-lift schemes were studied, the leading edge remained fixed. On the trailing edge the presence of a thickened section housing the main landing gear resulted in the decision to increase chord by reducing sweep inboard to zero and fit track-mounted slotted flaps of the largest chord possible, and driven by hydraulic ball-screw jacks to 50 deg for landing. A second flap section was added outboard, driven like the first but to only 30 deg. From the flap to the tip was a fully powered aileron made in inner and outer sections, the inner portion having a tab, moving as a single unit. There were no spoilers. The wingtips were extended forwards carrying anti-flutter masses and pitot heads.

Tupolev was proud of the extremely large circular-section fuselage of visibly area-ruled profile, with a ruling diameter of 2.5m (98.4in) from the cockpit to the tail. It was instrumental in achieving commendably low figures for overall supersonic drag. The first section comprised a pointed conical nose, of 15 deg angle, occupied by the radar, the entire lower half of the nose being the radome. Next came the pressurised crew capsule, made chiefly of thick magnesium alloy lagged with ATIM insulation. The navigator/bomb-aimer was seated behind the radar with an inclined bomb-aiming window and two superimposed windows on each side giving an undistorted view over the lower half of the forward hemisphere. At the upper level the pilot was seated on the centreline with the radio-operator/gunner behind, with side and roof windows. All three crew had K-22 seats, which could be lowered through the three ventral entrance hatches to winch the crew up to their cockpits. The three hatches were in a shallow gondola which faired the optical bombing window back into the fuselage.

Behind the rear pressure bulkhead was the longest section, measuring 8.52m (27ft 11½in), occupied by the electronics racking, nose landing gear, LAS-4M liferaft and first two protected fuselage tanks. At the point where the wing-root fairings terminated the inwards taper of the fuselage began. The next section, tapering throughout its length, was occupied entirely by fuel and by the wing carry-through bridge, of high-strength steel. The next section housed fuel above the centreline and the bomb bay below. Here the taper in width was sharply reversed, returning to the maximum fuselage width at the end frame a short distance behind the engine inlets. The long final section tapered to end at the width of the gun turret. This section housed fuel, including a trimming tank, the braking parachute, tailskid, tailplane/elevator mounts and actuators, and the gun magazines.

The tailplanes were pivoted for longitudinal trim and carried fully powered elevators working together for control in pitch only, the span being 10m (32ft 9¾in) and leading-edge sweep 59 deg. As in the 98 these surfaces were mounted as low as possible on the fuselage, the only apparent alternative being the top of the fin, with sharp dihedral. The vertical tail began with a dorsal fin swept at 80 deg and continued to the tip at 60 deg with a rounded top, carrying a powered tabbed one-piece rudder. Tail areas were 40sq m (430.5sq ft) horizontal and 28.93sq m (311.4sq ft) vertical.

Even in the Type 105 prototype the engines were of the VD-7M type, each rated at 12,800kg (28,220lb) dry and 16,000kg (35,273lb) with maximum afterburner. Their installation was perfect from the viewpoints of engine performance and also asymmetric (one engine inoperative) handling, but it necessitated high servicing trestles and special equipment for changing engines by withdrawing on rails to the rear. There were five distinct forms of cowling, but all had a circular-section inlet ring 0.8m (32in) long which, driven by four actuators, could be translated forwards to open up the whole duct immediately upstream of the engine whenever full power was needed at low airspeeds, particularly on take-off. Three large doors both above and below a central longeron gave excellent access to most parts of the engine and associated accessories. In all installations the variable multiflap nozzle projected behind the end of the cowling, the engine being removed on rails to the rear. Each engine was attached by three antivibration mounts picking up the strong fin-spar frames of the rear fuselage, which were machined forgings. In cross-section the top of each cowling curved in and down before curving sharply up at the junction with the fin (this shape was repeated in the 128 interceptor, where the engines were in the fuselage).

Each main landing gear superficially resembled a stronger version of that designed for Aeroplane 98 but retracting inwards. A single vertical main shock strut carried the rocking bogie beam inclined so that, restrained by a front snubber strut, the front wheels hit the runway first. This rotated the beam horizontal, the whole unit then pivoting slightly aft restrained by the forward drag strut. After take-off, a jack pulled the unit inwards, the bogie being housed in the thickened wing root. Tyre size was 1,100mm by 350mm (43.3in by 13.8in), and brakes

were of the hydraulic multi-disc type with anti-skid control.

The nose gear again resembled a strengthened version of that of Type 98. The extended leg was vertical, carrying a hydraulically steered and braked twin-wheel truck with tyres 1,000mm by 280mm. The whole unit retracted backwards hydraulically into a narrow box with one front and twin side doors. The rear fuselage was protected by a steel-shod skid, retracted by its shock strut. A large tube beneath the rudder housed twin braking parachutes.

The unobstructed square-section weapons bay behind the rear spar was configured for a normal load of 3,000kg (6,614lb) or a maximum of 9,000kg (19,840lb), including N (nuclear) or TN (thermonuclear) bombs. There was no provision for external loads. Despite the performance of this aircraft it was required to have a tail turret, with twin 23mm guns (originally planned to be NR-23), with 600 rounds total. Together with the integral wing tanks the total fuel capacity was 51,830 litres (11,401gal).

Aeroplane 105 was built at GAZ No. 156 and assembled at LII Zhukovskii from March 1958. It was first flown by Yu T Alasheyev on 21 June 1958, with no major problem. A static-test airframe was also built. Continued improvements led to Aeroplane 105A.

Span 23.745m (77ft 10⅞in); length 41.921m (137ft 6½in); height 10.995m (33ft 2⅜in); wing area 166.6sq m (1,793sq ft).
Empty weight 37,900kg (83,554lb); fuel 42,500kg (93,695lb); weight loaded (normal) 67,500kg (148,810lb), (maximum) 90,000kg (198,413lb).
Maximum speed (from 6,750–11,000m) 1,450km/h (901mph, Mach 1.36); service ceiling (dry power) 11,250m (36,910ft), (afterburner) 13,600m (44,620ft); range (3-tonne bomb load at 11,000m) 5,800km (3,605 miles); take-off run 2,250m (7,381ft); landing run 1,420m (4,658ft).

105A, Tu-22B

To save money this aircraft retained the same engines as the 105, and was held to similar weights. One of the principal changes was that, to improve the longitudinal distribution of cross-sectional areas according to transonic Area Rule, the main landing gears were arranged to retract backwards into faired containers projecting behind

the wings, as had previously been done with subsonic aircraft. This increased the track (measured between the bogie mid-points) from 7.98m to 9.12m (29ft 11in). CAHI tunnel testing showed the drag advantage at low supersonic Mach numbers, and the same arrangement was adopted on the long-range interceptor 128. Tyre sizes in production aircraft were (main) 1,160mm by 290mm and (nose) 1,000mm by 280mm.

As the wing root no longer needed to house the landing gear it was reduced in thickness, the trailing-edge fillet being eliminated and chord and area being reduced by sweeping the trailing edge inboard of the landing-gear fairings. Span was slightly reduced, the engine installations were modified, tailplane span increased to 10.09m (33ft 1in), the top of the fin made horizontal (reducing area to 22.01sq m), the fuel system improved (a drawing shows the seven tanks in the fuselage) and the bomb bay reconfigured for up to 12,000kg (26,455lb) of conventional bombs, as an alternative to eight types of N and TN. The remotely controlled tail barbette was changed to a lighter 261P pattern with two AM-23 guns each with 250 rounds, with a TSP-1 TV sight immediately above the engines linked to the PRS-3A *Argon* II radar immediately above the guns. The main radar was a *Rubin* I, optimised for anti-ship attack and linked to the new OPB-22 bombsight.

Other equipment included an RV-25 radar altimeter, DISS doppler, twin ARK-11 ADF, an NI-50BM navigation computer, a DAK-DB-5 astrocompass, dual hf/vhf radios, an SP-50M ILS, SRO-2 or 2M IFF, and newly developed warning receivers and Elint systems, including antennas in front of the tip pods (the pitot heads being mounted on the fuselage). Provision was made for overload take-offs with four SPRD solid a.t.o rockets.

From 1962 tests took place with a Yakovlev-type non-retractable in-flight-refuelling probe added above the nose, the external portion being detachable (later being fitted with a drogue deflector guard on the underside), all plumbing being internal. The tankers were the Tu-16N. The weapon bay was sealed by two double-hinged doors. Typical loads could include one FAB-9000, two nuclear bombs, three FAB-3000, eight FAB-1500s, twenty-

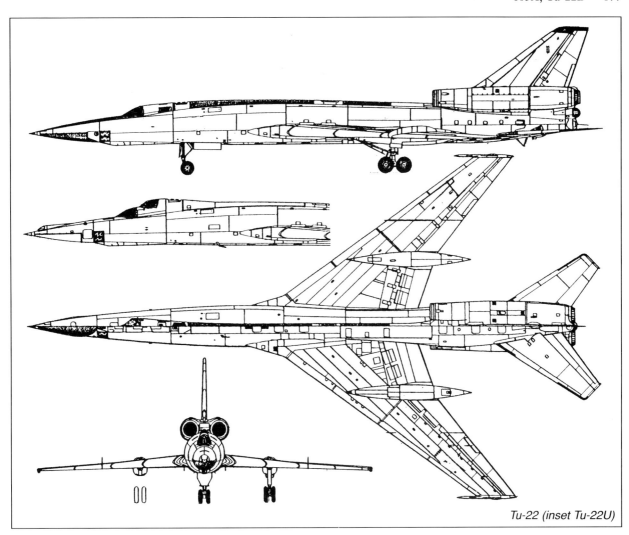

Tu-22 (inset Tu-22U)

four FAB-500, forty-two FAB-250s or -100s, or either eight or eighteen mines, depending on the pattern.

The first Aircraft 105A (the second Yu prototype) was first flown on 7 September 1959 by Alasheyev accompanied by navigator I Gavrilenko and operator K Shcherbakov. Testing generally went well until 21 December 1959 when, during high-Mach testing, elevator flutter resulted not only in loss of the aircraft but also of Alasheyev and Gavrilenko, only Shcherbakov managing to eject. The main result of this catastrophe was to redesign the horizontal tail entirely. The final result was to use one-piece 'slab' tailplanes, fully powered by irreversible electro-hydraulic actuators for both longitudinal control and trim. Each surface had main spars at 10 per cent and at 70 per cent chord, plus a steel spigot at 58 per cent about which it pivoted on an axis parallel with the rear spar. This demanded that the surface should abut against a kinked fixed portion, as shown in the

three-view drawing. To eliminate a recurrence of the flutter the tips were 'cut off' at an angle corresponding to Mach 1.5.

Just before this accident the 105A had been accepted for service as the Tu-22, a number used previously for an aircraft which never entered service (Type 82). Series production was quickly organized at Kazan. During July–August 1960 the first three series aircraft were intensively tested at LII Zhukovskii, and before the end of the year a complete production programme had been planned for five versions, the Tu-22B, Tu-22R, Tu-22K, Tu-22U and Tu-22P, in that order.

The VVS designation of initial series aircraft, the free-fall bomber, was Tu-22B. Only ten were built, VVS-DA acceptance being received in September 1962. Pilots initially approached this enormous supersonic aircraft in awe, but found it nice to fly. Its popularity was marred only by the excessive pilot workload and poor view through the sharp wedge windscreen.

The existence of this aircraft became known to the West at the 1961 Aviation Day, when all ten airworthy examples took part in the flypast. Western analysts agreed it was the work of V M Myasishchev, and the ASCC invented the reporting name 'Bullshot' (considered improper), changed to 'Beauty' (considered too favourable), and finally to 'Blinder' (which had originally been allotted to the Tu-28-80).

Span 23.646m (77ft 6⅛in); length 41.6m (136ft 5⅞in), (with probe) 42.75m (140ft 3in); height 10.15m (33ft 3½in); wing area 151.23sq m (1,628sq ft), (with full flaps) 162.25sq m (1,746.5sq ft).

Weight empty 40,940kg (90,168lb); fuel 42,900kg (94,578lb); loaded weight (normal) 85,000kg (187,390 lb); maximum landing weight 60,000kg (132,275lb).

Maximum speed (sea level) 890km/h (550mph), (12,000m, 39,370ft) 1,510km/h (938mph, Mach 1.42); service ceiling (dry power) 11,700m (38,390ft), (afterburner, weight 63,000kg) 14,700m (48,230ft); range with 3 tonnes bomb load at 950km/h 5,850km (3,635 miles); take-off run at 85,000kg 1,950m (6,397ft); landing run at 53,000kg 2,500m (8,202ft), or with parachutes 1,370m (4,494ft).

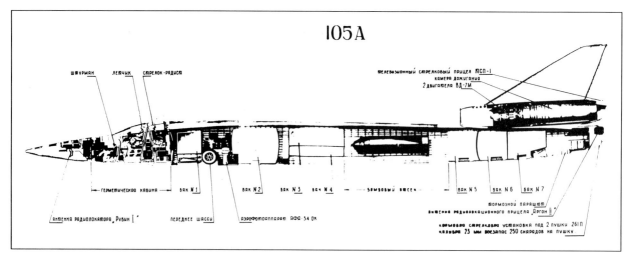

OKB drawing showing internal arrangement of 105A.

Tu-22R This dedicated reconnaissance version for the VVS and Aviatsiya-VMF was the first for a new mission to go into flight test. The weapon bay could be configured for bombs or for a choice of reconnaissance sensors mounted on a standard pallet, usually 5.55m (18ft 2in) long, with quickly plugged-in electrical and environmental-control connections. A typical pallet load would be seven cameras of types AFA-42/20, 42/100 and 41/20 and night NAFA-MK-75. Behind the pallet was a magazine ejecting photoflash cartridges. The main radar was the *Rubin* IA. Development flying occupied from 1960 until mid-1963. A total of about sixty aircraft of this type were built at Kazan. The ASCC name for all reconnaissance versions was 'Blinder-C'.

Tu-22RD From 1962 all reconnaissance versions were fitted with a flight-refuelling probe, the designation meaning *Dalnyi* (long-range). In 1965 the engine was changed to the RD-7M-2 (fitted from the outset to the second and subsequent RD aircraft) rated with maximum afterburner at 16,500kg (36,376lb), with a different convergent/divergent nozzle visible behind a further modified cowling. The greater power enabled the aircraft to be certificated at increased weights. As in other versions the tail gun installation was changed to a single AM-23, with a 600-round magazine.

Tu-22RDM This version was produced from 1975 by upgrading the

RD with 'BREO/REP' installations for improved cameras and IR (infrared) linescan, plus comprehensive Elint and Comint passive receivers and recorders and new defensive electronics. Like the RDK (next) this version usually has a 'canoe' container under the centreline.

Tu-22RDK This version was produced by installing the *Kub*-4 (cube) strategic SLAR (sideways-looking airborne radar), plus the same defensive electronic-warfare systems as the RDM. As in the RDM, equipment for visual reconnaissance was retained, including such cameras as the AFA-45 and AFA-54DS.

Dimensions unchanged.
Empty weight 40,900kg (90,168lb); maximum take-off weight 92,000kg (202,822lb); maximum landing weight 60,000kg (132,275lb).
Maximum speed (SL) 890km/h (550mph), (12,000m, 39,370ft) 1,610km/h (1,000mph, Mach 1.52); service ceiling 13,800m (45,275ft); combat radius (400km full-power dash, hi-lo-hi) 2,200km (1,367 miles); maximum range (normal fuel) 4,900km (3,050 miles); ferry range (bomb bay fuel) 5,650km (3,511 miles); take-off run 2,250m (7,380ft); landing speed/run 310km/h (193mph), 2,170m (7,120ft), with chute 1,650m (5,410ft).

Tu-22K The main production version, believed to number 150, was this missile carrier. The prototype was first flown in early 1961, and it was included in the ten displayed on the

A Tu-22KD, taking off, without missile. (G F Petrov/RART)

Tu-22R with bay doors open. (G F Petrov/RART)

1961 Aviation Day. The underside of the nose bulged with the radome of the *Rubin*-PN radar with a wider antenna for targeting and initial guidance of the Kh-22 cruise missile. This large rocket-propelled weapon, which had not then entered flight test, weighed 6,800kg (14,991lb), and had to be carried on the centreline recessed into the fuselage. Overnight the K version could be reconfigured to enable internal loads to be carried instead, the free-fall bombsight being the PSB-11. In theory three Kh-22s could be carried, with one under each wing, but this was not possible within weight limits. The aft bay of the main landing-gear pods housed chaff/flare cartridge dispensers and a strike camera. Clearing the complete electronics *Kompleks* for service proved to be a protracted task, not fully completed until 1967. Thus, though most Tu-2Ks had entered service by late 1964, they were not certified combat-ready for a further three years. The ASCC name was 'Blinder-B'.

Tu-22KD By the time the missile carrying version had been cleared for service most had been brought up to this (*Dalnyi*, long-range) standard by the addition of a flight-refuelling probe, together with RD-7M-2 engines, which were fitted to most from the outset. In the 1980s a total of twenty-five KD aircraft saw action in Afghanistan, seventeen had been painted in camouflage and exported to Libya (one being shot down by a SAM over Chad) and others were exported to Iraq (one shot down

by an Iranian F-4). About seventy, with upgraded electronics, served in the Smolyensk and Irkutsk air armies until 1992.

Tu-22KP Prototypes only were produced of this electronic-warfare version based on the airframe of the Tu-22KD. High-power jammers were installed in the weapon bay and elsewhere, with transmitter antennas mainly in the nose and wingtips.

Span 23.646m (77ft 6⅞in); length (excluding probe) 41.6m (136ft 5¼in); height 10.15m (33ft 3½in); wing area 162.25sq m (1,746sq ft).

Empty weight 38,300kg (84,436lb); maximum take-off weight 84,000kg (185,185lb) (with a.t.o.) 94,000kg (207,231lb); maximum landing weight 60,000kg (132,275lb).

Maximum speed at 12,000m 1,510km/h (938mph, Mach 1.42); service ceiling 14,700m (48,230ft); range 5,850km (3,635 miles); take-off run 2,200m (7,220ft); landing speed 310km/h (193mph), landing run 2,400m (7,874ft).

Tu-22U This pilot-conversion trainer seated a second (instructor) pilot cockpit above and behind the pupil in place of the usual rear cockpit. Though fuel capacity was hardly changed, the range and endurance

A Tu-22RD seen from a tanker. (G F Petrov/RART)

A standard Tu-22KP, with Tu-95s and Tu-160s in the background. (G F Petrov/RART)

developments are described under numbers 106, 125 and 145. Altogether the Tu-22, under chief designer Dmitri S Markov throughout its career, proved to be a most versatile and useful aircraft, with an operational life much longer than that of its only Western counterpart, the B-58 (which though four-engined was lighter and less-powerful). Like so many other Soviet aircraft it was consistently misidentified, misunderstood and underrated by Western analysts, who failed to identify most of the many variants yet were convinced that one version was an interceptor.

106

Since 1957 it had been the intention to replace Aeroplane Yu by a heavier type, carrying a cruise missile and powered by either two uprated VD-7M engines or two of Kuznetsov's larger NK-6s, in either case rated at 20,000kg (44,090lb). This was overtaken in May 1960 by an improved Aeroplane 106, with greater fuel capacity and wing area and powered by two NK-6F engines each with a take-off rating in afterburner of 23,000kg (50,705lb). It would have been a striking aircraft, with a high T-tail similar to that of the Tu-134 passenger transport. Span would have been 23.89m (78ft 4½in), length 40.2m

were reduced, as was maximum speed. The main navigation and bombing radar was replaced by a simpler mapping/weather radar, and the tail turret was replaced by a fairing to preserve CG position. Trainers fitted with a flight-refuelling probe were designated Tu-22UD. The ASCC name was 'Blinder-D'.

Tu-22P, PD This version was built new for Elint (A-VMF) and ECM jamming (CIS AF) duties. The former has the REB-K complex with a 'farm' of blade antennas covering fourteen selected wavelengths from metric to *c* 1cm, and twin 6.2m (20ft 4in) 'canoe' fairings with four (one at each

end) Chipthorn 'hockeystick' antennas. The Tu-22PD has a large Elint pod scabbed under the fuselage and receiver/jammer antennas on each side of the cockpit, the nosegear doors, the wing roots, the tip pods, the landing-gear pods, the engine nacelles and the rear fuselage, and a 2.4m (7ft 10½in) SPS receiver/jammer fairing replacing the tail turret. About forty of these electronic-warfare versions were still airworthy in 1994. The ASCC name is 'Blinder-E'.

At least nine Tu-22s were converted as testbeds for a range of items from supersonic propulsion and navigation systems to the main landing gear for the Tu-22M. Subsequent projected

A Tu-22U landing. (RART)

(131ft 10in), wing area 179.2sq m (1,929sq ft) and maximum weight 99,000kg (218,254lb). Maximum speed was estimated at over 2,000 km/h (1,242mph, Mach 1.88), service ceiling 18,000m and range with bomb load 6,750km (4,195 miles).

Much work was done to increase speed to as close to 3,000km/h (1,864 mph) as possible, whilst trying to achieve a lift/drag ratio of 4.88 (compared with 5.8 for the rival Type 125). Prolonged investigation was also done to reduce field length, at one time with a large group of RD-36-35 lift jets. Configuration changes included moving the engines to underwing pods and (with tail engines) fitting 'swing wings' able to pivot to 20 deg, 65 deg and 72 deg. Rival studies were the Tu-22/NK-144 and thin-wing Tu-22RTK. In the event, partly because of the enormous sums transferred from manned aircraft to the ICBM and space programmes, none of these aircraft, nor the next-generation 125 and 134, was built, work finally settling on the Tu-22-derived Tu-22M.

107

This number was allocated to the military assault version of the Tu-104. It was distinguished by a large door under the rear fuselage from which two short ramps could be hinged down for loading vehicles and other loads. This door was based on that of the 75, and so was the DK-7 tail turret fitted with two AM-23 cannon. The prototype was fitted with full military avionics and equipment. Data were similar to those of the Tu-104B.

108

Civil cargo version of Tu-104. Type 109 No information.

Tu-110

In June 1955 Tupolev assigned a brigade under Dmitri S Markov to

ASU-57 assault gun emerging from the 107 prototype. (RART)

produce a four-engined version of the Tu-104. Though the RD-3 was a mature engine, it was considered that a four-engined aircraft might have a better chance of selling in world markets, besides having increased all-round performance and improved engine-out qualities. The design task was considerable, the entire wing centre section being new. The four 6,500kg (14,330lb) Lyul'ka AL-7P engines were installed side-by-side behind the rear spar, though their inlets were staggered, that to each inner engine being further forward than the adjacent outer. The main wing panels, with unchanged anhedral, were thus attached further from the fuselage, increasing span (initially to 36.98m [121ft 4in], later to 37.5m [123ft]) and wing area, and increasing track from 11.325m to 13.730m and eventually to 15.746m (51ft 7in).

Two 110 aircraft were built. The first, No. 5600, was based on the Tu-104A. It made its first flight in the hands of D V Zyuzin on 11 March 1957. For some reason it bore a VVS-style red star on the tail, instead of the usual national flag. The second was based on the Tu-104B, with stretched forward fuselage and wide-chord flaps. The ASCC assigned the name 'Cooker', but no production orders were placed for what was in most respects a superior aircraft. A version was projected with Soloviev D-30 engines of similar power, as used in the twin-engined Tu-134. Data below are for the first aircraft.

Span 37.5m (123ft 0⅜in); length 38.3m (125ft 7⅞in) [No. 2, 40.06m (133ft 2in)]; wing area 182sq m (1,959sq ft) [No.2, 186sq m, 2,003sq ft].

Weight empty 44,250kg (97,553lb); weight loaded 75,000kg (165,344lb) normal, 79,300kg (174,824lb) maximum.

Maximum speed 1,000km/h (621mph) at 9,000m; economical cruising speed 890km/h (553mph); service ceiling 12,000m (39,370ft); range 3,450km (2,144 miles); take-off run 1,600m; landing speed 235km/h (146mph) landing run 1,430m (4,691ft) (1,000m with parachutes).

111

This 1955 project was for a twin turboprop twenty four-passenger civil or military transport.

112, 113

No information.

Tu-114

This famous transport derivative of the Tu-95 was for many years the largest, heaviest, most capacious and longest-ranged civil aircraft in the world. Moreover, despite having propellers, it also repeatedly demonstrated its ability to cruise at speeds

Tu-110 prototype, with civil number and red star. (G F Petrov/RART)

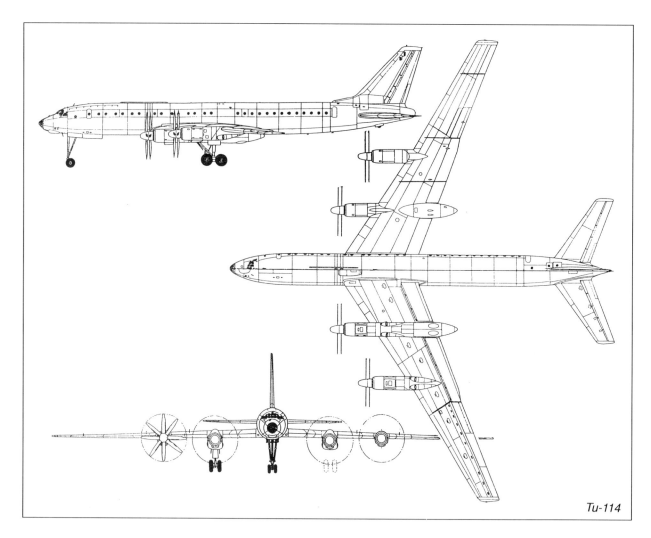

Tu-114

similar to those of swept-wing jets. Its use of propellers made it difficult for non-technical observers to realise that it was to a large degree a 'Jumbo Jet' more than a decade earlier than the 747, but 50 years after the Tu-114 first flew we may well see similar aircraft, powered by propfans.

Despite having a later number, the Tu-116 (described separately) flew first. This straightforward demilitarised bomber greatly assisted the much larger task of putting a completely new fuselage above a modified Tu-95 wing. At the time, May 1955, this was the largest pressurised fuse-

lage ever designed. The *Eskiznyi* project three-view, reproduced here, shows the original scheme for a true double-deck aircraft.

The ruling diameter of the fuselage was 4.2m (165.4in) and, in contrast to the 95 and 116, the entire volume back to the tailplane was heated and

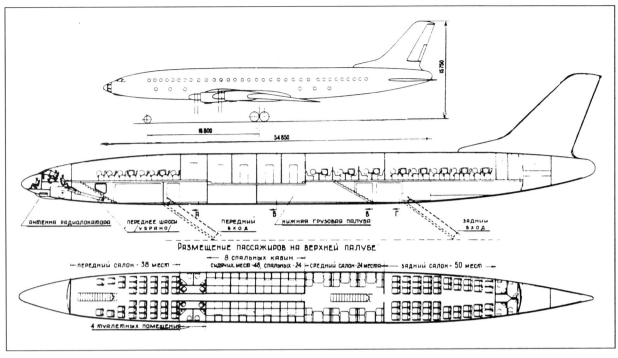

OKB drawings of TU-114.

pressurised to the high maximum dP of 0.59kg/sq cm (8.4lb/sq in). The environmental system included air-cycle machines under the wing centre section served by a ram inlet projecting under the fuselage, plus an auxiliary inlet further forward on the right. The cabin volume was 33.2cu m (11,732cu ft), plus large underfloor compartments. It was at first calculated that the latter could hold 90cu m of cargo and 35cu m of baggage, but when the 114 was actually designed it was found that large compartments would be taken up by the wing centre section, an under-floor kitchen and walk-in service bays, reducing the volume available for bag-gage/cargo holds to 24cu m (front) and 46cu m (rear).

Thanks to the large diameter of the fuselage, there was no difficulty in accommodating the wing centre section between the bottom of the fuselage and a flat level main floor. To fit the larger fuselage the width of the centre section had to be increased. The original three-view shows a span of 52.25m (171ft 5in), and a main-gear track increased from the Tu-95's 12.55m to 14.75m (48ft 5in), but in the event the span was extended by only 1.1m (43.3in), giving a track of 13.7m (44ft 11⅜in). The greater span increased the wing area, and, in order to keep down the take-off field length despite a modest increase in weight, the decision was taken to extend

the chord of the flaps. In turn, this involved extending the wing chord at the trailing edge from the root to the flap/aileron junction (not shown in the 1955 drawing). At the same time, a similar modification was made on the Tu-104. The variable-incidence tailplane was increased in span from 14.78m to 16.7m (54ft 9½in), its area being increased to 64.9sq m (699sq ft), and it was brought lower to near the mid-level on the fuselage. Area of the vertical tail was slightly increased to 41.4sq m (445.6sq ft). Otherwise the flight controls, like the engines, pro-pellers, main landing gear and twin

An unusual view of the Tu-114 prototype. (RART)

A Tu-114 series aircraft. (RART)

retractable tailwheels, were left almost unchanged.

In the prototype the engines were initially the 11,995hp NK-12, but the 14,795hp NK-12MV was retrofitted, and used in the series aircraft. The nose landing gear was entirely new, taller than any designed previously, with the hydraulically steerable twin-wheel truck carried by a narrow V of extremely long KhGSA tubes. The unit was pulled back by a single rear bracing strut into a long bay with four doors. De-icing was by jetpipe heat-exchangers along the wing, by raw AC current on the tailplane, windscreens and propellers, and by a rubber boot on the fin.

The prototype, L-5611 *Rossiya* (Russia), made its first flight in the hands of A P Yakimov on 15 November 1957, over six months after the first 116. It won the Grand Prix at the Brussels World Fair in 1958, was at the 1959 Paris airshow and completed NII-GVF testing in July 1960. The prototype was followed by thirty-one series Tu-114s built at GAZ No. 31 at Kuibyshev/Samara. All had identical fuselages, but there were various interior arrangements including night configurations with bunks. Day seating could be provided for up to 224 passengers in unique 4+4 rows. A more common arrangement seated 170, in three cabins for forty-two, forty-eight, and fifty-four, plus smaller cabins and a midships dining area linked to the underfloor kitchen by an electric lift. Flight crew usually comprised a navigator in the glazed

nose, two pilots, an engineer and the operator for the radio and *Rubidii* radar.

Apart from the sheer size and weight of the 114, and its rather long take-off run (compared with other 1950s aircraft) a major problem was that its door sills were 5m (16ft 5in) from the ground. Drawings show the planned arrangement of self-contained centreline staircases leading up to the passenger deck, but this attractive idea was eventually abandoned (though it would have been used on the military 115, and much later was adopted for the Ilyushin Il-86).

On 15 September 1959 the Tu-114 prototype flew General Secretary N Khrushchyev nonstop to New York. The GVF test crew was led by Ivan M Sukhomlin, and they established many of the world records set by these aircraft, including: 24 March 1960, 871.38km/h (541mph) over a 1,000km circuit with 25 tonnes payload; 1 April, 857.277km/h (533mph) round a 2,000km circuit with the same load; 9 April, 877.212km/h (545mph) round a 5,000km circuit with the same load; 21 April, a 10,000km circuit (Moscow – Sverdlovsk – Sevastopol – Moscow) with 10 tonnes payload; and 12 July 1961, a height of 12,073m (39,610ft) with 30 tonnes payload. Line service began on 24 April 1961 to Khabarovsk, 6,800km in a scheduled 8hr 15min. Aeroflot Tu-114 destinations included Delhi, Montreal, Conakry/Accra and (with Japan Air Lines titles) Tokyo.

Tu-114D From January 1963 luxurious 120-seaters with two crews and auxiliary tanks, giving a total fuel capacity of 83,000 litres (18,258gal), flew the world's longest route, from Moscow to Havana. The outward flight refuelled at Murmansk, but the 10,900km (6,773mile) return was flown nonstop.

The Tu-114 carried over six million passengers in 19 years of Aeroflot service. Two are preserved. The ASCC reporting name was 'Cleat.'

Span 51.1m (167ft 8in); length 54.1m (177ft 6in); wing area 311.1sq m (3,348sq ft).

Weight empty 85,800–88,200kg (189,153–194,444lb); fuel/oil normal 57,700kg (127,207 lb), maximum 60,800kg (134,041lb) weight loaded 179,000kg (394,621lb); maximum landing weight 128,000kg (282,187lb).

Maximum speed 880km/h (547mph) at 7,100m; economical cruising speed 750–770km/h (466–478mph); service ceiling 12,000m (39,370ft); range (15 tonnes load, 1hr reserve) 9,720km (6,040 miles); take-off run 1,600m (5,249ft); landing speed 205km/h (127mph), landing run 1,550m (5,085ft).

115

This 1955 project was for the military assault-transport version of the Tu-114. It would have had inbuilt staircases, a large rear ramp door for cargo and a DK-7 tail gun turret.

Tu-116

In 1955 the Kremlin asked Tupolev to build a long-range transport based on the Tu-95. The much larger fuselage and consequently reduced fuel of the Tu-114 made its range inadequate for the nonstop global missions the Soviet leaders wished to be able to make, and they issued a requirement for two or three special-purpose VIP passenger aircraft with the longest range possible.

Tupolev assigned the task to Arkhangel'skii. As far as possible the basic aircraft was unaltered, so that bombers on the production line could be modified during manufacture. A pressure cabin was fitted behind the wing, joined by a communication tunnel to the forward fuselage. Together the two cabins provided 70.5cu m (2,490cu ft) of pressurised and heated space for the flight crew of seven/eight and a normal complement of eighteen passengers, with a VIP cabin for the head of delegation. In

other configurations twenty or twenty-four passengers could be accommodated. A galley and two toilets were provided, access from the ground being via a new ventral airstairs door under the aft cabin.

Avionics were based on those of the bomber, but in some cases changed for civil counterparts. The mapping radar under the nose was related to the *Rubidii* MM-II of the Tu-95, the main communications complex was the *Gelii* (Helium) comprising RPS, 1-RSB-70M, US-8 and triple RSIU-4P. The usual ARK-5 radio compass and RV-2 (low) and RV-17 (high) radio altimeters were fitted, together with SP-50 ILS. Apart from intercom one of the traditional pneumatic-tube message systems linked the cockpit and cabin, and the passengers had a *Mir* (world) radiogram for relaxation. An LAS-5 emergency boat was carried, plus two SP-12 liferafts. A total of 77,800 litres (17,114gal) of fuel was accommodated in sixty-six flexible tanks.

GAZ No. 18 at Kuibyshev/Samara built two of these special passenger aircraft, bearing VVS registrations 7801

and 7802. The former carried out factory testing from 23 April to 4 October 1957. The second flew on 3 June 1957, and completed GOSNII State testing in March 1958. This test programme was in the hands of engineer N G Zhukovskii and test pilot V K Bobrikov. Appearance of the bigger Tu-114 in November 1957 took the limelight away from the 116. The second aircraft was then given civil registration 76463, and was operated subsequently by Aeroflot. After a nonstop flight each way to Havana, Bobrikov made headlines with an 8,600km flight Chkalovskaya – Irkutsk – Chkalovskaya. Later in 1958 an Aeroflot crew took twenty-four officials and journalists on three zigzag tours of the Soviet Union, in the course of which nonstop flights were made from Chkalovskaya to Vladivostok. The belief that these aircraft were designated Tu-114D was, for once, a mistake made not by Western but by Russian journalists!

Commanded by N N Kharitonov, No. 7801 also made many distinguished flights, besides making tests on new equipment for the Tu-95,

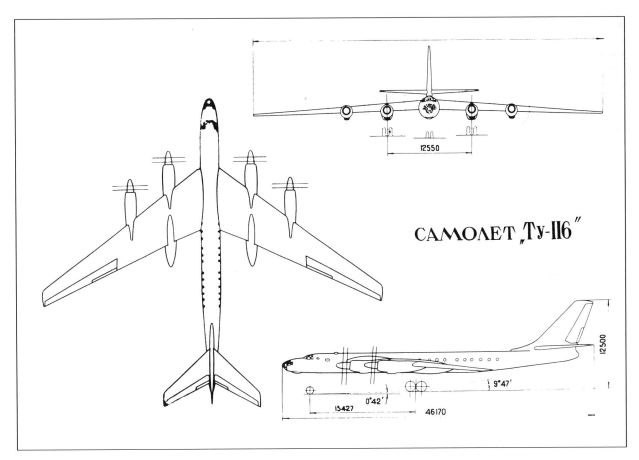

САМОЛЕТ „Ту-116"

OKB drawing of Tu-116. For aircraft as built see drawing of Tu-95 variants.

Two views of Tu-116 No. 76462. (RART)

starting with the NK-12MV/AV-60N powerplant with autofeathering and braking capability. No. 7802 was retired in April 1991, and both are preserved.

Span 50.1m (164ft 4⅛in); length 46.17m (151ft 5¾in); wing area 283.7sq m (3,054sq ft).

Weight empty 79,000kg (174,162lb); fuel/oil (maximum) 66,700kg (147,048lb); weight loaded (normal) 124,100kg (273,589lb), (maximum) 143,600kg (316,578lb).

Maximum speed 630km/h (391.5mph) at sea level, 870km/h (541mph) at 6,300m; climb to 10,000m in 19min; range (normal fuel) 11,190km (6,953 miles); take-off run 2,300m (7,545ft); landing speed 252km/h (157mph), landing run 1,450m (4,757ft).

117–118

Type 117 was a 1956 project for the military version of the 110, with a full-section rear ramp door for bulky cargo and a DK-7 tail gun turret. 118 no information.

119–120

In 1956 the Kremlin decided to fund a nuclear-powered strategic bomber, and in 1957 work began on the task of developing this aircraft. One of the many associated research programmes was to test-fly the proposed molten-salt reactor, and this contract was assigned to Tupolev, using an aircraft based on the Tu-95M. To this day it remains classified, though the accompanying photograph shows the associated underwing coolant tanks, the dorsal fairing over the core of the reactor and the kinked bottom line to the rear fuselage. Heat from the operating reactor was dissipated through large radiators. Aeroplane 119 was flown 'in the early 1960s', by which time the project had been abandoned. Type 120 no information.

121–122

The prototype of an unmanned strategic bomber, reconnaissance vehicle or cruise missile. It would have had intercontinental range, cruising at over Mach 2. Type 122, no information.

123

Extensive flight-testing from 1961 led to operational service with this unmanned supersonic reconnaissance vehicle, called a DBR (long-range without-pilot reconnaissance). Named *Jastreb* (Hawk), it was carried on a MAZ-537 cross-country vehicle, from which it was launched by two large solid motors. The delta wing, of 3.1 per cent thickness, had a leading-edge sweep of 57 deg and in some versions –4 deg anhedral. The tips were cut off

The Tu-119 parked at Zhukovskii. (G F Petrov/RART)

at the Mach angle, and under the trailing edge at mid-span were two small but very sharply swept fins. Around the tail end were three delta control surfaces spaced at 120 deg. A ventral semicircular multi-shock inlet served the KR-15 turbojet, derived from the Tumanskii R-15 designed for the MiG-25. The KR-15 was designed to cruise in afterburner, with a sea-level rating of 14,000kg (30,850lb). The autopilot was guided by an inertial platform updated by doppler radar.

At the conclusion of the mission the vehicle was brought over a recovery area where the nose, housing one oblique and three vertical/panoramic AFA-54 cameras, was separated and retrieved by parachute. The rest of the vehicle was expendable. Production Jastrebs were delivered from GAZ No. 18 at Voronezh, and served for 15 years until 1980. The Tu-123 was the basis of other surveillance vehicles with different sensors.

Span 7.94m (26ft 0⅝in); length 26.95m (88ft 5in); wing area 24.1sq m (259sq ft).

A DBR Type 123 cruise missile being prepared for launch. (Alfred Matusevich archive)

Weight empty (typical) 14,000kg (30,865lb); fuel 16,600kg (39,597lb); launch weight 38,500kg (84,877lb).
Cruising speed 2,715km/h (1,687mph, Mach 2.55); cruising altitude 22,000m (72,180ft); endurance 90min; range 3,000km (1,864 miles).

Tu-124

In 1957 this passenger transport, looking like a smaller version of the Tu-104, was launched to replace the Il-14 and Li-2 and bring jet speed to short-haul routes flown from secondary airfields. Despite its obvious similarity to the 104, hardly a single structural part was common to the two aircraft. The engine KB of P A Soloviev produced the 5,400kg (11,905lb) D-20P bypass engine (low-ratio turbofan) specifically for this aircraft. The Tu-124 thus became the first aircraft in the world specifically designed for short-haul services with turbofan engines to enter service.

The wing was almost a scale of that of the 104, with leading-edge sweep 40 deg to the kink at the end of the centre section and 36 deg from there to the tip. Major differences included a trailing edge inboard of the landing gears at 90 deg to the fuselage, double-slotted flaps with small hinged panels (called deflectors) immediately upstream on the underside of the wing to seal the slot gap with the flaps housed, large spoilers (called interceptors) immediately upstream on the upper surface, driven in unison as lift-dumpers after landing, and a large air-brake under the fuselage driven down 40 deg on the approach and for rapid letdown. Above the wing were two fences on each side, the outer fences being at the junction between the flaps and the manual spring-tabbed ailerons.

The fuselage retained the same 2.9m (114in) diameter as the B-29. Back to the convex bulkhead at the leading edge of the tailplane it was pressurised to 0.57kg/sq cm (8.1lb/sq in) by bleed-air cycle machinery drawing fresh air from a ram inlet under the belly. Tail control surfaces were again manual with spring tabs, the horizontal and vertical tail areas being respectively 26.55sq m and 10.90sq m (286sq ft and 117sq ft). The main landing gears were shorter and lighter than those of the 104, pushed by the front drag strut to lie inverted in twin-door pods. The wheels had anti-skid brakes, initially drum type but in the second production series changing to multi-disc type. Tyre size was 865mm by 280mm, inflated to 6.5kg/sq cm (92.5lb/sq in), suitable

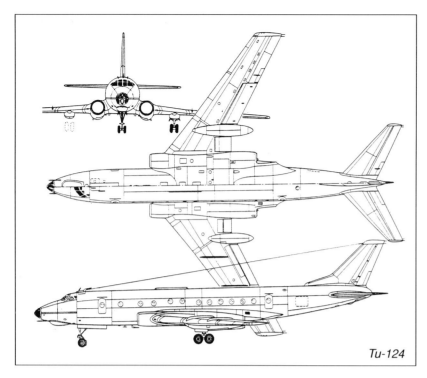

Tu-124

nose were two retractable landing/taxi lights, with FRS-200 lights added in the sides of the landing-gear pods from aircraft No. 45035. Type SPM-1 beacons were above and below the forward fuselage.

The prototype, L-45000, was first flown by A D Kalina on 24 March 1960. Development was exceptionally trouble-free, and Aeroflot line service began to Tallinn on 2 October 1962. The fuselage had two plug doors at front and rear, and 2+2 seating for forty-four passengers. Inevitably, the wing caused two annoying steps in the floor, though headroom in the small centre cabin was not seriously affected. Flight crew comprised a navigator in the glazed nose and two pilots, with a jump seat for a radio operator, but two-crew operation was shown to be practical using the navigator and one pilot. From the outset, furnishing decor was modern, with no mahogany or brass. Small changes were seen in early aircraft, and 45005 had a short nose and slightly taller fin.

Tu-124V This was the standard series aircraft, usually with cabins seating twelve/twelve/thirty-two for a total of fifty-six. Maximum payload was 6,000kg (13,228lb).

Tu-124K A few were built as *Konvertiruyemyi* passenger/cargo versions, some having a VIP interior. The K2 export version was sold to Interflug, the LSK (E German AF) and Indian and Iraqi Air Forces, but the only customers for the 124V were CSA (two) and Interflug (two). This was despite the type's many attractive features and low price ($1.45m in 1965). Total production was 154. The ASCC name was 'Cookpot'.

for unpaved surfaces. The steerable nose wheels had 660mm by 200mm tyres, inflated to the same pressure, with the option of deflectors to protect the engines from FOD (foreign-object damage). Track was 9.05m (29ft 8¼in) and wheelbase 10.036m (32ft 11⅛in).

Landing-gear retraction, and operation of flaps and spoilers, was hydraulic, at 210kg/sq cm (2,987lb/sq in). Almost all servicing could be done while standing on the ground. Two pressure fuelling sockets could fill the sixteen tanks in eight minutes, those in the outer wing being integral, housing 13,120 litres (2,886gal). Uniquely, following electrical failure a ram-air turbine could be extended under each wing to drive the tank booster pumps.

De-icing was by hot air on the wing and engine inlets, and electrothermal on the tail and windscreen. Under the rear fuselage was a steel bumper, though this did not scrape until the ground angle reached 12 deg. Immediately behind this was an emergency braking parachute. Avionics included standard radio with a wire antenna from fin to cockpit and two short but prominent antenna coupler strips above the mid-fuselage, RBP-4 radar, SRO-2 IFF with antennas above the nose and under the rear fuselage, dual radio altimeters, radio compasses and ADF, and VOR/ILS. The engines were mounted behind the rear spar in a fireproof bay with large access doors on the underside. Each carried an 18kW starter/generator. Under the

A Tu-124 of the Czech airline CSA. (RART)

Span 25.55m (83ft 9⅞in); length (except No. 45005) 30.58m (100ft 4in); wing area 119.37sq m (1,285sq ft).

Weight empty 22,900kg (50,486lb); fuel/oil 10,800kg (23,810lb); weight loaded 36,500kg (80,467lb) normal, 37,500kg (82,673lb) maximum.

Maximum speed 978km/h (608mph); cruising speed 780km/h (485mph); maximum rate of climb 12m/s (2,360ft/min); service ceiling 11,500m (37,730ft); range (maximum payload, normal gross weight) 1,700km (1,056 miles), (maximum weight) 2,100km (1,305 miles); take-off run (normal weight) 800m (2,624ft); landing speed 190 km/h (118mph), landing run 900m (2,952ft).

A model of the original DRLO based on the Tu-95. (Yefim Gordon archive)

125

In 1961 this project was raised as a next-generation supersonic bomber to follow the Tu-22 (105A). Somewhat influenced by the USAF B-70, it was a canard delta. The initial choice of engines was two Kuznetsov NK-6 augmented turbofans each rated at 23,000kg (50,705lb) thrust, but recent research makes it clear that these were later rejected in favour of as-yet unknown later engines. The requirement was to carry a cruise missile weighing 4 tonnes (to fly 600km after release), and the range was put at 4,500–4,800km cruising at 2,500km/h

(Mach 2.35). This assumed that, by using compression lift like the B-70, cruise lift/drag ratio would be as high as 5.8. After much study the 125 was shelved, though the same configuration was used for the bigger 135 strategic bomber project.

Tu-126

In 1958, eleven years before the USAF began to work on its AWACS

counterpart, the Soviet Union tested prototypes of its first large DRLO, (long-range airborne surveillance radar). At this time OKB-156, headed by Tupolev, was instructed to study aircraft to carry what became the AK-RLDN (aviation complex for radar for patrol and air control). The requirement was to keep watch on all airspace surrounding the USSR.

The radar, named *Liana* (the same word in English), was being developed by what became NP Vega. The main antenna was distributed across a beam 10.5m (34ft 5in) long. To give

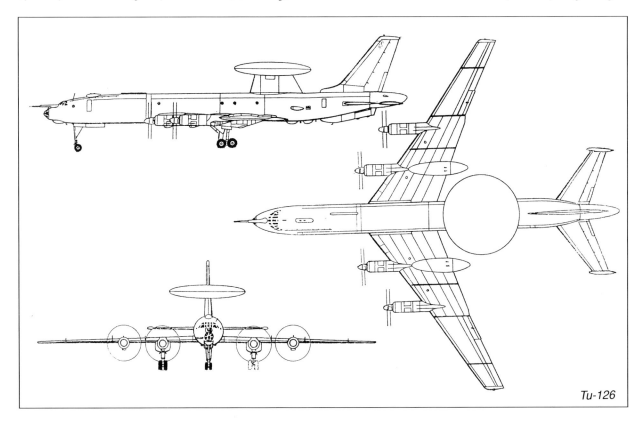

Tu-126

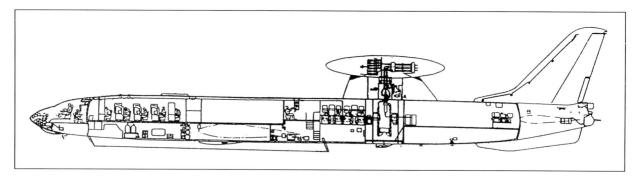

Interior layout of the Tu-126 as originally conceived, with twin AM-23 guns in the tail turret. The rest of the aircraft hardly changed, as described in the text.

The Tu-126 prototype, No. 618601. (Yefim Gordon archive)

the necessary 360 deg coverage in azimuth, this antenna, bigger than any previously designed for airspace surveillance, either had to rotate inside a radome or to have D-shaped fairings added in front and behind so that it formed the central diameter of a rotodome. The latter consistently proved to be the superior solution. Complete engineering design of the *Liana* installation took place in 1959, still ten years before the AWACS radar was designed by Westinghouse.

Work began at OKB-156 on studying how to fit this installation into a Tu-95M, but it was soon obvious that, even with a redesigned fully-pressurised fuselage, there was inadequate internal volume. In 1960 the decision was taken to base the carrier aircraft on the Tu-114, and this made the task at first seem straightforward. There was room for a total operating crew of twenty-four, of whom twelve would be resting off-duty at any one time.

Aft of the flight-crew cockpit in the nose the enormous pressurised volume above the floor was divided into six compartments. Compartment

1 housed the aircraft commander and eight operator stations, with a mass of underfloor equipment including liquid oxygen flasks, gaseous-nitrogen cylinders and the complete *Roza* (rose) station. No. 2, shown blank in the drawing, housed most of the main radar racking and an engineering workshop for repairs and servicing. No. 3 was for controlling the images sent to the operator stations, with a telescopic sight in the roof for viewing the antennas. No. 4 was a dining area. No. 5 housed the rotodome drive and cooling, with an air radiator under the fuselage. No. 6 housed beds for the off-duty crew, with the *Krystall* electronic complex at the extreme rear.

Each crew comprised two pilots, two navigators, a control officer for interceptors, three *Liana* operators, an operator for other sensors, an operator for air/ground radio, an engineer for equipment and an engineer to rectify faults. The main radar, an extremely powerful pulse-doppler equipment, used electronic scanning in elevation, while the antenna rotated at a normal operating speed of 10 rotations per minute. One of the principal

additional sensing systems was called *Lira* (Lyre), served by antennas in streamlined containers on the tips of the tailplane as in the Tu-95RTs. ECM receivers and jammers were distributed around the rear fuselage, and though the original intention was to have a Tu-95 tail turret, with twin AM-23 cannon, this was replaced in series aircraft by an SPS tailcone. Other antennas were mounted on a ventral underfin.

The primary AK-RLDN mission was long-range reconnaissance for the IA-PVO force of manned interceptors, working normally with the Tu-128S-4 complex to intercept all bombers at a distance of 1,000km (621 miles). The intention was to place the radar at an assigned patrol point 2,000km (1,242 miles) from its base, in emergency within three hours. Provided the Tu-126 was at a height between 2,000m (6,560ft) and its operating ceiling of 10,700m (35,100ft, compared with 29,000ft for the Boeing E-3), it was designed to detect fighters head-on at any height above 'treetop level' at a range of 100km (62 miles) or bombers at 200–300km, (124–186

A Tu-126 intercepted over the Mediterranean. (US Navy via Jean Alexander)

miles) or a warship at 400km (248 miles).

Development of the Tu-126 was assigned to a special brigade at GAZ No. 18 at Kuibyshev headed by veteran aircraft designer A I Putilov. By far the greatest problem to be overcome was EM (electro-magnetic) interference. EM effects extended throughout the aircraft, interactions between the numerous emitters and approximately twenty-five receivers proving difficult to eliminate. The problem was especially acute when everything was operative in the prototype aircraft, No. 6860, first flown in 1962. After three years of meticulous research and development it was possible to complete eight series aircraft, built at Kuibyshev in 1965–67.

After such difficulties it was a relief to find that initial integration into PVO with the first series aircraft in 1965 went without a hitch. The first operating base was Shyaulyai in Lithuania. The Tu-126 soon proved its ability to meet PVO requirements, no matter what the weather. Operating from this base, one of the first in service flew to the Kolski peninsula and thence round the entire Arctic shore of the USSR to land at Vladivostok 10 hours later. Soon flight-refuelling probes were added, the fuel pipe running externally along the right of the fuselage to the wing, extending normal patrols to 20 hours. For the first time the PVO had radar coverage of the entire Arctic, and ability quickly to respond to changed dispositions of NATO aircraft.

After 1970 the *Liana* radar was becoming inadequate to meet new threats. From 1969 development had been in progress on a replacement,

Shmel (bumblebee), and in 1974 this flew in the Tu-126 prototype. Reluctantly, Tupolev agreed that a completely new carrier aircraft was needed, and this was projected as the 156. The ASCC reporting name for the Tu-126 was 'Moss'.

Span 51.4m (168ft 7⅝in); length (excluding probe) 54.1m (177ft 6in); wing area 311.1sq m (3,349sq ft).
Weight empty 103,000kg (227,000lb); fuel 60,800km (134,041lb) loaded weight (normal) 171,000kg (376,984lb), (maximum) 175,000kg (385,800lb).
Maximum speed 790km/h (491mph); cruising speed to reach station 680km/h (423mph); patrol speed 530km/h at 10,000m (323mph at 32,800ft); unrefuelled range 7,000km (4,350 miles), (maximum fuel) 8,600km; unrefuelled endurance 10.2hr.

Tu-128

In 1953, long before the Yak-25 had entered service as the main radar-equipped night and all-weather fighter of the IA-PVO (manned interceptor force of the air-defence forces), it was

clear that this gun-armed subsonic twin-jet would be inadequate to defend Soviet airspace. The threat from new versions of B-52, supersonic B-58, improved V-bombers, Mirage IVA, XB-70 cruising at Mach 3, Mach 2 Vigilante flying from carriers in many locations, and Snark and Hound Dog cruise missiles posed terrible problems.

In 1955 it was decreed that the future IA-PVO interceptor would have to be capable of engaging targets flying at 2,000km/h (1,243mph) at a height of 21,000m (69,000ft); it also had to perform 2-hour escort missions. The task resolved itself into two parts, each of which posed a great challenge. One was to create an effective weapon system, in the USSR called a *Kompleks* (system), comprising the airborne interception radar and associated air-to-air missiles. The other task was to create the aircraft to carry it.

The Lavochkin OKB, having failed to get any of the three versions of its subsonic La-200 night fighter accepted, was accordingly awarded a contract for the La-250, dubbed

The Tu-28-80 prototype. (RART)

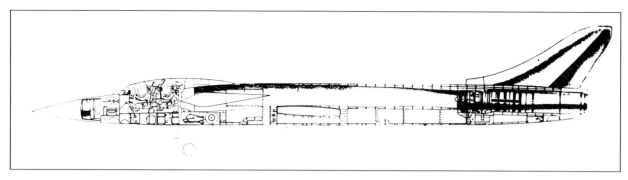

OKB drawing of internal arrangement of Tu-28 (later 128).

Anakonda. This enormous missile-armed supersonic interceptor was planned as the successor to the Yak-25, and it was hoped to meet all the expected threats from the USA and other hostile countries. Unfortunately, while on the one hand its K-15 *Kompleks* swiftly became out of date, the aircraft itself made slow progress, punctuated by serious incidents, and in 1958 the whole programme was cancelled.

By this time the threat was even greater. For short-range defence, even if only on a temporary basis, P O Sukhoi OKB was awarded a contract for the Su-9 with the K-51 *Kompleks* including the rather primitive RS-2US missile. To the astonishment of almost the whole Soviet industry, the big contract for the long-range interceptor was awarded to the Tupolev OKB. Factors influencing the decision were the fact that this was the biggest and most experienced of all design groups, it had extensive supersonic experience and could use possible close kinship between the proven Type 98 and the required interceptor, the fact that the interceptor would not be an agile dogfighter but a large aircraft firing missiles from a distance, and, above all, the reputation of the OKB for conservative reliability.

The principal contracts for the aircraft and its associated systems were placed in 1958. The *Kompleks* K-80 was based on the biggest and most powerful fighter radar in the world at that time, the RLS *Smerch* (Waterspout) developed by the F F Volkov KB, and the associated PR-S-80 sight. The missile was likewise to be large and long-ranged, the K-80, known as R-4 in production, by the M R Biesnovat KB. It was to be developed in

two forms, the R-4R with semi-active radar guidance and the R-4T with infra-red homing. The interceptor's engine was to be the VD-19, developed by the KB of V A Dobrynin. This was a derivative of the VD-7 turbojet produced for the Myasishchev 3M and Tupolev Tu-22 bombers, but with diameter slightly reduced and an afterburner and convergent/divergent variable nozzle added for flight at over Mach 1.6.

The new interceptor was given the *Izdelye* (article) number 128, but the Service designation was expected to be Tu-28. The prototype was designated Tu-28-80, indicating the *Kompleks* to be carried. It went ahead in late 1958 in Tupolev's Advanced Projects office directed by S Yeger. Chief designer was I F Nezval, and from the outset it was clear that this aircraft would be similar to the 98 bomber in size, weight, engine thrust and general performance, and that it should therefore be possible to reduce risk by making maximum use of 'read across' from the earlier programme. The single 98 was rebuilt as the *Komplex* K-80 test aircraft.

A N Tupolev personally oversaw the crucial aerodynamics of the wing and inlets. As often happens, the further the design progressed, the more did the similarity with the 98 evaporate. There was no requirement for internal carriage of weapons, so the wing was moved to the low/mid position. At first the wing was slightly reduced in span, but it was increased in area partly by increasing chord, to reduce thickness/chord ratio, and partly by putting a kink in the trailing edge so that the inboard trailing edge was at 90 deg to the fuselage. This enabled the root to be deeper and stronger.

The main wing structural box was bounded by heavy spars at 5 per cent and 40 per cent chord, with machined skins forming integral tankage. It was constructed in three sections, the centre section integral with the fuselage with massive girders bridging the body-width, and left and right panels extending to the tip. The leading edge was made detachable but fixed, with hot-air de-icing. It was to carry four missile pylons, a kink at the inboard pylon changing the leading-edge sweepback from 58 deg to 70 deg over the inboard portion. On the trailing edge were large-chord tracked slotted-flaps in single sections inboard and outboard of the main landing gear. Immediately ahead of each inboard flap was an air-brake, and ahead of each outer flap was a spoiler. From here to the tip was an aileron, driven by an electro-hydraulic power unit. Static anhedral was −4 deg, fractionally less than on the 98.

Though initially the tailplanes resembled those of the 98, eventually the tail was a fresh design, but retaining the leading-edge sweep of 58 deg. It was found that for the best longitudinal control both at supersonic speed and on the landing approach the tailplanes should be raised almost to the mid-position on the rear fuselage, and they were given 7 deg dihedral. Span was increased from 7.7m to 8m (26ft 3in), range of movement increased, and the elevators linked to the tailplanes purely to increase camber, with tabs omitted. The tailplanes always moved in unison, and each was mounted outboard of a quite large fixed fairing required by the more rounded profile of the fuselage. The vertical tail was completely new, with reduced chord giving higher aspect ratio and an air inlet at the front form-

An unusual view of a Tu-128. (G F Petrov/RART)

tor was slightly shorter, but it appeared to be longer because of the reduced cross-section of the fuselage. The fineness ratio (slenderness) of the fuselage was probably the highest of any fighter in history. The cross-section profile varied throughout. At the front was the giant radome, with a perfectly circular section. The pressurised tandem cockpits were again almost circular in section (the 98 being an upright oval), but the large air inlets exerted a profound effect, so that over the mid-fuselage Area Rule waisting almost halved the cross-section area, with a flat top and bulged underside below the wing. In contrast, the rear fuselage section was a broad flat oval.

Unlike the 98 the engines could be installed close together, parallel and horizontal. Inlet geometry finally differed from the bomber in that the cross-section was a D-shape, with the flat edge against (but standing away from) the fuselage, and in side view it was inclined instead of being vertical and the centrebody half-cone was increased in size and arranged to translate axially to match flight Mach number. A section drawing shows how the centrebody downstream of the cone tapers away inside the duct. As before, the underside was faired back to the underside of the wing, but the

ing a short dorsal fin. Leading-edge sweep of the fin finally settled at 60 deg, and the top was of Küchemann form, with a curved leading edge becoming horizontal at the top. The one-piece rudder carried a tab. Like the wing, all tail leading edges had thermal de-icing.

All flight-control surfaces were fully powered, with mechanical signalling to surface hydraulic power units. As eventually developed, manual rever-sion was possible on all axes – ailerons, tailplanes and rudder – but transfer back to powered control was possible only on the ailerons and rudder. The elevators served as a back-up, and to increase camber, while the spoilers were opened automatically to prevent any excursion beyond a critical value of AOA (angle of attack), and also to reduce wing lift after landing and thus improve braking.

Compared with the 98 the intercep-

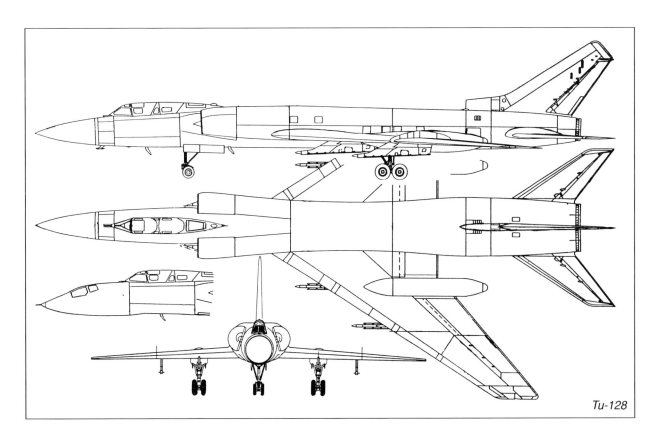

Tu-128

Tu-128UT prototype, with the original fin and dummy missiles. (G F Petrov/ RART)

inlet was extended further ahead of the wing. Internally, the duct curved up and inwards across the wing, in conformity with transonic Area Rule, before moving out again to become horizontal upstream of the engines.

To make the long fuselage stiff in the vertical plane a strong vertical keel ran down the centre above wing level. Fuel was housed in integral tanks filling the wing inter-spar box as far as the flap/aileron end-rib, in Tank No. 1 between the ducts and in a remarkable row of seven tanks along the lower part of the fuselage interrupted only by the small bay housing engine-driven accessories. No provision was made for inflight refuelling, and it was never to be thought necessary. Further details of the propulsion system are given later.

Geometry of the landing gears was based on that of the 98, except that, as in the 105A, drag was reduced by making the bogie main units retract backwards into fairings across and behind the wing. To achieve the desired smooth plot of cross-section area these fairings had to be larger than necessary; in the series aircraft they began at 20 per cent chord and ran parallel-sided for 4.2m (13ft 9½in)! Track, measured between the bogie mid-points, was 6.2m (20ft 4in). Wheelbase varied slightly, but in series aircraft was typically 10.69m (35ft 1in). The nose gear had a tall vertical leg with side V-braces, carrying a hydraulically steerable and braked twin-wheel truck with tyres 660mm by 200mm (26in by 7.9in). It retracted backwards hydraulically into a bay behind the rear pressure bulkhead with four doors. The main units,

pivoted behind the rear spar, each had a vertical shock strut with lateral V-braces and front and rear oleo snubber struts to control the attitude of the pivoted truck carrying the four wheels with anti-skid multi-disc brakes and tyres 900mm by 275mm (35.4in by 10.8in). As in the bomber, a jack pushing a link near the pivot folded the whole unit backwards, the truck rotating to lie inverted. A braking parachute, later of 50sq m (538sq ft), was housed in a thermally insulated compartment in the underside of the rear fuselage.

The enormous radome presented a severe challenge to the Volkov KB who produced the RP-5 radar as part of the *Smerch* fire-control system. Major problems included electronic distortion and erosion by rain and hail at supersonic speeds. The pressurised crew compartment contained tandem cockpits for the pilot and navigator/ systems operator, both with upward-firing KT-1 seats. Projecting above the pressure drum were the large birdproof V-windscreen and separate upward-hinged metal canopies with inset windows. The crew boarded via a large trolley with an upper platform reached by an eleven-rung ladder.

The mock-up was finally approved in early 1960. While the scale of design and development effort needed by this aircraft is obvious, Tupolev was increasingly concerned at the repeated failure of the VD-19 to pass its qualification tests. Eventually a panic redesign took place, to enable the prototype to fly on the same AL-7F engines as used on the 98. This involved considerable redesign of the fuselage, with less-pronounced Area

Rule waisting, giving greater width throughout, and a cross section at the rear like a letter B on its side with bulges each side of the fin and rudder. Other modifications included completely different access doors, engine pick-ups (including the rails along which engines were withdrawn to the rear) and pipe connections, repositioned fire bulkheads and a large red-painted bleed-air grille on each side below the dorsal fin. Flight performance suffered slightly from the reduction in thrust, and a reduction in range resulted from the higher specific fuel consumption.

The wing was considerably modified. Span was increased, so that the ailerons stopped well short of the tips, and the ailerons were extended in chord to project slightly behind the trailing edge (echoing Tu aircraft of 40 years earlier). The leading-edge kink was almost eliminated, and 0.5m beyond the outboard pylons a shallow fence was added round the leading edge and upper surface. The vertical tail was redesigned by adding a trianglular area at the top of the fin, increasing the height of the aircraft on the ground from 6.51m to 7.21m (23ft 8in), the purpose being both to improve directional stability and, by cutting off the tip at the Mach angle, increasing the flutter-critical speed. The tailplanes were not Mach-cropped, but were increased in span from 8m to 8.22m (26ft 11⅜in), and they were made horizontal.

At last the redesign and ground testing was completed, and the Tu-28–80 made its first flight on 18 March 1961, in the hands of OKB pilot Mikhail V Kozlov and observer K I

Malkhasyan. This aircraft had only the outboard pylons for dummy K–80R/T missiles, but under its belly was a giant fairing occupied by test equipment including a sensitive receiver antenna to test the reception of target echoes. This fairing required the addition of two inclined rear ventral fins. On 9 July 1961 the 28–80 was included in the impressive flypast of new types in the Aviation Day celebrations at Moscow–Tushino, where (like the 98 before it) it was identified by Western analysts as a design by Yakovlev. The ASCC assigned the fighter-style reporting name 'Fiddler', though there was a general belief that this impressive aircraft could fly attack-bomber roles.

Span 16.53m (54ft 2¼in); length 29.90m (98ft 1⅛in); wing area 90.3sq m (972sq ft).

Weight data not known.

Maximum speed 1,700km/h (1,056mph); range 2,000km (1,242 miles).

This impressive Tupolev interceptor was, in fact, one of only two new types displayed in 1961 to go into production. Instead of Tu-28 the Service designation was the same as the OKB number, Tu-128. Development problems were considerable. There were prolonged difficulties with the radar fire control, and once NII-VVS testing had started (on 20 March 1962) numerous criticisms were voiced by NII pilots V Ivanov and future Cosmonaut G Beregovoi. In subsonic flight the yaw (directional) stability was poor, and with dummy missiles it became unacceptable. This threatened to preclude carriage of the desired armament of two R-4R and two R-4T, on four large wing pylons. Moreover, at over Mach 1.45 pedal pressure resulted in yaw reversal, pressure on

the left pedal causing a yaw to the right.

The ventral fairing and underfins were removed, and the vertical tail modified. A gas-turbine APU (auxiliary power unit) was added under the dorsal inlet in the fin to make the aircraft independent of external power supplies. De-icing of the tail surfaces was by fifth-stage compressor bleed air. The fuel system was improved, capacity being standardised at 18,600 litres (4,085gal). There continued to be no provision for external fuel or for inflight refuelling. Planar-spiral ECM antennas were added at the tips of the landing-gear fairings. The wing geometry was modified, the leading edge being made straight, with sweep angle of 56 deg 22 min 24 sec, aileron chord was reduced (the trailing edge beyond the aileron having a chord-extending strip added) and the fences progressively increased in depth from under the leading edge to mid-chord, thereafter tapering to zero ahead of the flap. Nevertheless, handling remained tricky and liable to be dangerous with an inexperienced pilot, so in 1966 the Tupolev OKB flew the first Tu-128UT dual-control trainer, described later. Backseaters were trained with the Tu-124Sh.

As a result of the significant departures from the specified values reported in the NII testing, about a year was added to the development time, mainly in extended improvement to the electronic interception and flight-control systems. Initial series aircraft were powered by the AL-7F-1 engine, rated at 6,900kg (15,212lb) dry and 9,900kg (21,825lb) with maximum afterburner. The inlets were redesigned, extending further forward and almost symmetrical in side view

with a vertical lip and a more horizontal bottom line. Tyre pressures were increased, reducing tyre sizes to (main) 800mm by 225mm and (nose) 600mm by 155mm.

The fuselage was lengthened, and the radome and nose tilted down 3 deg. Among many new items of equipment were a new pattern of red/white cockpit lighting, an AP-7P autopilot, NVU-B1 navigation *Kompleks* and a Put'-4P flight-control system giving semi-automatic guidance in level flight with altitude/heading hold, airfield homing, runway approach and, especially, automatic return to a pre-programmed position and missile lock-on to aerial targets. Altogether, the Tu-128 had the most advanced interception system in the USSR at that time, and in many ways the most advanced in the world. Ironically, in the first air firing with a live missile against a target (an Il-28M), in September 1962, the Tu-128 was damaged.

Tu-128 This initial series version did not become available to the IA-PVO until 1966. It was powered by the AL-7F-1 engine, and was equipped with *Kompleks* Tu-128S-4 including *Smerch* radar and two R-4R and two R-4T missiles, cleared to fire at ranges up to 40km (25 miles) in the absence of any ground control. This was the first Soviet interceptor able to engage enemy (or believed enemy) targets beyond the range of ground radar surveillance or ground control. A total of 189 series aircraft were delivered from Voronezh. Later production blocks added *Chaika* Loran, with a towel-rail antenna, and improved the operator's external vision by adding a porthole in the fixed metal structure

The first Tu-128M, construction number 501101 (G F Petrov/RART)

between the canopies and two small roof windows in his own canopy.

Tu-128UT Eleven examples of this three-seat dual-control trainer version were built in two sub-types corresponding to different versions of Tu-128. All had a new forward fuselage with a pressure cabin extended forwards so that the main radar was replaced by a third KT-1 seat for an instructor. His V-profile clamshell canopy was flush with the top of the nose, so that drag was hardly altered even though the extra cockpit caused a significant bulge on the underside. The UT carried pylons for four dummy missiles to preserve the flying characteristics of the interceptor. Except for the prototype, this version had area added at the top of a flat-topped fin.

Tu-128A The Tupolev OKB had worked with the Lyul'ka KB to improve aircraft performance, and this prototype switched to the AL-7F-2 engine, rated at 10,000kg (22,282lb) thrust with afterburner. Other changes included the RP-SA (*Smerch-A*) radar, with improved look-down performance, and a rear canopy with no roof window, though the porthole in the fixed canopy section between the cockpits was replaced by a substantial triangular window.

Tu-128M First flown on 15 October 1970, this was for 10 years the standard PVO version. All were rebuilds of aircraft previously in service. Conversion slightly increased span and length. More importantly, the Tu-128M was fitted with the definitive Tu-128S-4M fire-control, incorporating the RP-5M *Smerch-M* radar, and two R-4RM and two R-4TM missiles. These, besides being exceptionally reliable, had outstanding performance at long range even against targets hugging the ground. The engines were the AL-7F-4, the final production series of AL-7 with an afterburning thrust of 10,700kg (23,590lb). The final (small) block of 128M conversions were at last powered by the VD-19 engine, rated at 10,200kg (22,480lb) with afterburner but whose main attraction was even better fuel economy, giving greater range and endurance.

NII-VVS testing was completed in the spring of 1974, and the Tu-128M was certified for PVO service in 1979. [Incidentally, a standard book on Soviet aircraft published in Britain in 1975 said 'Now obsolescent, the Tu-28P was from late 1974 being phased out of service in favour of an interceptor version of Tu-22'. No such aircraft existed, and when that was written the Tu-128M was still four years away from operational service.]

Tu-128B This projected bomber version, powered with VD-19 engines, would have carried a 4½ tonnes (9,921lb) bomb load, *Initsiativa-2* radar and an aft-hemisphere search radar. It was rejected in favour of the Su-24.

Span (Tu-128) 17.53m (57ft 6in), (Tu-128M) 17.67m (57ft 11⅜in); length (Tu-128) 30.06m (98ft 7½in), (Tu-128M) 30.49m (100ft 0⅜in); height (all) 7.24m (23ft 9in); wing area (Tu-128M) 96.94sq m (1,043.5sq ft).

Weights (first production series) empty 25,960kg (57,231lb); maximum take-off 43,000kg (94,797lb).

Maximum speed at 11,000m (36,090ft), first series, clean, 1,915km/h (1,190mph, Mach 1.8), with missiles, 1,665km/h (1,035mph, Mach 1.57), (F-4 engines) 2,085km/h (1,296mph, Mach 1.96); loiter speed Mach 0.85; service ceiling (F-1 engines) 15,600m (51,180ft); range (first series, with full combat allowances) 2,565km (1,595 miles); PVO interception line 600–965km (373–600 miles); take-off run (first series) 1,350m (4,429ft); landing run 1,050m (3,445ft)

129–133

Aircraft 129, no information. No. 130, small unmanned reusable space launcher (aerospace-plane), configuration slender lifting-body delta with upturned tips, design 1961, prototype tested. 131, projected long-range air-

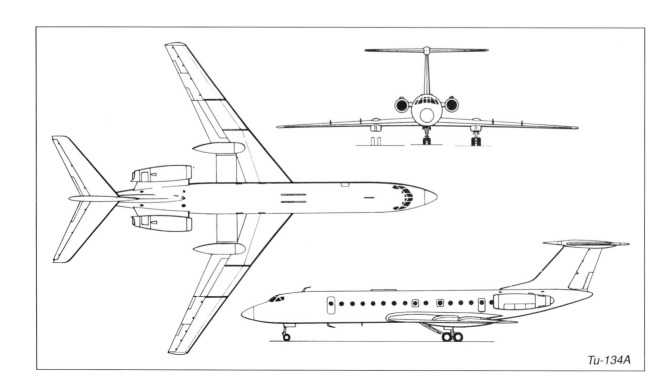

Tu-134A

defence missile system, dating from 1959. 132–133, no information.

134

This number was first used for a projected supersonic passenger transport derived from the 106. Powered by two Kuznetsov NK-6 duct-burning turbofans, each rated at 23,000kg (50,700lb) with maximum augmentation, it would have been a canard delta seating fifty to seventy passengers, payload being 5–8 tonnes. It was overtaken by the 135.

Span 23.66m (77ft 7½in); length 45.9m (150ft 7in); wing area 190sq m (2,045sq ft).
Weight empty 43,200kg (95,238lb); fuel 46,000kg maximum take-off weight 97,000kg (213,845lb).
Cruising speed 2,100km/h (1,305mph, Mach 1.98); range 3,000–3,500km (1,864–2,175 miles).

Tu-134

Originally designated 124A, this passenger aircraft was planned in late 1961 as a modified Tu-124, built to the then-fashionable formula with engines hung on the rear fuselage and a T-tail. By then the Tu-124 had shown it would be a success, capable of replacing the smaller piston-engined transports, but its wing-root engines slightly degraded installed efficiency, and caused cabin noise and vibration levels higher than those of Western rivals. Tupolev also believed a more capable aircraft could be built without much increase in weight, and the decision to stretch the fuselage was taken when P A Soloviev received a contract for the more powerful D-30 engine.

As far as possible the existing structure was retained, though stressed for increased weights, and with chem-milled panels used in the wing and fuselage, especially for the skins. The wing retained the same profile and sweep, though static anhedral was reduced to –1 deg 30 min. The centre section, no longer complicated by the engines, had increased span, thus giving a considerably greater effective lifting surface. The outer panels were also increased in span, and the ailerons were split into two equal-span

The first Tu-134A (65610, with extra '232' on the forward fuselage for the 1967 Paris Airshow). (RART)

sections, fitted with spring and trim tabs. The double-slotted flaps, of greater effective span, were driven in unison to 20 deg for take-off and to 38 deg for landing. The two spoiler/lift dumpers above each outer flap section were retained.

The fuselage diameter remained 2.9m (114in) but length was increased by 1.6m (63in), and by providing a shallow belly fairing the wing box was accommodated under a passenger floor entirely at one level, with constant 1.96m (77.2in) headroom along the aisle. A simplified design of door-type belly air-brake was retained. Two frames near the rear of the pressurised portion carried the engines on short stubs, a little above the fuselage axis. Each engine, in the prototype an interim D-20 version called D-20P-125, was installed in a neat pod with underside doors, without reverser or noise suppressor. Standard flight crew remained as on the 124, but the three cabins could accommodate (typically) sixteen + twenty + twenty-eight for a total of sixty-four passengers.

The tail was considerably enlarged. The area of the vertical surfaces was 21.25sq m (228.74sq ft), and together with the bullet fairing for the junction with the tailplane increased nominal height on the ground from 8.080m (26ft 6in) to 9.144m (30ft). The rudder was hydraulically boosted, and a yaw damper was fitted. The tailplane span was reduced slightly to 9.2m (30ft 2in), but this was entirely the aerodynamic

surface, driven by an irreversible jack for trim and carrying geared-tab elevators. The greater span of the wing centre section increased track from 9.05m to 9.45m (31ft), and wheelbase was 13.93m (45ft 8⅜in). The main landing gears were strengthened for greater weights and rough airstrips, the pressure of the 930mm by 305mm tyres accordingly being reduced to 6kg/sq cm (85lb/sq in). The anti-skid multi-disc brakes were greatly increased in capacity, and the braking parachute was removed. The nose landing gear was strengthened, and tyre size widened to 660mm by 220mm. Systems were essentially unchanged, though the three groups of fuel tanks in each wing housed the increased total of 16,500 litres (3,629gal), the single pressure-fuelling socket being under the right leading edge.

Project 124A was the first complete aircraft assigned to Leonid Selyakov, though with the far-senior Arkhangel'-skii keeping a watchful eye as his deputy. This curious arrangement was mentioned in the first public announcement, which did not take place until the prototype had made 100 test flights. Two aircraft on the 124 assembly line, 45075/76, had in early 1962 been earmarked for completion as 124As, and the basic design was accepted by the GVF in August that year. Aircraft 45075 was first flown at Kharkov by A D Kalina on 29 July 1963. There were no serious problems, and brief details were released on

A Tu-134Sh with bay doors open. (Robert J Ruffle)

29 September 1964, the OKB number being changed to 134. As previously noted, this number had already been assigned to a supersonic aircraft. When it was clear that this would not be built the number was transferred to the 124A, to follow Boeing's marketing of a 'family' of civil transports with similar numbers. The second prototype, 45076, was at the 1965 Paris airshow.

By this time the Kharkov line was in full production. All subsequent aircraft had the D-30 engine, handling an increased airflow and with greater pressure ratio, increasing take-off thrust to 6,800kg (14,991lb), with better fuel economy. After building five pre-production aircraft, seating was standardised at seventy-two (forty-four in a longer front cabin, twenty-eight in the rear), passengers entering via a plug door at left front and stowing their baggage behind the flight deck opposite the galley. Cargo was loaded at the rear of the cabin via a door under the right engine. The first production aircraft, 65600, was exhibited at Paris in June 1967. By this time extensive line experience had been gained with aircraft carrying cargo (if anything), prior to release to passenger service on 9 September 1967 to Sochi. The first international service, to Stockholm, followed three days later.

This aircraft attracted more foreign interest than had the Tu-104 or 124, and sales were made to Bulgaria, East Germany, Hungary, Iraq, Yugoslavia and Poland. During production many

improvements were made. By 1969 the D-30 Series II was available, with an STM-10 pneumatic starter and twin-clamshell target-type reverser, and as a customer option each engine could carry an English Electric constant-speed drive to power a 115V, 400Hz AC electrical system. A year later an APU was added in the tailcone for ground air-conditioning and main-engine starting. The ASCC assigned the reporting name 'Crusty', retained for all subsequent versions.

Span 29.01m (95ft 2in); length 34.95m (114ft 8in); wing area 127sq m (1,370sq ft).

Weight empty 26,500kg (58,422lb); fuel 10,530kg (23,219lb); maximum take-off weight 45,000kg (99,200lb); maximum landing weight 40,000kg (88,183lb).

Cruising speed 870km/h (541mph) maximum, 800km/h (497mph) normal; maximum rate of climb 14½m/s (2,855ft/min); service ceiling 11,500m (37,730ft); range (against 50km/h headwind, 1hr reserve) 2,400km (1,490 miles) at 850km/h (528mph) with 7 tonnes payload, 3,070km (1,908 miles) at 850km/h with 5,190kg, 3,500km (2,175 miles) at 800km/h with 3,000kg, take-off run 1,050m (3,444ft); landing speed 220km/h (137mph), landing run 880m (2,887ft).

Tu-134A Aircraft 65624 was completed as the prototype of this stretched version, and exhibited at Paris in 1969. The most obvious change was a 2.1m (83in) fuselage extension, increasing capacity by one seat row to seventy-six, but with capability of seating (in series aircraft) up to eighty-four, all 2+2, partly by reducing the baggage space at the front. The second new feature was a

long pointed VHF antenna projecting ahead of the fin tip. Less obvious was a considerable increase in size of the horizontal tail, span being 11.8m (38ft 8⅝in). The main landing gears were modified, with similar wheels, brakes and tyres to the Il-18 and stronger legs, but keeping tyre pressure unchanged. The wing was locally strengthened, and avionic fit was upgraded (though Arkhangel'skii's 1964 claim of automatic landing in 50m visibility was not certificated for service).

The Tu-134A also achieved foreign sales, including Bulgaria, Czechoslovakia, Hungary, Poland and Yugoslavia. The third for Aviogenex (Yugoslavia) was the first to have what became the preferred configuration, with the new *Groza* (Storm) weather radar mounted in the more pointed nose, the navigator being replaced by an engineer at a side panel. The Tu-134A-3 introduced the D-30 Series III engine with a compressor zero-stage to give the same ratings at reduced turbine temperature and up to ISA+25 deg C ambient temperature.

Tu-134B In 1980 production switched to this improved variant. It introduced spoilers modified for DLC (direct-lift control) in flight, especially on the approach, an advanced cockpit with updated instruments (but not multifunction displays) cleared for two-pilot operation but usually retaining the engineer, and with new internal furnishing options. Normal seating rose to eighty and with the galley

The Tu-134Skh, No. 65918. (RART)

removed from the 134B-1 the total could reach ninety, or even ninety-six using compact lightweight seating, which became an option on the 134A-3. With Series III engines the aircraft was a 134B-3. Production continued not only for Aeroflot and the VVS but for no fewer than twenty foreign customers. The total for all versions was 853, all from Kharkov. In Aeroflot service the Tu-134 had carried 360 million passengers by 1995.

Dimensions unchanged except length 37,322m (122ft 5⅜in).

Weight empty 28,600kg (63,052lb); fuel 14,400kg; weight loaded (normal) 44,000kg (97,000lb), maximum take-off weight (A) 47,500kg (104,720lb), (B) 47,600kg (104,940lb); maximum landing weight 43,000kg (94,797lb).

Maximum speed 900km/h (559mph); maximum cruising speed 885km/h (550mph) at 10,000m; service ceiling (MTO wt) 11,900m (39,040ft); range (800km/h, 1hr reserve) 1,890km (1,174 miles) with maximum (8,200kg) payload, 3,020km (1,877 miles) with 5,000kg; take-off run 1,200m (3,936kg); landing speed 240km/h (149mph), landing run 1,600m (5,249ft).

Aeroflot passenger aircraft began to be withdrawn in 1988. Before this, Tupolev and various users had discussed conversion of Tu-134s into several types of crew trainer, of which the following are in service:

Tu-134BU B-series conversions into trainers for civil and military flight crews, including Cat IIIA automatic landings.

Tu-134Sh A number have been converted into military trainers for navigators. The cabin is equipped with nine or ten student and instructor stations, and an optical bombsight is in the glazed nose for aiming bombs of up to 250kg from two large pylons under the fuselage.

Tu-134BSh From the mid-1970s about twelve aircraft were rebuilt as bomber navigation trainers for crews of the Tu-22M. The cabin is completely devoted to twelve crew stations for navigation and bombing, using the *Rubin* radar in the simulated Tu-22M nose and practice bombs released from carriers on two underwing pylons. Overall length is increased to 41.87m (137ft 4½in), requiring a reduction in maximum take-off weight and an augmented yaw-damping system. Alternate cabin windows are removed, and electrical power is augmented paralleled AC. These aircraft equip the Tambov conversion school.

Tu-134UBL This version (UBL = training combat pilot) entered service with the 184th Polk (regiment) at Priluki to assist conversion to the Tu-160. These aircraft have the nose of the heavy bomber, increasing length to 41.92m (137ft 6½in), the reduced yaw stability resulting in a reduction in MTO weight to 44,250kg (97,553lb), and limiting speed at 10,000m height to 860km/h (534mph).

Tu-134SKh Aircraft 65917, built as an A-3, is today equipped for agri-

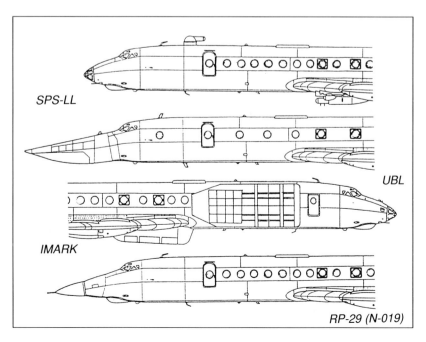

cultural and economic survey. The cabin contains ten operator stations for monitoring and recording the Earth's surface at optical, radio and IR (infrared) bands. A precision navigation system can control aircraft position anywhere in the world. A French multiband scanning system operates in four visual and two IR bands, data being taped and displayed in monochrome or colour. The two large pods are SLAR (side-looking airborne radar) operating in the 2cm band, with different sweep widths and scales from 100,000 to 500,000. Instrumentation weight is 6,000kg (13,228lb), and the crew numbers up to twenty-one. This aircraft has been extensively used in many countries. Data for Tu-134SKh.

Dimensions unchanged.
Weight empty 30,800kg (67,900lb); fuel 14,000kg (30,864lb); maximum take-off weight 49,000kg (108,025lb); maximum landing weight 43,000kg (94,797lb).
Cruising speed 450–850km/h (280–528mph); service ceiling 11,500m (37,730ft); range 3,600km (2,237 miles); field length 2,200m (7,217ft).

IMARK This aircraft (RA-65906) is developing the *Zemai* multifrequency polarised radar by NPO Vega-M, which also produced the radar of the Tu-126. The two longest-wavelength side-looking antennas are in a large unpressurised container under the mid- fuselage, while two further arrays

of flush antennas cover the surface of a giant part-drum scabbed on the right side of the fuselage. The radar operates at 4, 23, 68 or 230cm wavelength, and can carry out complete geological survey of a strip 6–12km from the aircraft track.

Tu-134LL At least fifteen aircraft have been converted into flying laboratories. One has been intensively used for Cosmonaut training, simulating weightless conditions. Another is being used by CAHI at Zhukovskii for deep-stall research (a problem with rear-engine T-tail aircraft) with a large emergency parachute in the tail. Aircraft 65908 was exhibited at Zhukovskii equipped for large-scale electronic and systems tests, with a ventral equipment container and a large side-looking antenna group on the right side of the fuselage.

135

Originally projected in 1959 as the 125, this long-range strategic bomber was almost a clone of the North American B-70. There were to be several versions, the drawing showing Variant No. 1, carrying a single Kh-22 or KSR-5 cruise missile recessed under the centreline. The large engine

compartment under the delta wing was to house four Kuznetsov NK-6 engines, each with sea-level rating under full augmentation of 23,500kg (50,705lb), giving a total thrust almost 20 per cent greater than the B-70. Despite this, the B-70's problems persuaded Tupolev to design for Mach 2.35, using aluminium alloy instead of stainless steel. The diversion of funds to strategic missiles made this expensive aircraft difficult to fund. When the USAF abandoned the B-70 the 135 was recast as a passenger transport, *see* later.

Span (wingtips horizontal) 28.0m (91ft 10⅜in); length 44.8m (146ft 11¼in); wing area (tips horizontal) 380.0sq m (4,090sq ft).
Weight empty, about 79,000kg (174,000lb); fuel about 108,000kg (238,098lb); loaded weight 190,000kg (418,871lb).
Cruising speed 2,500km/h (1,553mph, Mach 2.35); cruising altitude 22,500m (73,820ft); range on normal 2,500km/h mission 7,950km (4,940 miles); take-off speed 400km/h (249mph), take-off distance 2,300m (7,545ft).

The SST version was larger, seating 100–120 with a payload of 12,000kg (26,455lb), but retaining the same engines:

Span 34.8m (114ft 2in); length 50.7m (166ft 4in); wing area 417sq m (4,489sq ft).
Weight empty 87,000kg (191,800lb); take-off weight 175,000–205,000kg (385,800–451,940lb).
Cruising speed 2,100–2,200km/h (1,305–1,367mph, Mach 2.07); range 4,000–4,500km (2,486–2,796 miles) cruising at 15,000m rising to 22,000m (49,200–72,180ft).

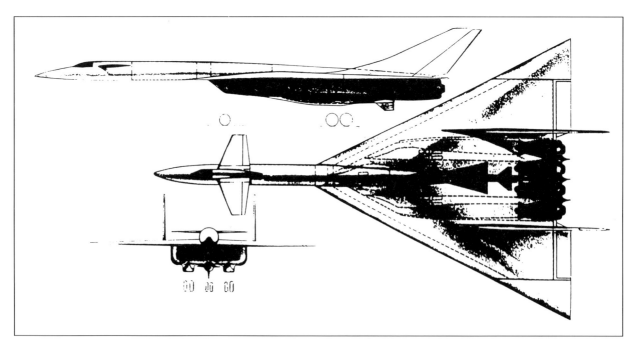

OKB drawing of Project 135.

This M-141 Strizh *has been put on display at Monino.* (RART)

136–143

Numbers 136–138, no information. Type 139 was a single prototype, tested in 1968, of a different form of unmanned reconnaissance vehicle derived from the 123 *Jastreb*. Type 140, no information. Type 141 also called M-141 and named *Strizh* (swift, the bird), was a multi-role unmanned vehicle scaled up from the VR-3 (Type 143). Differences included an R-9A cruise turbojet (related to the Tumanskii AM-9 family), folding wings and modified guidance and payload. First flown in 1975, the M-141's main role was reconnaissance, using cameras, side-looking radar, IR linescan or a radiation monitor. One bearing Ukrainian national markings was exhibited at Lvov in 1992. Data included overall length 10.05m (32ft 11½in) and range 1,000km (621 miles) cruising at 50–1,000m (164–3,280ft) at 950–1,100km/h (590–683mph).

Tu-142

This long-range naval aircraft was derived from the Tu-95RTs. It is featured as an integral part of the Tu-95 story because the two programmes became so interwoven, the later versions of Tu-95 having been based upon the Tu-142M-Z.

143

This major programme met a need for an unmanned multi-sensor tactical reconnaissance vehicle for repeated use by front-line troops. Tupolev OKB began work in 1972, and the resulting system, called VR-3 and named *Reis* (trip, journey), went into wide service five years later. It could operate as an RPV (remotely piloted vehicle) but was usually pre-programmed as a drone.

The simple metal/composite airframe has cropped delta foreplanes, delta rear wings with 58 deg sweep on the leading edge with ailerons, a circular-section fuselage with the payload carried in a detachable forward section, and a dorsal inlet feeding a Klimov TR-3-117 turbojet rated at 590kg (1,300lb) originally and 640kg (1,411lb) later. Above the engine was a compartment for a large recovery parachute, with a jettisonable rear fairing and a shallow fin with 40 deg sweep and a small rudder above. A requirement was minimum radar cross-section, peak reflective area being only 0.25–0.35 sq m (2.69–3.77 sq ft) depending on aspect.

Production VR-3s were transported either in two sealed drums on TZM-143 vehicles or fully assembled on their SPU-143 launcher, both being based on the standard BAZ-135 cross-country truck. After electrical and hydraulic systems had been checked the fuel tanks surrounding the engine and inlet duct would be filled with 190 litres (41.8gal). Almost the only short-

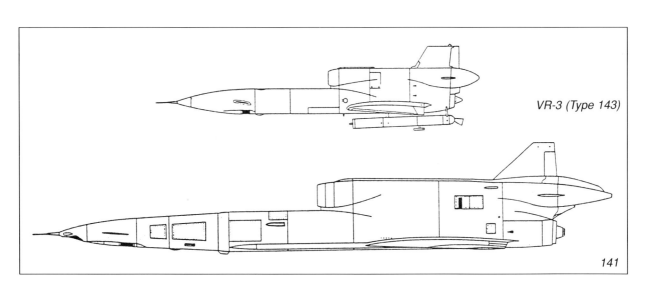

VR-3 (Type 143)

141

VR-3 Reis *being assembled on its launcher*. (Alfred Matusevich archive)

coming of *Reis* was that it could take an hour to programme its guidance system, depending on the flight profile and track. Then, with cruise engine at maximum-continuous thrust, the single launch rocket would be fired, burning out in 100m (330ft).

The trajectory could include four changes of altitude and two 270 deg turns. Guidance comprised an auto-pilot, ABSU-143 flight-control system, DISS-7 doppler radar and A-032 precision radar altimeter. Angular course error was within 6 min, giving a lateral error of 0.25 min per cent of distance (200m, 656ft, at 70km, 43½ miles, for example). Usual payload comprised either a PA-1 optical camera, a Chibis-B TV camera or Sigma radiation monitoring. Back at the recovery area the engine would be stopped, the parachute streamed to support the vehicle at its CG in a level attitude and, just before alighting, landing legs would be extended and two small rockets fired to reduce sink from 6m/s to 2m/s.

Production VR-3s were grey, with the nose, parachute fairing, inlet, all edges and fuselage lines in red. Each

was certified for six missions. Many squadrons were deployed by the USSR, and export customers included Czechoslovakia and Syria. Prototypes were tested of the long-range *Reis-D*, with a length of 8.25m (27ft 8in). *See* Tu-43.

Span 2.24m (7ft 4¼in); length 8.06m (26ft 5⅜in); wing area 2.9sq m (31.2sq ft).

Equipped empty and landing weight (optical payload) 1,012kg (2,231lb); fuel 150kg (331lb); launch weight 1,230kg (2,712lb).

Cruise speed 925km/h (575mph); cruise height 100–1,000m (330–3,300ft); radius of action 75km (46.6 miles); duration 13min.

Tu-144

This challenging programme for an SST (supersonic transport) began with studies spurred by similar work in Britain and the USA from 1956. By 1959 both OKB-23 (Myasishchev) and OKB-156 (Tupolev) were looking at possible configurations, some related to existing supersonic designs,

and undertaking supporting research, which was increasingly backed up by CAHI. On 14 May 1962 OKB-23 began testing a model of the M-53 SST from Mach 0.7 to Mach 3.

There was full recognition of the enormity of the task. Aircraft in the 100/150-seat size cruising at over Mach 2 could generate passenger-km so fast that few would be needed, while on the other hand the R&D expenditure would be colossal. The justification was that the USSR was the world's largest country, and use of seventy-five such aircraft could transform travel times on trunk routes. In 1962 Aeroflot calculated that it saved its passengers an average of 24.9hr on each journey, but that with seventy-five SSTs this could rise to over 36hr. Following the November 1962 decision by Britain and France to build Concorde, and the funding by the US Federal Aviation Agency of a nationwide SST design competition, the Council of Ministers ordered go-ahead on a Soviet SST on 16 July 1963. The Ministry of Aviation Industry issued order 276ss ten days later.

The 44-00 on its first flight, accompanied by the Mikoyan 211/1. (Tass)

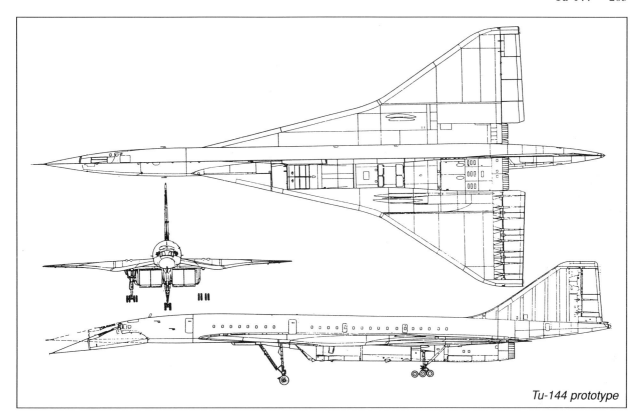

Tu-144 prototype

The outline requirement was for a slender-delta aircraft made primarily of light alloy to cruise at Mach 2.35 for 6,500km whilst carrying 121 passengers. A rough guess was that the programme cost would be three times that of the previous highest (Il-62 and M-50), and the underpinning R&D establishments, notably CAHI and CIAM, had to shelve some other work. The programme was assigned to OKB-156, which called it Aircraft K, with type number 144. Tupolev assigned Yuri N Popov as chief designer, with S M Yeger, D S Markov and Leonid Selyakov as deputies and Yu N Kashtanov as chief engineer. Propulsion was assigned to N D Kuznetsov. Initial contracts were received for two flight articles and a static-test airframe, all to be constructed at Zhukovskii. OKB-155 (MiG) was subcontracted to build one (later two) MiG-21 aircraft modified with a scaled wing of the 144. Designated 21I, Type 21–31, the first of these supporting aircraft did not fly until 18 April 1968, by which time it was too late to assist the original Tupolev design.

A desk model exists of a Tu-144 with serial 65000 with a delta canard mounted right at the bottom of the forward fuselage, a delta wing with a kink at the mid-point of each leading edge reducing sweep from 60 deg to 50 deg to increase span to the sharply pointed tips, and main landing gears with a row of four pairs of wheels retracting back between the pairs of underslung engines. This was not proceeded with, the wing actually adopted having a thickness/chord ratio of 2.5 per cent, a straight inboard leading edge swept at 76 deg with modest downwards camber at the kink, and an outer leading edge swept at 57 deg curving round to a broad tip aligned with the fuselage axis. In front view the upper surface was initially horizontal but sloped down gradually to give slight overall anhedral. Structurally the wing was a complex rectilinear assembly with ten spars joined to fuselage frames at 90 deg. Skins were chem-milled sheets filling the gaps between the parallel spars. Ruling material was VAD-23 aluminium alloy, with a titanium fixed leading edge and titanium or steel in regions of particularly high temperature or high stress. Manufacture was subcontracted to Antonov at Kiev.

The engine nozzles were in line with the trailing edge of the wing itself. Downstream were added four approximately square elevons on each side, each driven by two power units in external fairings on the underside.

The fuselage was basically of circular section, with a diameter of 3.4m (134in), with closely spaced forged frames, multiple stringers and thick chem-milled skin, integrally stiffened by crack-stoppers along the row of twenty-five small rectangular passenger windows on each side. The underside of the rear fuselage was titanium and stainless heat-resistant steel. Above it was added the large delta fin carrying a two-section rudder, the upper half having its two power units on the right and the lower having them on the left. All flight controls were given quadruple redundancy from the divided surfaces being driven by duplicate power units in separate systems, there being three 210kg/sq cm (2,987lb/sq in) hydraulic systems with 'majority-vote' automatic switching in the event of failure.

Kuznetsov was able to develop the NK-144 engine from the established NK-6 and NK-8. A duct-burning turbofan, it ran on the bench in 1965 at a thrust of 17,000kg (37,500lb), with a bypass ratio of about 1.6. It was cleared for flight in 1968 at 17,490kg (38,550lb), with bypass ratio closer to 1.8. It had been intended to install the engines in a single enormous ventral box, as in the '134' and Tu-135, but in 1966 it was decided to move them apart in two pairs to leave room on the

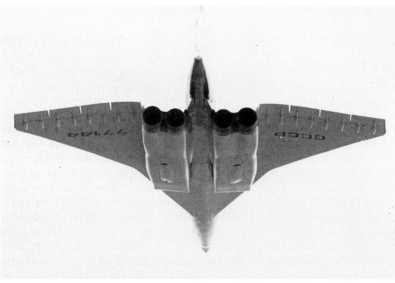

An unusual view showing the complete redesign between 44-00 (above) and 02-2 (below). (RART)

centreline for the nose landing gear. Unlike Concorde, where the engines are in short nacelles well outboard, the Tupolev choice was the longest possible inlet ducts, with parallel sides until near the engines when the width of each twin-engine box increased considerably. This put the four convergent/divergent variable nozzles all in a close group, overall length from the inlets being 23m (75ft). A total of 87,500 litres (19,247gal) of fuel could be housed in integral tanks in the outer wing, the inboard leading edge and the bottom of the fuselage, with fore and aft trim tanks, the pipe

to the tail trim tank running externally on the right side.

Separating the pairs of engines made it more difficult to retract the main landing gears into the wing, which outboard of the nacelles was so thin that, despite a magnificent design of landing gear with an overall retracted depth of only 620mm (24.4in), modest blisters were still needed both below and (two) above the wing. Each unit was a gateleg, with an upper frame pivoted to the top of the shock strut carrying the pivoted beam on which were mounted the three pairs of twin-tyred wheels with quadruple

steel plate brakes. Tyre pressure was 13kg/sq cm (185lb/sq in). Each unit was retracted forwards by rotating the upper frame to the rear, the long rear bracing strut thus pushing the bottom of the leg round an arc while the bogie truck rotated through 180 deg to lie inverted in its shallow bay. The very tall nose unit, with two hydraulically steered and braked wheels, was of typical Tupolev V-form, similar to that of the Tu-114, retracting to the rear and carrying three landing/taxi lights. A large braking parachute was housed in a thermally insulated bay in the underside of the rear fuselage.

The pointed nose could be drooped (hinged down) 12 deg for take-off and landing. Normal flight crew was to comprise two pilots and an engineer, but the first flight aircraft, 44–00, with registration 68001, had a crew of four all with upward ejection seats. The passenger cabin was 26.5m (87ft) long, with a width of 3.05m (120in) and aisle height of 2.16m (85in). The intention was to stow baggage at the front and rear in the usual Aeroflot manner, which with a galley and toilets left room for a maximum of 126 passenger seats, arranged mainly 3+2. There were two doors on each side forward and two emergency exits on each side over the wing.

With the MiG 21I/1 flying chase, 68001 made the world's first SST flight on 31 December 1968, the crew comprising Eduard Yelyan, co-pilot Mikhail Kozlov, test director V N Benderov and engineer Yu Seliverstov. There were few snags, and subsequent flying was devoted to perfecting the duplicated environmental system, with a remarkably large discharge duct between the engine nacelles, and the interlinked autopilot and inertial/ doppler navigation system. On 21 May 1969 a landing was made at Moscow Sheremetyevo, the first time the aircraft had been seen in public. Mach 1 was exceeded on 5 June 1969, Mach 2 on Flight 45 on 15 June 1970 and later in 1970 Mach 2.4 was achieved. To improve lift at low speeds and high angles of attack the leading edge was rebuilt with even more pronounced camber, the leading edge having sharp anhedral from the root to a maximum droop just inboard of the change in sweep. A more serious problem was that cruise drag was higher than predicted, continuous afterburning being necessary. Aircraft 44–00 was

01-2 displayed as No. 451 at Paris in 1973. (Jean Alexander/RART)

displayed as aircraft No. 826 at the 1971 Paris airshow. No. 044 was the static-test airframe.

Span 27m, later 27.65m (90ft 8½in); length 58.15, later 59.4m (194ft 10½in); wing area 469.8sq m (5,057sq ft).

Weight empty (typical) 79,000kg (174,162lb); fuel 70,000kg; maximum take-off weight 150,000kg (330,688lb); maximum landing weight 105,000kg (231,481lb).

Cruising speed 2,430km/h (1,510mph, Mach 2.285) at 16,000m rising to 18,000m (52,490ft to 59,055ft) with fuel burn of 26kg/km; range not stated but about 2,500km (1,553 miles); landing speed 292km/h (181mph).

After much study it was decided to effect modifications which almost amounted to the Tu-144 becoming a different aircraft, though the designation remained unchanged. The wing was increased in camber throughout, that on the leading edge extending from the root, so that the anhedral on the inboard leading edge was greatly reduced, and the downward camber at the rear being increased. In plan the change of sweep was moved forward, in effect adding chord outboard so that the wing was extended beyond the original tip, without the original gentle curve, to increase span and area. Whereas the prototype skins were flush-riveted, on this redesigned aircraft they were spot-welded. The four elevons on each wing were redesigned, though the only external changes were that the power-unit fairings started much further forward and the inner and outer edges of each surface were tapered to leave an increasing gap; thus, on a trailing edge with progressive anhedral, the elevons could be depressed to a greater angle than before without jamming against each other.

The fuselage was extended ahead of the wing by 6.3m (20ft 8in), with modified structure and materials. As in the wing, there was greatly increased use of honeycomb sandwich, and much of the titanium was of 6Al-4V alloy. There were now thirty-four passenger windows on each side, and a large cargo/baggage door at the rear on the right.

A completely new feature was the addition of retractable canards to the top of the fuselage behind the cockpit. Each surface was almost rectangular, with an extended span of 6.1m (20ft), with a fixed double slat and double-slotted flap. After retracting their flaps these surfaces pivoted aft to lie in recesses in the top of the fuselage. When extended, with 20 deg anhedral, they generated sufficient lift near the nose for the elevons to be depressed to their maximum angle to increase lift; previously, as in Concorde, the elevons had to be raised, pushing the aircraft downwards, on take-off and landing.

The engine installations and landing gear were totally redesigned. The nose landing gear was moved forward no less than 9.6m (31ft 6in), increasing wheelbase to 19.6m (64ft 3½in). In turn, this meant that the leg had to be even longer, and it carried six landing/taxi lights. It was redesigned to retract forwards, into an unpressurised underfloor box. Each main landing gear was redesigned to lie under the centreline of the engine nacelle, reducing track to 6.05m (19ft 10½in). It now comprised a vertical shock strut carrying a pivoted bogie beam with only two axles, each with inner and outer pairs of wheels with tyres 950mm by 400mm, the spread across each four tyres being almost double the previous distance. On retraction, the side bracing strut rotated the bogie truck about the beam through 90 deg, the axles being nearly vertical, so that it could then swing forward into a narrow thermally insulated bay between the engine ducts. The variable double-shock inlets were angled slightly inwards in plan, the nacelle downstream having an almost straight inner wall but sloping outwards on the outer side near the wing leading edge to accommodate the landing-gear bay. The engines were moved to the rear to balance the extended forward fuselage, the nozzles forming two pairs aligned behind the elevons. A retractable sprung skid was added at the extreme tail.

Among many other changes the vertical tail was modified, the nose-cone was given a larger glazed area, the fuel capacity was increased to 118,750 litres (26,121gal), the nose/tail trimming capability was increased (pumped aft at Mach 0.7 and back forward when decelerating through Mach 1.5) but with the piping all inside the fuselage, and the ejection seats were replaced by a strong forward entry door opened under hydraulic power. The engine was the Kuznetsov NK-144F, with ratings increased, the take-off thrust being 20,000kg (44,092lb), and the fuel was changed to thermally stable T-6, designed not to degrade or produce solid 'coke' even after soaking at high temperatures.

The first redesigned aircraft was 01–1, with registration 77101. First flown on 1 July 1971, it was in effect a different aircraft, on the whole much more capable. On 20 September 1972 it flew to Tashkent, 3,000km (1,864 miles) in 1hr 50min, reaching a peak speed of 2,500km/h. A N Tupolev died on 23 December 1972, his son Dr Alexei A Tupolev being appointed chief designer of the Tu-144. At this time there was every expectancy that line service would begin in the coming year. In the same month a French delegation visiting GAZ No. 18 at Voronezh was shown eight series aircraft on the assembly line. Aircraft 01–2, with registration 77102, was displayed as aircraft 451 at the 1973 Paris airshow. During that show, on 3 June it made a violent pull-up (to avoid a rogue Mirage or for some other reason) and suffered catastrophic wing failure, crashing at

Goussainville, killing Kozlov, Benderov and others.

Production continued with 01–3 static-test specimen, 01–4 system-test specimen, 02–1 registration 77103 and 02–2 (77144, not 77104). Aircraft 02–2 participated as aircraft 361 in the 1975 Paris airshow. Aircraft 03–1 (77105) was the first to fly with a completely new engine, P A Kolesov's RD-36-51A. Instead of being a low-ratio turbofan (bypass engine) this was a single-shaft high-compression turbojet, giving proportionately higher thrust in cruising flight and thus rendering continuous afterburning unnecessary. A novel feature was the use of a Laval propelling nozzle with a central adjustable spike (bullet). Another change was that all aircraft auxiliary power services were mounted on a remote gearbox, carried on the airframe and driven by a long shaft. Weighing 4,125kg (9,194lb), the RD-36-51A was lighter than the NK-144F, and gave similar thrust for better fuel economy, specific fuel consumption at Mach 2.2 at high altitude being 24.93mg/Ns (0.88lb/h/lb). This reduced cruise fuel burn from 26kg/km to 11.2kg/km.

Production continued with 04–1 (77106), 04–2 (77107), 05–1 (77108), 05–2 (77109), 05–3 (unregistered, used for various non-flying test programmes) and 06–1 (77110). On 20 August 1975 04–2 began trials of the auto navigation system, joined on 12 December 1975 by 05–1 equipped with the definitive NPK-144 *Kompleks* incorporating the ABSU-144 autoland system. On 26 December 1975, more than a year after the Tu-144D had flown (described later), aircraft 04–1 began regular cargo services in Aeroflot service between Moscow and Alma Ata. It flew the 3,260km (2,026 mile) sector regularly in less than the scheduled 2hr total block time. Most of the remaining series aircraft were fully furnished for 140 passengers, eleven in the front 1st-class cabin seated 2+1, thirty in the centre cabin seated 2+3, seventy-five in the next cabin seated 2+3 and twenty-four at the back seated 2+2. Maximum payload was 15,000kg (33,070lb). In March 1976 a preliminary certificate was awarded, and the Tu-144 appeared in Aeroflot timetables. Aircraft 06–1 appeared as No. 345 at the 1977 Paris airshow.

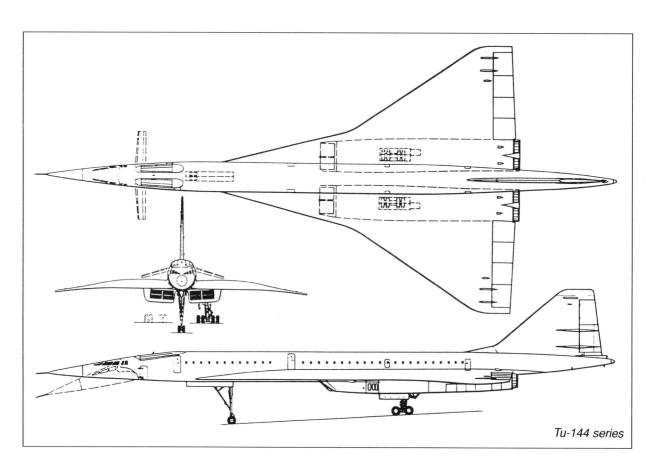

Tu-144 series

Span 28m (91ft 10⅜in); length 65.7m (215ft 6⅝in); wing area 503sq m (5,414sq ft).

Weight empty 85,150kg (187,720lb); fuel 92,000kg (202,821lb); maximum take-off weight 195,000kg (429,894lb); maximum landing weight 120,000kg (264,550lb).

Cruising speed 2,000–2,350km/h (1,243–1,460mph, Mach 2.2) at 16,000m rising to 18,000m; range with 13,000kg payload with commercial allowances 3,080km (1,914 miles); landing speed 280km/h (174mph).

Production continued with 06–2 (77111), the first Tu-144D, flown on 30 November 1974. This combined the Kolesov engine with a refined and slightly enlarged airframe with greater fuel capacity and cleared to greater weights. The new engine enabled the 144D to cruise without afterburner, at last extending the range closer to the requirement. An extra section in the fuselage enabled passenger capacity to be increased to 155, and the definitive avionics were installed. On 22 February 1977 06–2 flew to Khabarovsk, a great-circle distance of 6,280km (3,902 miles), far beyond the capability of the Tu-144.

Production continued with 07–1 (77112), 08–1 (77113), 08–2 (77114), 09–1 (77115) and terminated ahead of the intended total at 09–2 (77116). Of these, two, 08–1, not flown until 2 October 1981, and 08–2, flown on 13 April 1981, were delivered ready for service, the remainder being in some way incomplete or unfurnished, and the final two never left the Voronezh plant. In September 1977 route proving began ready for. Aeroflot scheduled services, and these were actually opened with a 144, 05–2, on 1 November 1977. Service SU499 was flown to Alma–Ata, a 3,520km (2,187 mile) sector which was all the 144 could handle, with a payload of about 10 tonnes, representing 100 passengers and baggage. The fare was R167, compared with Aeroflot's usual 110 or 130.

On 23 May 1978 06–2 suffered a relatively minor failure in part of the fuel system. This should have been survivable, but in fact by ill-fortune it led to catastrophic loss of the aircraft. This resulted in immediate cessation of scheduled services, the 102nd and last flight on SU499 being flown on 30 May. They were never resumed, in the opinion of Tupolev OKB because of a lack of enthusiasm by Aeroflot. None of the scheduled flights had been flown by the 144D, data for which appear below.

In July 1983 it was announced that fourteen records had been set by 'Aircraft 101', including a speed of 2,031.546km/h (1,262.344mph) round a 1,000km circuit with a payload of 30 tonnes, 2,012.257km/h (1,250.358 mph) round a 2,000km circuit with the same load, and a sustained altitude of 18,200m (59,711ft) again with a 30 tonnes payload. Not for ten years was it realised that 101, a genuine type designation, was in fact Tu-144D 08–2. All versions of the 144 family were given the ASCC name 'Charger'.

Span 28.8m (94ft 57⅞in); length 67.5m (221ft 5½in); wing area 506.9sq m (5,456sq ft).

Weight empty 86,900kg (191,578lb) fuel 92,000kg or 95,000kg; (202,825lb or 209,439lb) maximum take-off weight 207,000kg or 217,000kg (456,349lb or 478,395lb); maximum landing weight 120,000kg (264,550lb).

Cruising speed 2,124km/h (1,320mph, Mach 2.0) at 16,000m rising to 18,000m as before; range with 13,000kg payload 6,000–6,500km (3,730– 4,040 miles); landing speed 280km/h (174mph), landing run 2,600m (8,530ft).

Most Tu-144s were scrapped or used for instructional purposes at the Kazan or Kharkov Aviation Institutes. The better specimens, including the 144Ds, form a dwindling fleet based mainly at LII Zhukovskii, with the 144F, 36–51A and NK-32 and NK-321 engine. By cannibalising the others, a few are kept airworthy, for atmospheric research (in particular studying the ozone layer on a global basis) and to assist Buran pilot training. Aircraft 04–1 is parked at Monino and 06–1 at Ulyanovsk (Simbirsk). To the regret of ANTK Tupolev, no funds were available to build the Tu-144DA, dating from 1976, and powered by four Project-61 engines each rated at 22,950kg (50,595lb):

Span and length, slightly larger than 144D; wing area 544sq m (5,856sq ft).

Maximum take-off weight 235,000kg (518,078lb); payload 13–16 tonnes.

Cruising speed Mach 2–2.2; range with 160 passengers 7,500m (4,660 miles).

Such a capable basic platform could hardly escape being considered for other roles. One studied for years was a long-range naval interceptor, with *Zaslon* radar, comprehensive electronic-warfare systems and eight R-33E/R-33T air-ro-air missiles. Another was a strategic bomber and air-to-ground missile carrier, but this was abandoned in favour of the Tu-160.

145, Tu-22M

As early as 1962 the VVS, CAHI and Tupolev OKB were studying the fitting of pivoted 'swing wings' to the Tu-22. This very successful aircraft had triggered numerous projected improvements, as described earlier, and the 106 was studied with sweep angles of 20 deg, 65 deg and 72 deg. By 1965 this had been replaced by the 145, with no guns, powered by NK-144 engines rated at 22,000kg (48,500lb) thrust each, giving dash speeds up to 2,700km/h (1,677mph, Mach 2.54) at 14,500m, and cleared to fly at 1,100km/h 684mph) at heights down to 50m. Using engines already developed for the Tu-144, this was so impressive that the 145, and the 145R reconnaissance version, were nearly built.

Span (20 deg) 36.7m (120ft 5in), (60 deg, max) 23.66m (77ft 7½in); length 41m (134ft 6in); wing area (20 deg) 199sq m (2,142sq ft).

Weight loaded 105,000kg (231,481lb).

Maximum speed (maximum weight, with missile) 2,300km/h (1,429mph), (clean) 2,700km/h (1,678mph, Mach 2.54); range (subsonic) 6,900km (4,288 miles), (supersonic) 4,000km (2,485 miles); service ceiling 17,000m (55,775ft).

In early 1967 the decision was taken to go ahead with an aircraft with a smaller and more heavily loaded wing pivoted from 20 deg to 60 deg, and with NK-144–22 engines derated to 20,000kg (44,090lb). This was not given a new *Izdelye* number, and the Service designation, which was the only designation published, remained Tu-22M (*Modernizirovannyi*). Dr Alexei A Tupolev said 'Disguise as a mere modification was political'.

Tu-22M-0 In December 1967 the first metal was cut on the prototype, also called Tu-22KM and Article 45-01. This aircraft not only had a new wing but also a largely new fuselage. The engine KB of N D Kuznetsov, which had unsuccessfully competed to power the Tu-22 and had later produced a family of NK-144 engines, succeeded in getting the NK-144–22 accepted for the Tu-22M because it was available in the prototype timescale and was installationally interchangeable with the redesigned NK-22 intended for series aircraft.

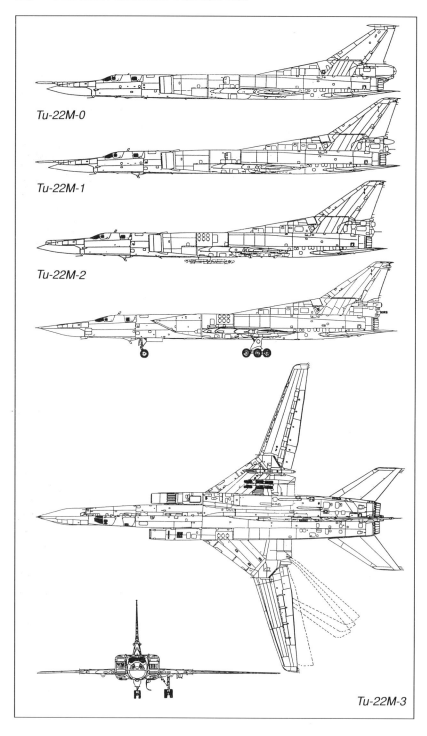

Tu-22M-0

Tu-22M-1

Tu-22M-2

Tu-22M-3

steel central wing box permitting the outer wings to pivot to four settings: 20 deg/30 deg/50 deg/60 deg. The wings were driven by duplex hydraulic motors in separate systems, so that any failure merely halved the drive power. A fence was added across the tip of the fixed inboard wing, extending round the leading edge. Maximum dash Mach number at high altitude was set at 1.89. The structural load factor remained 2.5, and was unchanged in subsequent versions. The fixed inboard leading edge was swept at a constant 55 deg to the root, the trailing edge being set at 90 deg to the fuselage axis. The main wing structural box required four spars, almost parallel, with two subsidiary spars further back, No. 6 at 90 deg to the fuselage.

On the back of No. 4 spar was fitted the forward bearing for the main landing gear by Hydromash, based on the Tu-22 bogie but with two extra wheels added closer together at the rear and with the centre pair splayed apart on gear extension, to give better 'flotation' with three sets of ruts on soft ground or snow. The top of the main leg was curved sharply out towards the wingtip; thus, as the breaker strut pulled the gear inwards, the whole assembly rose into the wing with the bogie itself housed between bridging spars 4 and 5 inside the fuselage. The doors to the bay were hinged on each side of the bomb bay in line with the side of the fuselage, and arranged to close after gear extension, the leg having its own door. Tyre size was set at 1,030mm by 350mm (40.55in by 13.8in). A fixed version of the new landing gear was tested on a Tu-22.

This was the first time Tupolev OKB had been able to achieve the optimum transonic Area Rule distribution of cross-section with a neat inwards-retracting landing gear. Nevertheless, tunnel testing showed that for the best possible compliance with the rule long shallow fairings should be added under the wings, extending beyond the trailing edge. These showed in satellite imagery, making Western observers believe the landing gears were the of Tu-22 type.

As before, the steerable levered-suspension twin-wheel nose gear retracted backwards into a bay aft of the crew compartment, the front door being linked to the leg and the left and right bay doors remaining open on the

The 45-01 was rolled out at Kazan (the old GAZ numbers by this time falling into disuse) in early April 1969, most systems testing having been completed indoors before the spring thaw. By this time it had been decided that progress would be speeded by producing not only the 45-01 and the requisite airframes for static and fatigue testing but also eight further prototypes. All nine aircraft received the VVS designation Tu-22M-0, though all differed in detail.

The new engines were placed inside a modified rear fuselage, fed by long ducts from fully variable inlets extending to the forward fuselage. This arrangement promised about 15 per cent better pressure recovery at Mach 1.8. The inboard wing remained as on the Tu-22 aerodynamically, but structurally it was redesigned. At one-third of the semispan, the outer-wing pivots were located far back at almost 50 per cent chord. Each was a plain steel bearing assembly in the titanium/

Tu-22M-0. (N A Eastaway/RART)

ground. All three landing gears were fitted with anti-skid brakes. The track was 7.3m (23ft 11½in) and the wheel-base 13.67m (44ft 10¼in). Tyre sizes were (main) 1,030mm and (nose) 1,000mm by 280mm.

The new inlets were outboard of an almost vertical splitter plate, standing away from the fuselage with a gap nowhere less than 150mm (5.9in) to divert sluggish boundary-layer air above and below. It was unperforated but vertical boundary-layer slits were provided in the pivoted inner wall, which was positioned by multiple rams to vary the profile and throat area according to flight Mach number. The sharp upper and lower lips (the upper being slightly ahead of the lower) were swept back in plan at 60 deg. At high power at low airspeeds extra air was admitted through a large suck-in door in the outer wall just behind the wing leading edge.

Despite being derated to 20,000kg (44,090lb) thrust, the NK-144-22 engines were the most powerful in any military aircraft up to that time. They were installed horizontally but toed inwards to bring the nozzles close together, with the afterburner petals visible beyond the fuselage. Large doors under the engines gave access to almost all the powered accessories, but the engines were arranged to be removed on rails to the rear. The rear fuselage was stressed for quadruple SPRD a.t.o. rockets (not normally used). In contrast to the Tu-22, and to a small degree influenced by the Mirage IVA and TSR.2, the decision

was taken to dispense with defensive guns in the tail. Accordingly, the single braking parachute was installed in a large container immediately below the rudder, which also housed the REP countermeasures and formed the aft end of the fuselage.

The area-ruling did not waist the fuselage, so the inlet ducts ran straight in plan, though the inner wall curved outwards (the opposite of what might be expected) as it rose over the wing. The enlarged crew compartment remained pressurised at a maximum dP (pressure differential) of 0.59kg/sq cm (8.39lb/sq in), but it was completely redesigned for a first pilot (who was usually the aircraft commander) on the left, a co-pilot on his right, the navigator/bomb-aimer behind on the left, and a *radist* electronic-systems officer on the right. All four crew were seated in forward-facing upward-firing KT-1 seats, with operating limits of 250m (820ft) and at low level 130km/h (81mph), or 300km/h (186mph) for forced simultaneous ejection. Each crew-member had an individual gull-wing roof hatch with a window and anti-flash blinds. For medium/high-level bombing the navigator could use the optical facility of the OPB-15 sight, looking through a flat ventral window (later in a blister under his cockpit), but he was normally seated at radar and multifunction electronic displays. High on the right behind the pressure cabin was a compartment for the LAS-5M five-man liferaft.

Compared with the Tu-22, the wing

area was increased by 10sq m (108sq ft). The outer wings were unchanged in profile, thickness/chord ratio at the root varying from 12 per cent at 20 deg sweep to 6 per cent at 60 deg. At high weights there was pronounced inflight flexure. On the leading edge were arranged three sections of powered slat, while the trailing edge carried three sections of double-slotted flap each running out on two tracks and preceded by a hinged cove strip. On the underside six removable panels provided access to the tracks and ballscrew drives. The outboard sections served as flaperons, with differential operation for roll control, backed up by spoilers ahead of the flaps operating differentially in conjunction with differential tailplanes. With weight on the main landing gears all six spoilers, pre-armed on the approach, flicked open to serve as air-brakes/lift dumpers.

The tailplanes were larger than those of the Tu-22 but set at the same low level, with a reflex (upturned leading edge) aerofoil profile. Leading-edge sweep was set at 60 deg, the tips being broad and parallel with the longitudinal axis. Each tailplane was driven independently by twin irreversible jacks over the remarkable angular range of +25 deg/–40 deg. The vertical tail was not greatly modified, leading-edge sweep remaining 80 deg on the dorsal fin and 60 deg on the fin proper, though the powered rudder was pivoted at its extremities and at two inset hinges with the leading edge set back between them. A slim antenna

The area-ruled hinge/main-gear fairing on a Tu-22M-0, also showing the underside of the flaps and air-brakes. (N A Eastaway/RART)

fairing was added above the fin. As already noted, the turret was replaced by the parachute compartment. The underside of the nose was occupied by the RBP-series main radar. The upper half of the nose was covered in access hatches, with a light on each side to assist night-time use of the flight-refuelling probe at the upper centreline.

The first Tu-22M-0 was rolled out from the enormous manufacturing plant at Kazan (originally called GAZ No. 22, named for S P Gorbunov) in May 1969. It was first flown by a test crew comprising B I Veremey, V P Borisov and A Bessonov on 30 August 1969. It was immediately obvious that the new aircraft solved one of the principal shortcomings of its predecessor by having two pilots with a good forward and downward view.

This prototype was apparently not noticed on US spy-satellite imagery until July 1970. Its appearance was then proclaimed in Washington to pose a serious strategic threat to the United States. This claim – whether it was actually believed or not – was to be a sticking point in the SALT and START treaty negotiations. In fact, the Tu-22M remained an in-theatre aircraft, the new wings being used only to reduce field length and increase weapon load. No aircraft of this family has even approached half the strategic radius of the Tu-95 or B-52. The new bomber was given the ASCC reporting name 'Backfire', later

changed to 'Backfire-A'. Someone in the West invented the designation 'Tu-26', which persisted for many years.

The nine M-0 prototypes all differed, sometimes visibly. A major modification introduced on the ninth aircraft was to fit a remotely controlled tail barbette with two GSh-23 guns, displacing the braking parachute to a ventral compartment. This aircraft also had the production engine, the more fuel-efficient NK-22, with an unchanged takeoff rating of 22,000kg (48,500lb). After a most successful factory and LII test programme one prototype, bearing callsign 33, is preserved at Monino and another, tail number 156, is at the Kiev VVS Institute.

Span (20 deg) 31.6m (103ft 8in), (60 deg) 22.75m (74ft 7⅝in); length 41.42m (135ft 10¼); wing area (20 deg) 173sq m (1,862sq ft).
Weight loaded 121,000kg (266,755lb).
Maximum speed 1,530km/h (951mph, Mach 1.44); range (maximum bomb load) 4,140km (2,573 miles); service ceiling 13,000m (42,650ft).

Tu-22M-1 A pre-series of ten aircraft for service test followed, with minor changes. All were powered by the NK-22 engine. The most obvious modification was to omit the under-wing area-rule fairings, which meant covering the hinge of each main leg by a fairing of the same depth terminating at the trailing edge of the wing. The landing gears themselves were also modified so that the centre wheels

no longer splayed apart on gear extension, the rear pair remaining close together. Another major modification was that, to counter incessant growth in weight, the outer wing panels were extended in span. The slats were also extended, continuing to the tips, but the trailing-edge flaps were unchanged and thus terminated well inboard. The rudder balance was modified, and there were numerous visible changes in antennas. The operational equipment is described under the M-2, the first true production version, commonality with which was estimated at over 95 per cent.

The M-1s were tested for four years, several of them by the Aviatsiya VMF (naval airforce). No ASCC name was assigned to the M-1.

Span (20 deg) 34.28m (112ft 5⅝in), (60 deg) 25m (82ft 0¼in); length 41.5m (136ft 1⅞in); wing area (60 deg) 179sq m (1,927sq ft), (20 deg) 183.58sq m (1,976sq ft).
Weight loaded 122,000kg (268,959lb).
Maximum speed 1,660km/h (1,032mph); range (maximum bomb load) 5,000km (3,110 miles); service ceiling 13,000m (42,650ft).

Tu-22M-2 This was the first series version, developed as Article 45-02 at Moscow and Kazan from 1971 by a team led by Boris Levanovich. This was the first version produced in series to meet the requirements of both the VVS and A–VMF.

To improve pressure-recovery at all Mach numbers the inlets were given large aft-facing boundary-layer ejectors above and below. A TA-series gas-turbine APU (auxiliary power unit) was installed in the dorsal fin, with the inlet on the right and exhaust on the left. Improved paralleled AC and 27V 500A DC electrics were served by four engine-driven generators and two on the APU. Improvements were made to the 210kg/sq cm (2,987lb/sq in) hydraulics. The fuel system was augmented, a new model of removable probe being fitted above the nose, with internal plumbing.

The required electronics *Kompleks* was the largest designed up to that time, with thirteen subsystems in nearly eighty LRUs (line-replaceable units). For navigation it included an updated triplex inertial system, plus an A-322Z doppler and retaining the RSBN-2S Shoran and twin ARK-15 radio compasses with blade antennas above and below the cabin, and a new automatic flight-control system for a

A Tu-22M-1 during its flight development. (G F Petrov/RART)

smoother low-level ride, manoeuvre limiting and bad-weather landing. A PNA *Rubin* navigation and bombing radar was installed, with an autopilot link to dual RV-18 radar altimeters for sustained flight at 150m (492ft). In 1975, a small formation, led by VVS Commander Marshal P S Deinikin, practised auto terrain-following at an altitude of 40–60m (131–197ft).

Other sensors included a blister fairing ahead of the nose landing gear for the combined OPB-15 electro-optical (TV) and visual bomb-aiming sight. Further forward, immediately behind the radar, were fitted left/right air-data probes, while on the underside between them were twin retractable landing lights. Steering (taxi) lights were installed under the inlets, white beacons above and below the centre fuselage, and a red air-refuelling signalling light in the leading edge of the fin.

The braking parachute, of cruciform type, was housed under the rear fuselage, in a compartment with twin doors, to permit installation of the specially tailored UKU-9K-502 turret. Its armament comprised two GSh-23 twin-barrel cannon, each fed from a 600-round magazine behind the APU under the fin. The turret was controlled by a PRS-4 fire control using a TP-1 EO (TV) sight served by the gun-direction radar immediately beneath it. These and other electronic complexes were cooled by projecting ram inlets on each side. The dorsal fin incorporated three large communications antennas, but the fin-tip fairing was deleted.

A great effort was required to integrate the computers controlling the navigation and weapon-delivery system. Most of the task had already been done with the prolonged testing of the Tu-22M-1, and with the M-2 simulators and test rigs used to shorten the time. Equally challenging was perfecting the complex defensive avionics with eighteen antennas mainly in the inboard wing leading edge, above and below the fuselage and on the fin, using a C-VU-10-022 computer for energy management. The main passive radar-warning receivers were located on each side of the nose and centre fuselage, along the inner leading edge (six), to the front and rear near the bottom of the fin, facing aft inboard of the tailplanes and at the top of the fin (raised above the rudder to house four antennas). An LO-82 multi-facet IRWR (infra-red warning receiver) was added above and below the forward fuselage. Ejectors for 192 chaff/flare cartridges in strips of eight triplets were recessed along each side of the rear fuselage near the tailplane leading edge, firing up and down. An AFA-15 camera was installed aft of the nose landing gear.

The normal free-fall bomb load was 12 tonnes (26,455lb), with a potential maximum of 24 tonnes (52,910lb), carried partly internally and partly on four bolt-on external MBDZ-U9M racks under the inlet ducts, each with a nominal capacity of up to three triplets of FAB-250 (551lb) bombs or a single Kh-31 cruise missile weighing 655kg (1,444lb). Free-fall conventional loads could include two FAB-3000s, eight FAB-1500s, forty-two FAB-500s (twenty-four external), or sixty-nine FAB-250s (thirty-six external).

Thanks to the sweep angles of the spars, the internal bay comprised a shallow front section with depth limited by the wing, normally occupied only by the recessed nose of a Kh-22 cruise missile, and deep centre and rear sections, each having left/right doors hinged in the conventional manner. The main bay provided an almost square cross-section, limited in width by the main-gear bays, large enough for all Soviet N (nuclear) or TN (thermonuclear) bombs or (in the later M-3 version) a rotary launcher for six cruise missiles of new types as described later. An alternative (A-VMF) option is three Kh-22M cruise missiles, one recessed on the centreline, the bomb bay having special doors with removable panels to fit round the missile, and the others on D2M (AERT-150) pylons just outboard of the main landing gears, each pylon housing the missile conditioning system, and the aircraft having the missile guidance transmitter.

Flight development of the M-2 began in April 1972. Again all problems were overcome reasonably quickly, and the new 'rocket carrier' was certified for combat duty with both the DA and the A-VMF. Service testing was done by the 185th Guards Regiment of the 13th TBAD (heavy bomber division), commanded by Col P S Deinikin, a future chief of the air staff. Total production at Kazan was about 220, split equally between the Air Armies and the A-VMF. All crews were trained at Tambov, using a fleet of Tu-134UBLs, but the regiment was initially based at Ryazan.

During production a succession of modifications were incorporated. The

A late series Tu-22M-2 with external bomb racks. (G F Petrov/RART)

inlet ducts were redesigned with the auxiliary suck-in side door replaced by nine smaller doors arranged in a 3 by 3 pattern, the tail radar was changed (the radome becoming blunt to reduce signal distortion), the avionics were updated, and the air-refuelling probe was removed and the socket faired over to comply with the SALT-2 treaty. A small force operated over Afghanistan from Poltava and Maryi (Turkmenistan) from December 1987, using free-fall bombs of various calibres against targets usually located only by geographic position. The ASCC name was 'Backfire-B'.

Engines, dimensions and weights, as M-2 except length 41.46m (136ft 2⅜in).
Maximum speed 1,800km/h (1,118mph, Mach 1.695); combat radius (maximum missile/bomb load) 2,200km (1,367 miles); service ceiling as M-1.

Tu-22M-3 Throughout the 1970s the engine KB of N D Kuznetsov had been developing an engine of unchanged frame size (thus fitting the Tu-22M engine bay) but with increased airflow and higher pressure ratio for greater power and improved fuel economy. This engine, the NK-25, was an augmented turbofan, with a maximum dry thrust of 19,000kg (41,887lb) and a maximum augmented thrust of 25,000kg (55,115lb), maintaining the Tu-22M's record of having the most powerful supersonic engine in service in the world. The NK-25 was flight-tested under a Tu-142 testbed, and then in a modified Tu-22M called the 22M-2Ye (the fastest aircraft of Tupolev design ever to fly), and qualified in 1976.

In 1980 the Tupolev OKB at Kazan accordingly planned a range of improvements in a new version, 45-03, powered by this impressive engine. Its power enabled the fuel capacity to be slightly increased, whilst reducing field length, raising combat ceiling and significantly improving engine-out performance at maximum weight.

As the engine inlet duct had to be modified to handle increased airflows, the inlets were redesigned for higher efficiency at peak Mach number. Helped by using Tu-144 testbeds, the inlets were changed to the horizontal wedge type, with a variable upper wall and hinged lower lip. The centre fuselage was structurally refined, eliminating the original auxiliary-inlet frame and moving the group of nine smaller auxiliary inlets further aft, where their induced airflow is slightly greater. Secondary airflows were modified, and the larger nozzles required some reprofiling of the rear fuselage. Another change was to redesign the fence at the tip of each inboard wing to have reduced chord above the wing and the same chord below. The electric generating and supply system was completely redesigned. Four parallel AC circuits were served by brushless generators driven at constant speed by hydraulic transmissions. Other subsystems were improved, and perfected on many rigs and, in the case of the navigation complex, in a Tu-104LL.

A basic change was the customer demand for operations to be undertaken in any weather at a height of 50–60m (164–192ft). As a result the outer wings were modified to oversweep to 65 deg. Another result, combined with severe maritime roles, was to fit an advanced multimode main radar with DBS (doppler beam sharpening) in an extended and slightly upturned nose. The socket for an air-refuelling probe was no longer fitted, but the associated lights and plumbing were retained as a prudent insurance against future need. The cockpits were improved in many ways and the

A Tu-22M-3 landing after a training flight. (Andreev via Petrov/RART)

seats changed to the KT-1M model, the minimum height being reduced to 60m (197ft). One aircraft tested K-36DM seats, specified for the Tu-160, but funds were not available for fitting this to the Tu-22M force. Fuel capacity was again increased, the resulting higher tyre pressure demanding Class I runways only. The new UK-9A-802 defensive turret was installed, with a single GSh-23M with superimposed barrels and case ejection chutes.

Though the Tu-22M3 could be equipped for internal and external free-fall bombs, it is officially referred to as a rocket launcher, qualified on five missile types. These are the Kh-22, weighing 6,800kg (14,991lb); the RKV-500B (about 2,500kg, 5,511lb); the Kh-15P (about 3,000kg, 6,614lb); Kh-31A/-31P (605/655kg, 1,334/1,444lb); and Kh-35 (610kg, 1,345lb). Prolonged Service testing by the 185th Guards Regiment was completed in 1986, and several early deliveries joined the M2 in operations over Afghanistan in October 1988. About 120 were delivered, effectively ending in December 1992. The ASCC name is 'Backfire-C'.

Tu-22MR No details have yet become available of this multi-sensor reconnaissance version. A modest number have been produced, starting in 1989, by modifying existing aircraft.

As noted earlier, all versions were said by the US Department of Defense to pose a direct threat to the USA, and were thus to justify major expenditure on counter systems. The unrefuelled combat radius was said to be '3,000 nautical miles', or 5,470km, an exaggeration of 130 per cent. The SALT-2 treaty demanded the removal of flight-refuelling probes and a ceiling on the production rate at thirty annually. Alarm is now expressed in Washington at the CIS wish to find export customers for new and used versions. Not least of Moscow's troubles was that, like the first Tu-160 unit, many of these aircraft have been taken over at Priluki by the government of Ukraine.

In 1994 the Russian VVS had 100 Tu-22M aircraft (virtually all M-3s) in service, and the naval air force had 165, of which twenty were in Ukraine.

Span (20 deg) 34.28m as before, (65 deg) 23.3m (73ft 2in); length 42.46m (139ft 3⅜in); height 11.05m (36ft 3in); wing area (20 deg) 183.58sq m as before, (65 deg) 175.8sq m (1,892sq ft).
Weights empty equipped 54,000kg (119,050lb); maximum fuel 53,550kg (118,060lb); maximum weapons 24,000kg (52,910lb); maximum take-off 124,000kg (273,370lb), (with a.t.o. rockets) 126,400kg (278,660lb); landing weight (normal limit) 78,000kg, (maximum) 88,000kg (194,000lb).
Performance maximum speed (at sea level) 1,050km/h (652mph), (at 11,000m, 36,090ft) 1,900km/h (1,181mph, Mach 1.79), (at light weight) 2,090km/h (1,300mph, Mach 1.97); cruising speed (at 11,000m) 900km/h (559mph); mission radius (all at minimum altitude, maxi-

mum bomb load) 1,500km (932 miles), (high altitude, maximum bombload or three Kh-22 missiles) 2,000km (1,242 miles), (internal bomb load only) 2,200km (1,367 miles); service ceiling (maximum weight) 13,300m (43,650ft), (lighter weight, speed not under Mach 1.3) 14,000m (45,930ft); take-off speed 370km/h (230mph), take-off run 2,050m (6,730ft) (with a.t.o. rockets, 1,920m, 6,300ft); landing speed 285km/h (177mph), landing run 1,450m (4,760ft) (1,200m, 3,940ft, with chute).

148

This advanced interceptor was designed to replace the Tu-128. Features included a design Mach number of 3.0, a variable-sweep wing, two Kolesov RD-36–41 or RD-36–51 engines (derived from the Dobrynin VD-19), a *Zaslon* multi-mode radar, a large internal bay for six or eight R-40 missiles, and range adequate for the design IA–PVO interception line to be set at a distance of 1,200km (746 miles). Estimated weight was 25 per cent greater than that of the Tu-128M. The project was launched in 1962, but PVO commander A L Kardomtsev rejected the 148 in favour of the MiG-25P followed by the MiG-31, and funding was stopped in 1969.

Tu-154

In the early 1990s this rear-engined three-jet transport was – say ANTK Tupolev – 'providing about half of all the passenger conveyance in the USSR' (what was meant was half the passenger-kilometres generated by Aeroflot). Even since 1991 the fragmented Republics with their own airlines have hardly disposed of any Tu-154s.

The basic design was a natural extrapolation beyond the Tu-134, resulting in an aircraft resembling a Boeing 727 but with a bigger airframe, lower footprint pressure and considerably more powerful engines in order to meet the severe demands for operation from short unpaved runways. The GVF (civil air fleet) requirement was issued in 1965. It was demanding,

calling for an aircraft to replace the An-10, Il-18 and Tu-104, using similar airfields and repair facilities, but cruising at 900km/h (560mph) for up to 6,000km (3,730 miles). Tupolev appointed S M Yeger chief designer, but he was nearing retirement and was replaced in 1970 by Dmitri S Markov.

A logical basis was to scale up the proven Tu-134, but the available engines virtually dictated that three engines would be necessary. In turn, this suggested triplexed flight-control and auxiliary power systems, giving duplex power after any single failure, and making it possible to eliminate the need for emergency manual flight control. CAHI conducted extensive testing to establish the optimum high-lift system, aiming at a lift coefficient of at least 2.7.

The wing was designed with a conventional three-spar structure with machined skins. Leading-edge sweep settled at 40 deg to a structural break

at 90 degrees to the rear spar at the main landing gear, and thence 38 deg to the tip. Mean sweep along the ¼-chord line was 35 deg, with zero dihedral and incidence +2.5 deg, decreasing to zero at the tip. Along the outer 80 per cent of each leading edge were arranged five sections of electrically powered slat, while along the trailing edge were fitted the largest possible triple-slotted flaps, driven hydraulically, the section inboard of the landing gear being at 90 deg to the fuselage. From the outer flap to the tip were fully-powered ailerons. Shallow fences were added at the flap/aileron junction and outboard of the landing-gear fairings, the inner fences terminating ahead of the flaps. Outboard of the inner fence were arranged three sections of hydraulically powered spoiler, the arrangement finally adopted being to use the outer spoiler to increase roll authority at low indicated airspeeds and the inner two as

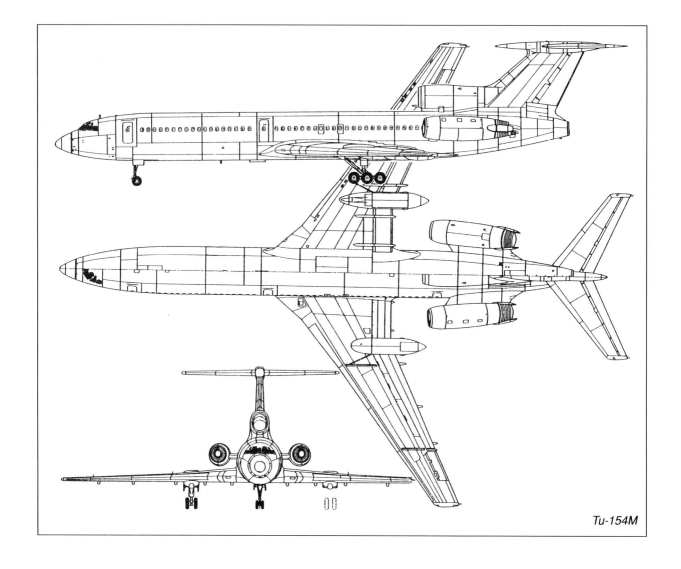

Tu-154M

A Tu-154B-1 of Tarom Romanian Airlines landing. (RART)

symmetric air-brakes. Inboard of the landing gear was a fourth spoiler used as an air-brake and, after landing, as a lift dumper.

The fuselage was structurally almost an enlarged version of the Tu-134, retaining the circular cross-section but enlarged to a diameter of 3.8m (149.6in), and with the crack-stop side panels perforated by windows changed to rounded rectangular shape. Maximum pressure differential was set at 0.63kg/sq cm (9lb/sq in). Cabin width and length were respectively 3.58m (141in) and 27.45m (90ft), with two doors ahead of the wing on the left and two service doors opposite, all plug-type 1,650mm by 762mm (65in by 30in). Despite the adoption of circular frames, instead of the pear-shape favoured in the West, the diameter was sufficient for shallow underfloor holds for cargo and baggage, two ahead of the wing providing 38cu m (1,342cu ft) and a small unpressurised compartment behind the wing. The design passenger accommodation was for 128–167, all in 3+3 seat rows.

The tail was again almost a scale of the Tu-134, though sweep on the tailplane was increased to 45 deg on the leading edge and 40 deg at ¼-chord. Pivoted to the top of the fin, with the fairing bullet extended forwards into a long vhf antenna, the tailplane was from the outset the primary control in pitch, the elevators adding camber. Span was 13.4m (43ft 11½in), dihedral zero and range of movement −20 deg/+ 25 deg. The fin

grew out of the inlet fairing for the tail engine, again with 45 deg sweep on the leading edge and carrying a one-piece rudder. As noted, the flight controls had triplexed drive without manual reversion, and their skins were honeycomb sandwich retaining a near-perfect exterior profile.

Landing gear again followed Tu-134 geometry, though to spread the load on unpaved surfaces each main unit had three fixed axles carrying pairs of wheels, all the same distance apart, with 930mm by 305mm tyres inflated to 8kg/sq cm (114lb/sq in). Off the ground the bogie beam tilted sharply nose-down, this rotation being continued on hydraulic retraction to the rear to stow the unit inverted in the typical Tu-style wing pod. The nose gear had twin wheels with hydraulic steering, carrying 800mm by 225mm tyres, again retracting to the rear, the leg shortening to fit in a small bay with four doors, the front pair remaining open on the ground.

Tupolev's long collaboration with N D Kuznetsov resulted in his NK-8-2 turbofan being selected, at a take-off rating of 9,500kg (20,945lb). The side engines were fitted with twin-clamshell reversers, while the centre engine was fed via an S-duct from a vertical-oval inlet above the fuselage and exhausted via a minimum-length jetpipe. The inter-spar wing box was sealed to form four integral tanks with a capacity of 41,140 litres (9,050gal), with single-point refuelling at 2,500 litres (550gal)/min. An additional group of four flexible cells could be

installed in the centre section to increase capacity to 46,825 litres (10,300gal).

Each engine drove a 40kVA alternator via a constant-speed drive to generate current at 200/115V at 400Hz, with converters supplying 27V DC. Electrothermal elements de-iced the slats and windscreens, all other de-icing being by bleed air. The three hydraulic systems operated at 210kg/sq cm (2,987lb/sq in). Above the centre-engine jetpipe, and exhausting into it, was installed an APU (auxiliary power unit), driving a fourth alternator and a 12kW starter/generator. Avionics included all the expected ICAO-standard radios and navaids, with VOR/DME, a moving-map ground-position indicator and Cat, II ILS. Standard crew comprised two pilots and an engineer, with up to three jump seats for supernumeraries.

The prototype, 85000, was first flown by Yu V Sukhanov on 3 October 1968. Bearing in mind the size and complexity of the aircraft, development was trouble-free, and in 1970 very large orders were placed at the Kuibyshev factory. As early as August of that year six development aircraft were delivered, and in July 1971 scheduled cargo services were started, together with occasional passenger services to Tbilisi. Scheduled passenger services began to Mineralnye Vody on 9 February 1972, followed by the first international service (to Prague) on 1 August 1972. The ASCC reporting name was 'Careless'.

Span 37.55m (123ft 2½in); length 47.9m (157ft 1¼in); wing area 201.45sq m (2,168sq ft).

Weight empty 43,500kg (95,900lb); fuel 33,150kg (73,083lb); maximum payload 16,000kg (35,275lb); maximum take-off weight 90,000kg (198,416lb); maximum landing weight 80,000kg (176,367lb).

Maximum speed (10,000m) 955km/h (593mph), limited to 575km/h (357mph) IAS or Mach 0.9; cruising speed 900km/h (559mph); service ceiling 10,500m (34,450ft); range (maximum payload) 2,520km (1,566 miles); take-off run 1,140m (3,740ft); balanced field length 2,100m (6,890ft); landing speed 235km/h (146mph).

Tu-154A

This introduced the NK-8-2U engine, with a take-off rating of 10,500kg (23,150lb). It also added a centre-section tank with a capacity of 8,250 litres (1,815gal), but with the contents transferable to the aircraft system only on the ground. Another change was to interconnect the flaps and tailplane to maintain trim, with a pilot over-ride should the movement exceed the desired +3 deg on flap extension. Other upgrades included two extra emergency exits, a separate jetpipe for the APU, and refinements to systems and avionics. Production switched to this version in 1973, all aircraft for Aeroflot being numbered in the 85000 series.

Dimensions unchanged.

Weight empty 49,200kg (171,960lb) (108,466lb); fuel 39,750kg (87,634lb); maximum payload 18,000kg (39,683lb); maximum take-off weight 94,000kg (207,231lb); maximum landing weight (normal) 78,000kg (emergency) 94,000kg (207,231lb)

Maximum speed (10,000m) 990km/h (615mph); cruising speed 910km/h (565mph); range (16,000kg payload) 3,200km (1,988 miles); balanced field length 2,200m (7,217ft).

Tu-154B

In 1977 this became the standard series version, with many refinements. Lateral control was improved by increasing spoiler authority, the ailerons being reduced in span, stopping 2m (80in) short of the wingtips (seldom shown on Western drawings). The flight-control system introduced French Thomson-CSF/SFIM avionics and autopilot, and the interlinked ILS receiver was qualified for Cat, II. Weights were increased, and the cabin was extended aft, with an emergency door on each side, to increase maximum seating to 180. Maximum pressure differential was reduced to 5.9kg/sq cm (8.4lb/sq in).

In 1980 production switched to the Tu-154B-2, which was to become

the most numerous version. In this the centre-section tank was made part of the aircraft fuel system, some Tu-154B aircraft being thus modified. A new *Groza* (Storm) weather radar was installed, other avionics being again upgraded, and to reduce tyre scrubbing in crosswinds the front bogie axles were pivoted to be aligned with the landing runway. In 1982 the all-cargo Tu-154S was announced, with a maximum payload of 20,000kg (44,090lb). It has a powered cargo door 2.8m by 1.87m (110.25in by 73.5in) and a tough floor equipped with ball mats and rollers for handling up to nine ISO 88in by 108in pallets.

Dimensions and engines unchanged.

Weight empty 50,775kg (111,398lb); fuel 39,750kg (87,633lb); maximum payload 19,000kg (41,887lb); maximum take-off weight 96,000kg (211,640lb), later 98,000kg (216,050lb); maximum landing weight 94,000kg (207,234lb).

Maximum speed (10,000m) 1,000km/h (621mph); cruising speed 915km/h (569mph); service ceiling 12,800m (41,995ft); range (maximum payload) 2,750km (1,710 miles); balanced field length, 2,200m (7,217ft).

Tu-154M

Originally announced as the Tu-164, this superseded the B-2 in production in 1984. Three years earlier a B-2 (85317) had been returned to Kuibyshev and its engines replaced by Soloviev D-30K turbofans, which in the series aircraft are D-30KU-154-II. These are basically considerably more powerful than the NK-8-2, but for this application were derated to 10,600kg (23,370lb) up to ISA+15 deg C, giving long life and expected high reliability. In addition, the specific fuel consumption was reduced (in typical cruise from 0.76lb/h/lb to 0.70), giving greater

range for the same fuel burn. The side engines were fitted with twin-bucket type reversers, as on the Il-62.

Many other refinements were added. The APU was replaced by the TA-92 pattern, and it was moved to below the mid-fuselage. The spoilers were enlarged but the slats made smaller, triple INS (inertial navigation systems) were fitted, enclosed overhead hand-baggage bins fitted along the cabin and all furnishing materials made non-inflammable. The Tu-154M was certificated for operations in ambient temperatures from +50 deg to –50 deg C, still in Cat II minima.

The Tu-154 was exported to seventeen foreign airlines, recipient countries including Afghanistan, Bulgaria, China, Cuba, Czechoslovakia, Egypt, East Germany, Hungary, Iraq, North Korea, Latvia, Mongolia, Poland, Romania and Syria. Production continued to 1994, by which time (according to marketing firm Aviacor) the total number of all versions was about 1,015. Of these, eleven were bought by government organisations, thirty-two by the VVS (Soviet Air Force) and 146 were exported. The total includes 537 Tu–154B and 292 Tu–154M versions. In 1995 marketable M–versions were being upgraded at customer request with an autopilot/navaid *Kompleks* called *Jasmin*, cleared to Cat II autolanding, British 32–place liferafts and optional extras Omega, Orion, TCAS and a linking block called ABSU–154.

Span 37.55m as before; length 48m (157ft 5¼in); wing area 202.45sq m (2,179sq ft).

Weight empty 54,800kg (120,811lb); fuel 39,750kg (87,633lb); maximum payload

The Tu-154 prototype landing. (Jean Alexander/RART)

18,000kg (39,683lb); maximum take-off weight 100,000kg (220,460lb); maximum landing weight 80,000kg (176,370lb).

Maximum speed (10,000m) 990km/h (615mph); cruising speed 850km/h (528mph); service ceiling 12,800m (41,995ft); range (maximum payload) 3,700km (2,300 miles); balanced field length 2,500m (8,202ft).

Tu-154M2 This further-modernised version was originally projected to fly in 1995, and remains an active project. Apart from a careful structural audit, to clear the airframe to a further 15,000+ flight cycles, it would have the three engines replaced by two Aviadvigatel (originally Soloviev) PS-90A. These high-bypass-ratio turbo-fans would each be rated at 16,000kg (35,275lb) take-off thrust. Their main advantages would be a dramatic reduction in noise and fuel burn, the fuel per passenger being only 62 per cent as high as that of the Tu-154M. A further upgrade would be an area-navigation system, and possibly satellite communications.

Tu 154S This designation has been repeated for a projected version to gain experience with cryogenic fuel (*see* 155, next). Whilst retaining 10t of kerosene, 13t (28,660lb) of LNG would be housed in a lagged tank filling the rear fuselage behind a shortened cabin, supplying three NK–89 engines each rated at 10.5t (23,148lb). Up to 130 passengers or 14.5t cargo would be carried 2,700km (1,677 miles).

155

In 1990 Tupolev announced that a team led by Vladimir Andreyev was developing an 'ecologically clean'

passenger aircraft designated Tu-156 (this number had previously been assigned to a totally different aircraft, *see* below). To support this project, Tu-154 No 85035 was bailed back from Aeroflot and converted to burn cryogenic (very low temperature) liquefied-gas fuels in an experimental NK-88 engine mounted on the right side. This engine, another version of the NK-8 turbofan family, is part of a strategic research programme into the use of cryogenic fuels by the Samara (Trud) enterprise, which took over the Kuznetsov design bureau. Without necessarily changing the ratings, the engine was modified with a fuel and combustion system able to accept fuels at temperatures as low as -255 deg C.

The three principal fuels are LNG (liquefied natural gas), liquid CH_4 (methane) and, coldest of all, LH_2 (hydrogen). Apart from careful insulation and fire-protection systems throughout, the Tu-155 had its passenger cabin largely given over to the special fuel and feed system. Behind the flight deck are three seats for experimental technicians, followed by a cabin where two engineers control the cryogenic supply. Behind this, the front part of the cabin, with blue-tinted windows, contains eighteen high-pressure helium gas spheres along the ceiling to provide the pressure feed, together with ninety hydrogen cylinders on the floor. Under the floor are high-pressure nitrogen bottles for purging and fire prevention. Next is a cabin for six guests, followed by a buffer zone. Behind this is the hermetically sealed aft compartment housing the main cryogenic fuel tank, with a capacity of over 1 tonne (2,205lb), sufficient for over 3 hours. An emergency inflight drain mast was added above the fin.

The Tu-155 first flew in April 1988. Subsequently it visited the Hanover airshow, Bratislava and a Congress on LNG at Nice. Work continues, now aimed at the projected Tu–156 and Tu–204SPG. In mid-1995 all concerned were awaiting Government funds to convert three 154M transports to 156 standard and supply twelve NK–89 engines, which are similar to the NK–88 but tailored to LNG fuel. Each would have three NK–89 engines and an LNG tank behind the cabin. As noted in the introduction, ANTK Tupolev is also collaborating with Daimler-Benz Aerospace on a proposed cryogenic test aircraft based on the A310.

156

In 1976 urgent studies were put in hand to find a carrier aircraft for the new DRLO radar *Shmel* (bumblebee). A photograph of a model in the story of the Tu-126 shows the installation in that aircraft, but it was concluded that a newer alternative had to be found. The first possible answer studied was to use newly built Tu-142M airframes, but these clearly would have been too cramped internally. Accordingly, OKB-156 quickly schemed an entirely new aircraft, the 156.

This had a configuration similar to the Boeing AWACS. The radar would have been carried on twin narrow pylons, instead of a single wider pylon on the centreline as in the Tu-126. Careful design made it possible to reduce the overall size of the aircraft considerably, the cabin width being 3.8m (150in), little more than that of the Tu-95, and overall dimensions

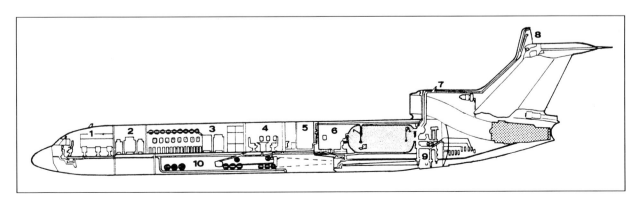

Internal arrangement of Tu-155; 1, technicians; 2, control engineers; 3, hydrogen and helium tanks; 4, guest cabin; 5, buffer zone; 6, hermetically-sealed fuel cabin; 7, auxiliary drain/vent; 8, main drain and vent; 9, main control complex; 10, nitrogen bottles.

The Tu-155 on show at Zhukovskii in 1993. (RART)

smaller than the Tu-126. The four Soloviev D-30KP turbofans, each of 12,000kg (26,455lb) thrust, would have been hung in widely spaced underwing pods with reversers. The photograph of a model does not show that there would have been three four-wheel main landing gears, one being on the centreline. Consideration was given to installing a tail turret.

Unfortunately for Tupolev, it was concluded that the need was too urgent for a completely new design, which was estimated to take seven years to get into PVO service. Attention therefore focussed on modifications of existing aircraft, Tupolev offering the Tu-154. This was soon decreed to be inadequate, because even when re-engined with two D-18 turbofans the range and endurance was too short. The obvious choice proved to be the Ilyushin Il-76.

Novozhilov declined, and the job went to the eager Beriev, who produced the A-50. Data for 156:

Span 45.8m (150ft 3in); length (no refuelling probe) 52.5m (172ft 3in); wing area 307sq m (3,305sq ft).

Weight loaded 182,000kg (401,235lb).

Maximum speed, restricted to 850km/h (528mph); cruising speed 720km/h (447mph); range 5,200–7,500km (3,231–4,660 miles) depending on fuel capacity; unrefuelled endurance 8.1hr; operating height 10,000m (32,800ft).

Tu-156

In 1995 number 156 was repeated for two cryogenic-fuel aircraft derived from the Tu-154. The Tu-156, also called the 156S, would use almost the same airframe as the Tu-154S.

Kerosene capacity would be 11t (24,250lb) and the lagged tank in the rear fuslage would hold 13,100kg (28,880lb) of LNG. To gain service experience, it would carry cargo only. Engines would be three NK-89, of 10,500kg (23,150lb) thrust each. Operating weight would be 57,400kg (126,543lb), maximum take-off weight 100t and payload 14,600kg (32,190lb). Using a 2,500m runway, range with this payload would be 3,450km (2,144 miles). A prototype could fly in late 1996, but it would be another six years before LNG could be in widespread use.

A Tu-156M passenger version could follow soon after. Passenger capacity would be 135. The Tu-156M2, again for availability in about 2001, would carry 20t (44,090lb) of LNG in two large saddle tanks above the cabin, fed to two (not three)

Model of the original project 156. (Yefim Gordon)

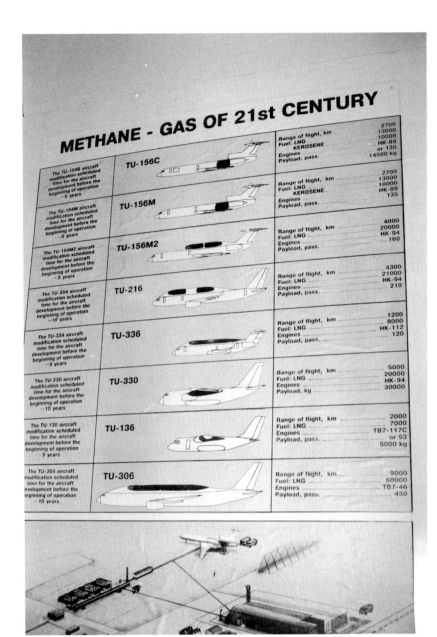

		METHANE - GAS OF 21st CENTURY		
The TU-154B aircraft modification scheduled time for the aircraft development before the beginning of operation – 6 years	TU-156C	Range of flight, km Fuel: LNG KEROSENE Engines Payload, pass.	2700 13000 10000 HK-89 or 130 14500 kg	
The TU-154M aircraft modification scheduled time for the aircraft development before the beginning of operation – 6 years	TU-156M	Range of flight, km Fuel: LNG KEROSENE Engines Payload, pass.	2700 13000 10000 HK-89 135	
The TU-154M2 aircraft modification scheduled time for the aircraft development before the beginning of operation – 8 years	TU-156M2	Range of flight, km Fuel: LNG Engines Payload, pass.	4000 20000 HK-94 160	
The TU-204 aircraft modification scheduled time for the aircraft development before the beginning of operation – 10 years	TU-216	Range of flight, km Fuel: LNG Engines Payload, pass.	4300 21000 HK-94 210	
The TU-334 aircraft modification scheduled time for the aircraft development before the beginning of operation – 9 years	TU-336	Range of flight, km Fuel: LNG Engines Payload, pass.	1200 8000 HK-112 120	
The TU-330 aircraft modification scheduled time for the aircraft development before the beginning of operation – 10 years	TU-330	Range of flight, km Fuel: LNG Engines Payload, kg	5000 20000 HK-94 30000	
The TU-130 aircraft modification scheduled time for the aircraft development before the beginning of operation – 9 years	TU-136	Range of flight, km Fuel: LNG Engines Payload, pass.	2000 7000 TB7-117C or 53 5000 kg	
The TU-304 aircraft modification scheduled time for the aircraft development before the beginning of operation – 10 years	TU-306	Range of flight, km Fuel: LNG Engines Payload, pass.	9000 60000 TB7-46 450	

Wall chart showing Tupolev's cryogenic fuel (LNG, methane) projects. (Mike Vines)

NK-94 turbofans each rated at 18t (39,683lb). It would carry 160 passengers 4,000km (2,486 miles).

70, Tu-160

The heaviest and most powerful combat aircraft of all time, this strategic bomber was built to a programme which began in 1967 when DA (long-range aviation) Gen-Col V V Reshetnikov studied the Sukhoi T-4MS (so-called '200') and Myasishchev M-20. VVS chief P S Kutakhov assigned

Sukhoi smaller aircraft, but the excellence of the M-20 led to its adoption, after modification as the M-18, with a horizontal tail instead of a canard. With so big a project CAHI's top men, G S Byushgens and G P Svishchyev, led the aerodynamic backing. It was finally decided only Tupolev was big enough to tackle the job.

In January 1975 design went ahead at Tupolev as Aeroplane K, with the out-of-sequence *Izdelye* number 70. The Service designation, Tu-160, is what might have been expected as the OKB number. Though VVS Col Evgeni Vlasov called the Tu-160 'an expensive countermeasure to the B-1',

in fact it would probably have gone ahead even if the USAF bomber had never existed.

Authorisation to create this bomber specified an optimised aircraft regardless of cost, and the Kremlin demand continued after cancellation of B-1 on 30 June 1977. Indeed it was exactly at this time that the OKB received an order for prototypes. Dr Alexei A Tupolev, then OKB titular head, appointed as team leader Valentin I Bliznyuk. Prototype assembly was to be in the Tupolev facility at LII Zhukovskii, with any subsequent series production to be at Kazan.

According to Vlasov 'Looked at from the outside, the Tu-160 and B-1B are similar. This is explained by the fact both aircraft have similar objectives: a long radius of action despite a heavy load of equipment and stores, the capability of deceiving enemy defence systems at low or high altitudes, and of having minimum radar, IR, optical and acoustic signatures. However, they bear resemblance from the outside only; the differences are fundamental'.

Perhaps the most basic single difference is that the Tu-160 has 79 per cent more installed engine power. Last of the engines to bear the initials of KB director Nikolai Kuznetsov, the NK-32 is one of the most powerful military engines in history, with a take-off rating of 14,000kg (30,843lb) and maximum augmented thrust of 23,100kg (50,926lb). A second difference is that, though significantly larger than the USAF bomber, the Tu-160 has smaller radar cross-sections and lower aerodynamic drag. Dr Tupolev said 'We believe that this is the most aerodynamically efficient supersonic aircraft ever built'.

Col-Gen Boris F Korolkov commented that there is a marked difference between the radar cross-sections, even ignoring the B-1B's external carriage of missiles and the claimed much better performance of the Tu-160's EW (electronic-warfare) systems, though in fairness to the US aircraft it has got better. Finally, partly thanks to idealised variable-geometry inlets, the Tu-160 is faster at sea level and almost twice as fast at altitude.

The four engines are installed in paired nacelles under the enormous fixed inner wing. This has a leading edge which in plan view curves

The Tu-160 first prototype. (RART)

continuously from a sweep angle of 90 deg where it blends imperceptibly into the fuselage near the cockpit. Thanks to this acute sweep, the leading-edge radius is large, affording both space and good antenna sizes for powerful internal ECM (electronic counter-measures). Two giant beams link the pivots for the outer wings, 19.2m (63ft) apart, with the sweep angle selected by control-stick buttons offering 20 deg, 35 deg or 65 deg.

Each outer wing is straight-tapered from root to tip, with full-span four-section hydraulically driven slotted leading-edge flaps (essentially slats) and full-span four-section double-slotted trailing-edge flaps. Outboard of the trailing-edge flaps, but stopping 3.5m (11ft 6in) short of the wingtip, are powered ailerons. These are all-speed primary roll-control surfaces, but backed by spoilers and tailerons. With flaps extended, they droop 20 deg. After prolonged research the

decision was taken to retain the optimum engine installation, leaving nowhere for the wing trailing edge to penetrate at 65 deg sweep. The problem was solved by making the inboard trailing edge hinge progressively upwards to form a large vertical fence at maximum sweep.

The tail looks deceptively conventional. The fixed fin is actually very sharply swept, with large chord but little height. On it are mounted the swept and sharply tapered slab tailplanes and the one-piece unswept but sharply tapered rudder (there is no fin above the tailplanes). The tail-planes are more strictly tailerons, because they can operate together for pitch authority or in opposition for control in roll. Their span is 13.25m (43ft 5⅝in) and area 55.6sq m (598.5sq ft); leading-edge sweep is 44 deg and they are without dihedral or anhedral. All flight-control surfaces are driven by electro-hydraulic power units with

dual FBW (fly by wire) and standby mechanical signalling.

The fuselage cross-section is the minimum necessary for crew, fuel and payload, and is significantly less than that of the B-1B or Tu-22M. Drag and radar cross-section are further reduced by the acute angle of the conical nose, and by the use of special computer routines to achieve optimum shape and control machine tools in production. Of over forty-five antennas, only three project as blades or spikes. Apart from the previously mentioned hinged 'flap fence', no fences or vortex generators were needed anywhere.

Structurally, this aircraft broke new ground with Tupolev in its very extensive use of honeycomb sandwich skin and precision-controlled RAM (radar-absorbent material) covering. Each engine-pair is installed in a rectangular-section nacelle hung under the inboard wing, with nacelle width

Aircraft 01 landing at Engels in early 1993. (Andreev via Petrov/RART)

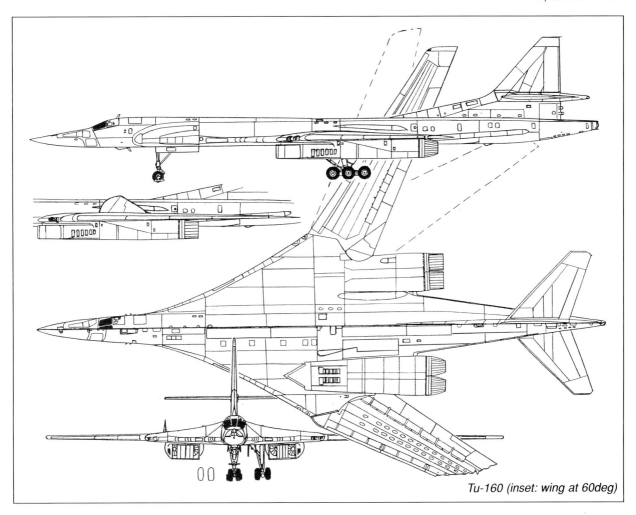

Tu-160 (inset: wing at 60deg)

increasing progressively from front to rear and the rear upper portions projecting above the wing as separate jet-pipe fairings. Despite experience of horizontal-wedge inlets with the Tu-144 and Tu-22M-3, for minimum radar cross-section the inlets were designed similar to those of the B–1A (not the fixed-geometry B–1B) with a vertical splitter leading back to an inner wall variable in profile and throat area. ANTK Tupolev believe this inlet achieves higher pressure recovery and lower drag from 0 to Mach 1.9 than any other inlet flying. The outer wall is also vertical, incorporating five inward suction-relief doors of progressively reduced height, matching the profile of the duct which at the throat is tall and narrow. For 'stealth' reasons, consideration is being given to modifying these doors with zigzag edges. As in Tupolev's previous supersonic bombers, the complete variable nozzle of each engine projects behind the nacelle, the engine being withdrawn on rails to the rear.

To minimise radar cross-section, the nose landing gear is installed behind, not under, the pressurised crew compartment. It carries landing and taxi lights, and spray/slush deflectors are mounted behind the twin wheels. Tyre size is 1,080mm by 400mm, the truck being hydraulically steerable through ±55 deg and retracted pneumo–hydraulically to the rear into a bay with left/right doors which remain open when the gear is extended. Each main landing gear has a six-wheel bogie, the tyre size being 1,260mm by 425mm. Unlike the Tu-22M, all three pairs of wheels are in line, because this aircraft operates from uncontaminated paved runways. The bogie supports a massive oleo leg made by Hydromash, installed inboard of the engine nacelle, which restricts the track to 5.4m (17ft 8⅝in). Wheelbase is 17.875m (58ft 7¾in). The main drag strut incorporates the retraction jack which pushes the gear to the rear. The leg pivots back and the lower portions also move inwards, while the bogie somersaults to lie inverted in a box which projects upwards to cause

'canoe' blisters between the engines and the fuselage. These blisters taper at the rear into shallow pipe/cable-loom fairings carried externally on each side of the rear-fuselage integral tanks to the closure bulkhead at the aft end.

A very large braking parachute can be streamed from the fairing between the horizontal and vertical tails. Internal fuel capacity (*see* data) is 50 per cent greater than that of the B-1B. Tankage is integral throughout, the main refuelling doors being in the front face of the inboard leading edge. Each aircraft was built with provision for a retractable inflight-refuelling probe above the nose, but in conformity with START treaties this capability remains unused. There are four autonomous hydraulic systems, each energised by an engine-driven pump to the high pressure of 280kg/sq cm (3,983lb/sq in).

The Tu-160 is entered via a pull-down ladder to the forward weapon bay, and walk forward through the main avionics compartment. As in the

The Tu-160 is the most powerful military aircraft of all time. This example has no number. (Andreev via Petrov/RART)

Tu-22M, the crew number four, but they have forward-facing K-36DM or K-36LM seats with zero height/zero airspeed capability, under jettisonable roof hatches. The large front windows have the same birdstrike strength as those of the Su-24. All windows have nuclear flash blinds, and the pilot-cockpit side windows are openable on the ground.

The FBW flight controls are moved by dual fighter-type sticks. Despite the fact that strategic bombers inevitably are restricted by severe flight limitations, so-called 'carefree' manoeuvres are possible under the protection of avionics. The navigator/bomb-aimer has the use of the *Obzor-K* multimode radar with terrain-following capability, and an OPB-15T optical sight plus video looking ahead from a Tu-22M-type ventral blister. The usual electronics suite includes vlf to uhf communications, dual ADF (auto direction finding), triple INS (inertial navigation system), Glonass satellite-navigation receiver and dual ILS (instrument landing system).

The *radist*, in this aircraft having no gun to manage, has become the full-time defensive-systems operator. According to Vlasov 'The EW systems fully meet their purpose. The inter-linked subsystems are easily removable and maintainable, and all modules are quickly replaceable. The navigator's panels are provided with eight computers, complex bomb and missile control systems, radar displays and long-range autonomous and non-emitting navigation systems linked with Glonass satellites. A computer generates the actual achieved flight route on a topographic chart. There are more than 100 computers aboard each Tu-160. The automatic air-

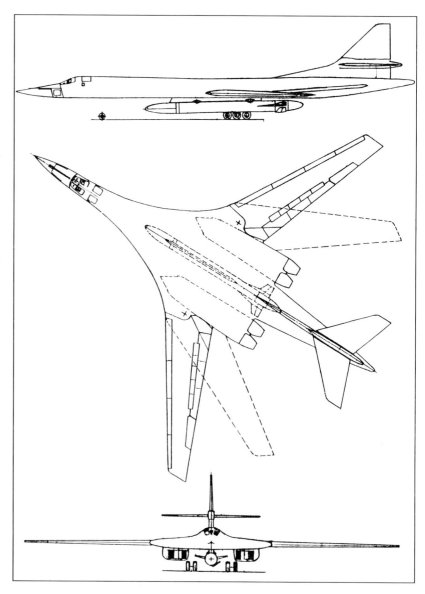

Tupolev and partners are marketing the Burlak *space launcher.*

combat and defence systems can interrogate many targets simultaneously'. The system is named *Baikal*, after the lake.

The passive radar and IR warning systems are new, and through computers serve what are claimed to be the most comprehensive ECM/IRCM active defence systems ever created. The main batteries of twenty chaff/flare dispensers are arranged in flush triplet groups surrounding the rear fuselage. The defensive-systems operator has a complete tactical display, and can at any time take over manual control in situations where an automatic response might betray the aircraft's presence. All weapons are carried internally in two enormous bays extending from the nose gear to the engine nozzles, one ahead of the wing bridge and the second behind. The bays are 1.92m (76in) wide and measure 11.28m (37ft) overall, with simple downward-hinged doors. Nominal capacity is 30,000kg (66,139lb), the maximum being 40,000kg (88,183lb). The 30 tonnes load was used in setting various speed/altitude/load records, but even this would seldom be approached in normal missions. The forward bay is equipped with an MKU-6-5U rotary launcher for six RK-55 or Kh-15P cruise missiles. The rear bay can have two rotary launchers each for twelve smaller RKV-500B cruise missiles. ANTK Tupolev state that there would be no problem in carrying loads even heavier than 40 tonnes, but there is no requirement. All development effort continues to be applied to mission radius and radar/IR cross-section.

The 70-01 test team was headed by L V Kozlov and B I Veremey. The first flight took place at Zhukovskii on 19 December 1981. The development fleet comprised ten aircraft, all slightly different. Production of 100 at Kazan was authorised in 1985. Aircraft No. 12 was presented for inspection by a US delegation on 2 August 1988, its unblemished white finish being admired. The third prototype, 70–03, bearing tail number 14, was used for 1,000km and 2,000km circuit records with a 25-tonne load at Mach 1.58 and 1.63, and for heights to 13,894m (45,584ft) with 30 tonnes. Instructor training was undertaken from 1985 at Dolon AB Siberia, with development and early production aircraft. Later crews were qualified with sixteen converted Aeroflot aircraft comprising Tu-134UBL pilot trainers and Tu-134BSh navigator/bomb-aimer trainers. The latter is fitted with Tu-160 radar and cleared for low terrain-following, with underwing pylons for practice bombs. This type serves at the Tambov school which also trains Tu-22M-3 crews. The simpler UBL serves alongside the first operational unit of Strategic Aviation, the 184th Regiment at Priluki, Ukraine. This unit was declared operational in April 1987, later qualifying with RK-55 missiles. By early 1988 the regiment kept aircraft on 24hr nuclear alert, but this has not been required since 1991. In 1995 the unit was still in Ukraine because construction work at Engels (their planned permanent home) was not yet completed. The regiment has two squadrons of ten available aircraft each, with others in reserve. As usual, nuclear weapons are under the control of a special sub-unit, made more difficult by the two nationalities involved.

One pre-production aircraft exhibited by the 184th in 1992 showed the original cockpit with a single multi-function display for each pilot, the rest of the area being mainly dial instruments. The front-line aircraft have a newer (classified) cockpit. A major problem is that no two aircraft are exactly alike, some even differing in inlets, quite apart from avionics standards. The 184th's CO, Col Valerii Gorgol, said 'The aircraft is complex and costly to maintain, and crews have complained of the cockpit environment, escape system and navigation system'! In the pre-1994 era there were also severe shortages, extending to such items as aircrew helmets and ground-crew ear defenders. Worse, in 1992 Ukraine laid claim to the aircraft. In early Ukraine had nineteen, compared with Russia's twelve (excluding non-operational aircraft) of the 1096th TBAP at Engels. Production was halted at about No. 38 in late 1992, though subsequently a little work has been resumed. In August 1994 seven of the development fleet were still at Zhukovskii, five being considered not worth bringing up to operational standard and being cannibalised. The ASCC reporting name is 'Blackjack'.

Tu-160P This was a 1979 project for a long-range heavy interceptor with missiles matched to over-the-horizon radar.

Tu-160 SK In 1993 this version was proposed as the carrier aircraft in the *Burlak* space transport system. The Tu-160SK *Burlak* (a barge-puller or carrier vehicle) would carry under its centreline the *Diana* rocket vehicle. This would be 22.5m (73ft 10in) long, and weigh 32,000kg (70,550lb), and be capable of placing a payload of up to 1,100kg (2,425lb) in low Earth orbit. The carrier aircraft, under control from an Il-76 command aircraft, would accelerate to 500m/s (about Mach 1.7), at a height of 13,500m (44,290ft) before releasing the space booster. The system involves ANTK im Tupoleva with booster builder MKB Raduga and with six other major companies.

Span (20 deg) 55.7m (182ft 9in), (35 deg) 50.7m (166ft 4in), (65 deg) 35.6m (116ft 9¾in); length 54.1m (177ft 6in); wing area 340sq m (3,660sq ft).

Empty weight c118,000kg (259,000lb); maximum fuel 171,000kg (376,984lb); loaded weight 275,000kg (606,260lb); maximum landing weight 165,000kg (363,762lb)

Maximum speed (SL) 1,250km/h (777mph, Mach 1.02), (13,000m) 2,200km/h (1,367mph, Mach 2.07), (light load) 2,500km/h (1,553mph, Mach 2.35); cruising speed 850km/h (528mph); maximum rate of climb 70m/s (13,780ft/min); service ceiling 15,000m (49,200ft); maximum unrefuelled combat range 12,300km (7,643 miles); endurance 15hr.

Tu-204

Launched in 1982 as the most important future Soviet passenger aircraft for Aeroflot trunk routes, replacing the Tu-154, the Tu-204 has seen the vast structure of that operator dismantled and replaced by a profusion of smaller airlines, many of whom are importing Western equipment. Despite this, the Tu-204 remains the biggest single programme by ANTK im Tupoleva, and is likely to support the production of several hundred aircraft in many versions.

It was designed by a team led by Lev Aronovich Lanovskii. Its size and shape bear a very close resemblance to the unbuilt Hawker Siddeley 134, and some similarity to the Boeing 757, though its detail engineering is entirely different. The basic airframe, largely advanced light alloys but 18 per

Tu-204

cent composites, is designed for a life of 60,000 hours, accumulated in 45,000 cycles over 20 years. At the outset it was decided to use a conventional tail, and hang the two engines under the wings. For this aircraft and the Il-96 the Soloviev engine KB (now Aviadvigatel, led by Yuri Reshetnikov) created the PS-90, an advanced turbofan of high bypass ratio with a take-off rating of 16,140kg (35,582lb) and competitive specific fuel consumption. Early in the programme it was decided to forge links with Western suppliers of engines, avionics and possibly other items. This has been done, and customers can be offered aircraft which not only vary in fuel capacity, fuselage length and passenger/cargo provisions but also in the proportion of Western content.

The CAHI-developed wing has a supercritical profile with a thickness/chord ratio of 14 per cent at the root and 9.5 per cent at the tip. Modern wings need less sweep, and the angle on the leading edge is 28 deg. Structurally the wing is a three-spar box with one-piece skins on each side 22m (72ft) long, sealed to form integral tanks. Much of this structure is aluminium/lithium alloy. The skins aft of the rear spar, and the large wing/body fairing, are glassfibre hon-

eycomb. On each leading edge are four sections of hydraulically-powered slat. On the trailing edge are double-slotted flaps, normally driven hydraulically to 18 deg for take-off and 37 deg for landing. The inboard section at 90 deg to the fuselage is in two parts, the outer having reduced landing angle to avoid the jet, each running on a single track and preceded by two air-brake/lift dumpers. The swept-back outer section runs on two tracks and is preceded by five sections of spoiler. Further out are the ailerons, which stop inboard of the washed-out tip with a winglet. Like all the flight-control surfaces, the ailerons are made in one piece of carbon-fibre composite, tabless and driven by a triplexed power unit with FBW (fly by wire) control.

The parallel section of fuselage is 3.8m (149.6in) wide and 4.1m (161.4in) high, the frame/stringer structure including aluminium/lithium and titanium alloys. There are main doors on each side at the extreme front and rear and Class-1 exits on both sides ahead of and just behind the wing. Length of the passenger cabin is 30.18m (99ft 0in), typical width over the interior trim is 3.57m (140.5in), and maximum pressure differential 0.6kg/sq cm (8.53lb/sq in). The powered tailplane, with a span of 15m (49ft 2½in), carries

camber-increasing elevators, and the fin torsion box forms a 2,820 litre (620gal) trim tank.

The engines, supplied by what is now Perm Motors company, are hung on pylons placed between the two inboard slat sections. They are housed in full-length cowls incorporating reversers with fan-duct blocker doors and all-round cascades. Normal fuel capacity is 30,000 litres (6,600gal), not including the fin. All tanks are filled at 3,000 litres (660gal)/min from a socket under the fuselage. Each main landing gear has a bogie with four 1,070mm by 390mm tyres, retracting inwards, the fuselage-bay doors closing after gear extension. The nose gear has a twin-wheel truck with 840mm by 290mm tyres, with FBW steering through ±70 deg, retracting forwards, the front pair of doors closing after gear extension. Track is 7.82m (25ft 8in) and wheelbase 17m (55ft 9¼in).

In the tailcone is a TA-12-60 APU (auxiliary power unit). This is used for main-engine starting, ground environmental control and inflight emergency electric power to drive one of the hydraulic pumps. The three hydraulic systems operate at 210kg/sq cm (2,987lb/sq in), and all three serve the flight controls, including spoilers and air-brake lift dumpers, and land-

The first Tu-204-120. (Rolls-Royce)

ing gear. Two systems operate the flaps, slats, wheel brakes and nose-wheel steering. Electric power is generated at 200/115V 400Hz, with converters for 27V DC. Liquid-oxygen capacity is 4,160 litres.

This is the first Russian aircraft certificated for ICAO Cat IIIa. The cockpit is equipped for two pilots, but Aeroflot also specified a flight engineer and seat for a supernumerary. The uncluttered modern cockpit has traditional yokes (adopted after flying a Tu-154 with sidesticks) and six large EFIS (electronic flight instrument system) multifunction displays, the standard suite being by Rockwell Collins (but see later). Flight management is by a triple autopilot with integrated flight director and thrust management. Avionics integration was done in partnership with Sextant Avionique and Honeywell. The expected communications, navigation, radar and blind-landing are supplemented by optional Glonass or Inmarsat satellite receivers, Acars (auto control and reporting), windshear detection and autoland, to be cleared later to Cat IIIb.

Many passenger arrangements are available, typical one-class being 214 in 3+3 rows. Another scheme is three 2+2 rows first class (total twelve) and 184 tourist. In each case a buffet/galley and toilet are behind the flight deck and a larger buffet/galley, two toilets and a service compartment at the rear. Underfloor holds are equipped for Russian 2AK-0,4 or 2AK-0,7 containers, five in the 11cu m (388.5cu ft) forward hold

(limit 3,625kg, 7,992lb) and seven in the 15.4cu m (544cu ft) rear hold (limit 5,075kg, 11,188lb).

Aeroflot announced a requirement for 500 Tu-204s, and the GVF initially ordered two static/fatigue test airframes and four flying prototypes powered by PS-90AT engines. The first, 64001, was rolled out from the Ulyanovsk plant in November 1988 and flown by OKB test pilot A I Talavkin on 2 January 1989. At that time Aeroflot had placed an initial order for eighty-eight aircraft. Following the break-up of the Soviet Union, this contract became void, and individual Republics gained authority to place their own orders. In 1992, following reorganisation of the former Soviet industry, the Ulyanovsk plant was privatised under the name Aviastar. In the same year ANTK Tupolev, Aviastar and FRIC (Fleming Russia Investment Corporation, a branch of a merchant bank) formed a Russian company called Bravia (British Russian Aviation) to market and support Russian aircraft worldwide, with varying degrees of Western content.

Span 42m (137ft 9½in); length 46.22m (151ft 7¼in); wing area 168.62sq m (1,815sq ft).
Weight empty 53,800kg (118,607lb); operating weight 53,800kg (128,527lb); fuel 24,000kg; payload 21,000kg; maximum take-off weight 93,500kg (206,125lb); maximum landing weight 86,000kg (189,594lb); zero-fuel weight 79,300kg (174,824lb).
Cruising speed 830km/h (516mph, Mach 0.78) at 10,650–12,200m (34,950–40,000ft) reached in 22–25min; range with maximum payload 2,900km (1,802 miles), with 196 passengers and maximum fuel 3,850km (2,392 miles); take-

off speed 269km/h (167mph), take-off run 1,230m; (4,035ft); landing speed 245km/h (152mph), landing run 850m (2,788ft)

Tu-204-100 Initial series aircraft, in production by Aviastar from 1992, centre-section tank for 7,000 litres (1,540gal), 1992 basic price (no Western content) US$38m. ARMAC (CIS) certification December 1994.

Dimensions, engines, unchanged.
Empty weight 54,600kg (120,370lb); fuel 29,500kg; payload 24,000kg; maximum take-off weight 99,500kg (219,356lb).
Range with maximum payload 5,300km (3,293 miles).

Tu-204-120 Sixth aircraft, completed with Rolls-Royce 535E4 engines flexibly rated at up to 19,550kg (43,100lb) thrust, first flown 14 August 1992.

Tu-204-122 As Dash-120 but fitted with Rockwell Collins avionics.

Tu-204C Cargo version of any previous sub-type with powered side door 3,405mm by 2,190mm (134in by 86in) and new sandwich floor with ball mat and rollers for pallet or container payload up to 25,000kg (55,115lb).

Tu-204-200 Additional fuel not only in centre section but also in adjacent underfloor hold, otherwise as Dash-100; deliveries from 1995. All Dash-200 family to be built by KAPO at Kazan.

Weight empty 56,500kg (124,559lb); operating weight 59,000kg (130,070lb); fuel 32,400kg (71,430lb); payload weight-limited to 25,200kg;

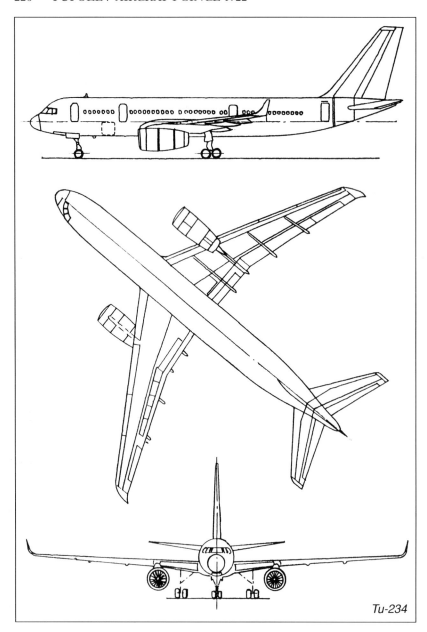

Tu-234

Tu-224 Westernised version developed jointly by ANTK Tupolev and AlliedSignal of the USA, for future series production by KAPO (Kazan Aviation Production Association) for foreign operators. Dimensionally unchanged, features include RR 535 engines, a Garrett GTCP-31-200/260 APU, Aviaagregat uprated landing gear with Rubin wheels fitted with Carbenix-4000 brakes and a choice of either a new type of NIIShP or Michelin tyres, Teplo-obmennik/ Allied Signal/Liebherr environmental-control system, Litton or Honeywell inertial navigation system, and a totally integrated avionic fit by Allied-Sig-nal or by ARIA, a joint venture by AlliedSignal and the Russian NIIAO (state avionics research institute). The programme is managed for ANTK Tupolev by Yuri Vorobjov, previously Chief Designer of the Tu-404.

Tu-234 Projected destretched version, with PS-90P engines each rated at 16,140kg (35,583lb), seating 99-166 passengers depending on range required. Probably also to be offered by Bravia with RR engines. Span reduced by curving winglets vertical.

Span 40.88m (134ft 1½in); length overall 40.2m (13ft 10⅝in).

Weight empty about 54t; operating weight about 58t; payload 16t; maximum take-off weight 84.8-103t (186,949-227,072lb) depending on range required; zero-fuel weight 76t (167,549lb).

Cruising speed 830-850km/h (516-528mph); range (maximum passengers) 9,250km (5,750 miles); take-off distance (hot/high) 2,050m (6,725ft); landing distance required 1,800m (5,900ft).

Tu-243

In 1995 Tupolev is marketing an upgraded version of the 143 as a mobile all-weather reconnaissance system. The air vehicle has been modified only in detail, but is offered as a package complete with SPU-243 launch vehicle, TZM-243 refueller, KRK-243 testing vehicle and POD-3D data-processing centre. As before, various optical, IR or TV sensors may be fitted, and missions may be flown repeatedly from any site.

Span 2.25m (7ft 4½in); length 8.29m (27ft 2½in).

Take-off weight (excluding boost rocket) 1,400kg (3,086lb).

Speed, programmable 850-940km/h (528-584mph); range 360km (224 miles); operating height 50-5,000m (164-16,400ft).

maximum take-off weight originally 107,900kg, later upgraded to 110,750kg (244, 158lb); maximum landing weight 89,500kg (197,310lb); zero-fuel weight 84,200kg (185,626lb).

Cruising speed 840km/h (522mph); range with maximum payload 6,330km (3,933 miles), with 196 passengers (payload 19,565kg, 43,133lb) and maximum fuel 7,500km (4,660 miles).

Tu-204-220 As Dash-200 but with RR 535E4 or 535F5 engines, marketed by Bravia with varying Western content. Specification close to previous.

Tu-204-22 As Dash-220 but with Rockwell Collins avionics.

Tu-204-230 Number reserved for possible version with PW2240 engines.

Tu-204-300 Destretched version, see Tu-234.

Tu-204-400 Number reserved for possible stretched version, length about 52m (170ft), passenger capcity up to 250.

Tu-214 Projected version burning cryogenic fuel, see Tu-216.

Tu-216 Projected version burning cryogenic fuel, with 21t (46,296lb) of LNG carried in front and rear lagged saddle tanks above the fuselage, fed to NK-94 engines. It would carry 210 passengers 4,300km (2,672 miles). Development is estimated to require up to 10 years.

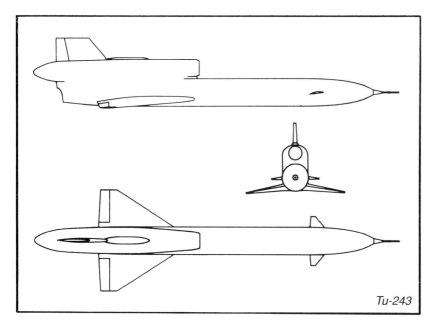

Tu-243

Tu-244

As the company with more experience of large supersonic aircraft than any other in the world, besides being the first to fly an SST (supersonic trans-port), it was natural that ANTK Tupolev should never have lost sight of the obvious need for a second-generation SST, when the airlines flying long sectors are ready to buy one. The CAHI, LII, CIAM and many other national organisations combined with Tupolev to fund large research pro-grammes, notably between 1979 and 1993, which included extensive flying of Tu-144 and other laboratory aircraft to investigate aerodynamics, propul-sion, structures, ecological factors (not only noise) and systems.

The results, certainly more exten-sive than anything done by a Western company, underpin the design of the project 244. A basic design objective was a lift/drag ratio of 15 at Mach 0.9 and at least 10 at Mach 2. The airframe would be a mixture of light alloys and composites, with only small amounts of titanium or steel.

Wing profile is broadly symmetric but with a slightly drooped fixed inboard leading edge swept at 75 deg (drawings show 80 deg) and a hinged outboard leading edge swept at 35 deg. The trailing edge would be made up of eight elevons on each side divided 2–1–5 by the engine nacelles. The individual engine installations would to some degree be simpler versions of those of the Tu-160. The engines themselves would be Samara aug-mented turbofans derived from the bomber's NK-321, with a take-off rating (with ideal nozzles) of 33,000kg (72,750lb).

All four landing gears would retract forwards. The three main units, each

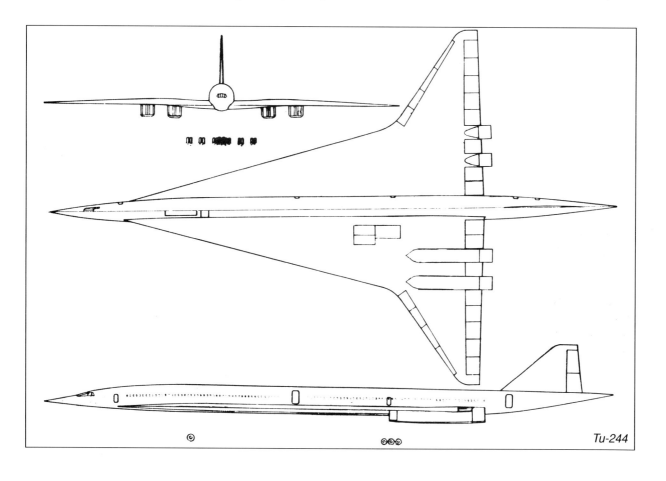

Tu-244

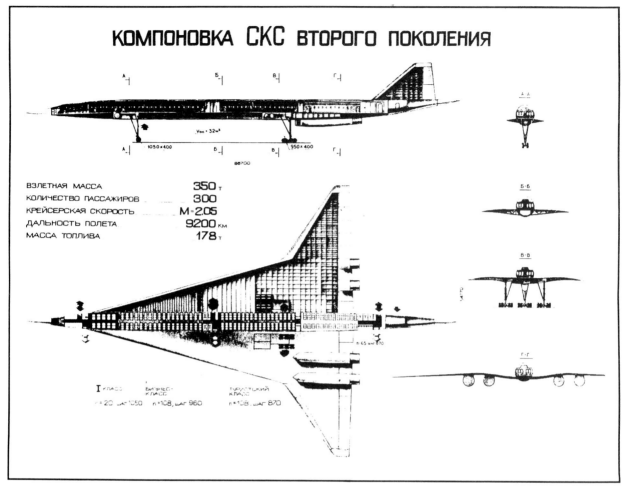

КОМПОНОВКА СКС ВТОРОГО ПОКОЛЕНИЯ

ВЗЛЕТНАЯ МАССА	350 т
КОЛИЧЕСТВО ПАССАЖИРОВ	300
КРЕЙСЕРСКАЯ СКОРОСТЬ	М=2.05
ДАЛЬНОСТЬ ПОЛЕТА	9200 км
МАССА ТОПЛИВА	178 т

Tu-244 project drawings.

The Tu-24Skh mock-up. (RART)

with twelve 950mm by 400mm tyres, would somersault to lie flat in the wing or an underfloor bay. The steerable nose gear, with twin tyres 1,051mm by 400mm, would shorten to fit a bay under the forward cabin. Like the elevons, the twin rudders would have FBW control and be powered by quad-redundant electro-hydraulic systems. Typical seating would be twenty 1st-class (1,050mm pitch), 108 business (960mm) and 108 tourist (870mm), but one-class accommodation would be for about 300. Initial certification would be to ICAO Cat IIIa. In the author's opinion it would be the height of folly for any group to try to build a new SST without the partnership of ANTK Tupolev.

Span 54.47m (178ft 8½in); length 88.7m (291ft); wing area 1,200sq m (12,917sq ft).
Weight empty 172,000kg (379,189lb); fuel 178,000kg (392,416lb); maximum take-off weight 350,000kg (771,605lb).
Cruising at 18,000–20,000m (59,055–65,600ft) 2,180km/h (1,355mph, Mach 2.05); range (300 passengers) 9,200km (5,717 miles).

Tu-24Skh

Searching for new markets to maintain employment, ANTK Tupolev exhibited a mockup of this utility and agricultural aircraft in 1993. The metal/composite airframe would have untapered wings with slotted-flaps, large ailerons, fixed outboard slats and down-curved tips. The elevators have large horn-balances and the rudder a nose balance over its mid-section. Main landing gears are simple pin-jointed tripods, and the tailwheel castors. A door on the left opens the cockpit with one or two front seats, behind which is a dusting hopper or chemical tank. The engine is a 355hp M14PS driving a 2.5m (8ft 2½in) three-blade propeller. The mock-up had a two-bladed propeller and was fitted with a ram-air driven dusting distributor. In 1995 Tupolev expected to deliver the first of 400 in 1998, at a base price of US $100,000. Performance figures are estimates.

Span 13m (42ft 7¾in); length 9.25m (30ft 4⅛in); wing area 28.04sq m (302sq ft).
Weight empty 990kg (2,183lb); fuel 70kg (154lb); chemicals 900kg (1,984lb), maximum loaded weight 2,100kg (4,630lb).

Maximum speed 235km/h (146mph); working speed 120–140km/h (74.5–87mph); rate of climb 4.5m/s (885ft/min); range with maximum (means auxiliary) fuel 2,000km (1,242 miles); take-off run 180m (590ft); landing run 100m (328ft).

Other projected versions are: Tu–24P patrol, R marine life survey, S ambulance, ST sport and parachuting, T five passengers or 800kg (1,764lb) cargo, and V trainer.

Tu-34

Also announced in 1993, this attractive multirole light transport was planned around the use of two pusher piston engines (Russian or American), each rated at 220hp and driving a constant-speed four-blade propeller. Since then it has grown in dimensions, the wing being twice redesigned, and the Tu-34 is now offered in two sizes with different weights and with a choice of piston engines or turboprops. In either case, the metal/composite airframe has wide-span track-mounted high-lift flaps, full-span leading-edge slats, outboard tabbed ailerons and mid-span spoilers, a T-type tail, and tri-cycle landing gear retracting into the finely streamlined fuselage.

Piston engines This version has a slightly sweptback tailplane and an unpressurised cabin 2.75m (108in) long, 1.6m (63in) wide and 1.34m (52.8in) high, arranged for a pilot and passenger in front and a row of three seats behind. Alternatively 450kg (992lb) of cargo can be carried, or the interior can be equipped for surveillance, ambulance or other roles. The leading edge of the wing, originally tapered, is now at 90 deg to the longitudinal axis, the tips are raked, and the adoption of foreplanes

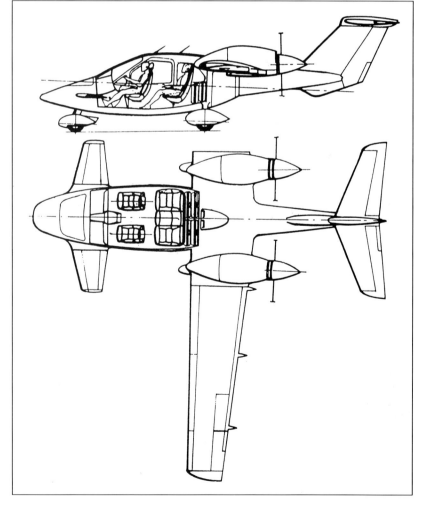

Passenger version of Tu-34.

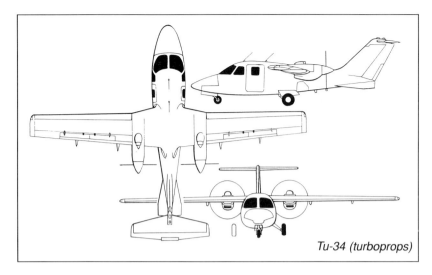

Tu-34 (turboprops)

be carried or a front passenger, stretcher and attendant. With the piston-engined model on hold, it is planned to fly the prototype turboprop Tu-34 in 1998.

Span 13.4m (43ft 11⅝in); length 9.9m (32ft 5¼in); wing area not stated.
Weight loaded 2,520kg (5,556lb).
Cruising speed 450km/h (280mph) at 5,000m; service ceiling 7,500m (24,600ft) with cabin altitude 2,600m; range 580km (360 miles) with 700kg (1,543lb) payload, 2,100km (1,305 miles) with 440kg; required runway length 400m (1,312ft).

is being studied. Engine power has risen to 230hp each, giving a cruise fuel consumption of 45–50 litres (10–11gal)/h.

Span 13.2m (43ft 3⅜in); length 9.44m (30ft 11⅝in); wing area not stated.
Weight loaded 1,900kg (4,189lb).
Maximum speed 360km/h (224mph); cruising speed 280–329km/h (174–199mph); range 1,800km (1,118 miles); required runway length 400m (1,312ft).

Turboprop version This aircraft would be powered by two pusher turboprops, such as the Allison B17, or Turbomeca Arrius, of 420hp each. The wing would have slight forward sweep and straight tips, and the horizontal tail would have equal taper, the elevator being a single surface with a central tab. The circular-section fuselage would be pressurised, and house a cabin 3.0m (118in) long, 1.55m (61in) wide and 1.34m (52¾in) high, able to carry a pilot and a passenger in front and up to six more passengers in two staggered bench seats behind. Alternatively, 700kg (1,543lb) or cargo could

Tu-130

Another civil project is this twin-turboprop convertible passenger/cargo transport. The fuselage would be pressurised, but behind the wing the cross-section would change from a circle to a broad oval, the beaver tail incorporating a full-width rear ramp door 3.5m (138in) long. Aft of the flight deck the cabin would be 8.5m (27ft 10⅝in) long, 2.8m (110in) wide and 2.2m (86.7in) high, typical seating in the all-passenger role being for fifty-three. Engines would be the Klimov

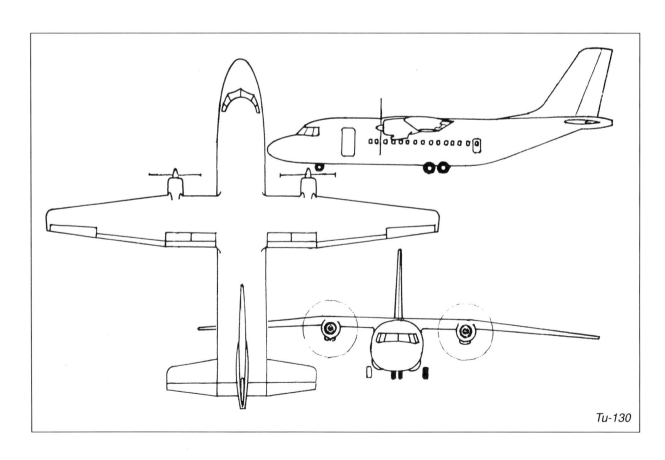

Tu-130

TV7–117S, as already produced for the Il-114, with provision for burning LNG or propane. Landing gear would be as on the C-130, with side-by-side nosewheels and tandem main wheels. In 1994 ANTK Tupolev was planning to fly a Tu-130 prototype in 1996, but as this book went to press construction had not begun.

Span 26.54m (87ft 0⅞in); length 22.75m (74ft 7⅝in); wing area not stated.
Maximum payload 5,000kg (11,020lb); maximum take-off weight 21,000kg (46,296lb).
Cruising speed 500km/h (311mph) at 7,000m (22,966ft); range with maximum payload 2,000km (1,242 miles); required runway length 1,800m (5,905ft).

Tu-136 Projected cryogenic-fuel version with 7t (15,432lb) of LNG in two external tanks under the wings, retaining TV7-117S engines. It would carry fifty-three passengers or 5t (11,020lb) of cargo 2,000km (1,242 miles).

Tu-304

This number was ANTK Tupolev's internal designation for a 'mega-carrier' in the 800–1,200-seat class. Around 1990 the project had span and length of 95m and 90m (312ft and 295ft), respectively, and an MTO weight of some 670,000kg (1,477,000lb). In 1992 this aircraft was sharply scaled down to the 300-seat level, using two Samara/KKBM NK-44 or Rolls-Royce Trent turbofans in the 90,000lb class. Like all their new projects, Tupolev would like to avoid building a prototype, and in 1992 were predicting possible production aircraft from 1998, but this is now clearly unlikely.

Tu-324

Tu-324

In 1995 ANTK Tupolev announced this new twin-jet transport. Yet a further scale-down of the very efficient wing, this aircraft would have a smaller body cross-section than the Tu-334. It is projected in two versions. The basic Tu-324A business and VIP version, which could replace the Tu-414, would be furnished for up to ten. In mid 1995 it was hoped to fly the first 324A in late 1998. The derived Tu-324 regional transport would have reduced fuel capacity and a longer fuselage seating up to 50, and economics '30 to 40 per cent better than competitors'. Both would be powered by two of the previously unknown Soyuz R-126-300 turbofans. Data for Tu–324:

Span 24.7m (81ft 0in); length 26.2m (85ft 11½in); height 7.3m (23ft 11½in).
Maximum payload 5t (11,020lb); maximum take-off weight 24,630kg (54,299lb).
Cruising speed 800km/h (497mph); range (50 passengers) 2,500km (1,533 miles), (30 passengers) 4,500km (2,796 miles); runway length 1,800m (5,900ft).

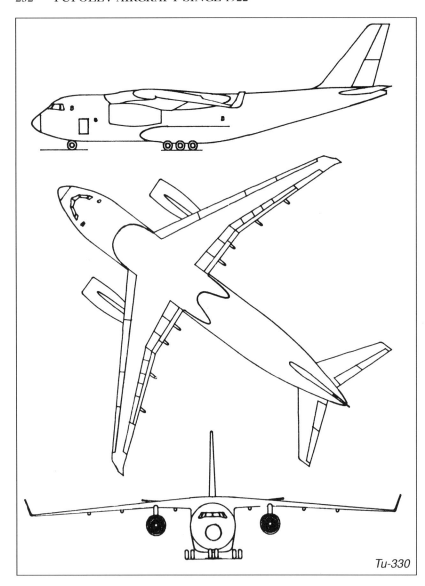

Tu-330

Data for Tu–324A:

Dimensions unchanged except length 23.2m (76ft 1½in).
Payload, normal 1t, maximum 3t (6,614lb); maximum take-off weight 25,400kg (555,996lb).
Cruising speed 800km/h (497mph); range (3t payload) 5,900km (3,666 miles), (1t payload) 7,900km (4,910 miles); runway length 1,950m (6,400ft).

304, 306

Number 304 was ANTK Tupolev's internal designation for a 'mega-carrier' in the 800–1,200-seat class. Around 1990 the project had span and length of 95m and 90m (312ft and 295ft), respectively, and an MTO weight of some 670t (1,477,000lb). In 1992 this aircraft was sharply scaled down to the 300-seat level, using two Samara/KKBM NK-44 or Rolls-Royce Trent turbofans in the 90,000lb class. Like all their new projects, Tupolev would like to avoid building a prototype, and in 1992 were predicting possible production aircraft from 1998, but this is now impossible.

Instead attention is now focussed on the 306, which would be fuelled by 60t (132,280lb of LNG (cryogenic liquefied methane) housed in a giant lagged saddle tank above the fuselage. Powered by two of the previously unknown TV7-46 engines, it would carry 450 passengers 9,000km (5,590 miles). Partners, additional to the existing partner on cryogenic fuels, Daimler-Benz Aerospace, are sought to enable this to go ahead.

Tu–330 model (September 1994). (RART)

Tu-330

In 1991 ANTK Tupolev, looking for *konversiya* projects and noting its success in using major parts of the 88 in order to produce the 104, and of the 95 to produce the 114, decided to scale the efficient wing of the 204 up and down to serve as the basis for a range of derived transport aircraft. It had already announced the smaller 334, and it followed with this important cargo aircraft, using almost the same wing, engine pylons, cockpit and avionics as the 204 but with some 334 features and with three inboard spoilers, and scaled up to a span of 47.47m (155ft 9in). In 1993 the 330 was scaled down again to 204 size, the span being increased only by the wider body. The circular-section fuselage would be fully pressurised, interior available length being 19.5m (63ft 11¾in), minimum (floor/ceiling) width being 4.0m (157in) and height also 4.0m (157in) except towards the rear where it tapers to 3.55m (140in). The engines would be PS-90A turbofans in the 16t (35,273lb) class, though for export various Western engines would be alternatives. Fuel burn would be 140 g/t-km (7.9oz/t-mile). By 1999 it is intended to switch to NK-94 engines burning 20t of LNG in lagged fuel pods hung under the wings. This would increase range to 5,000km (3,100 miles).

Tupolev hope this aircraft will be built in substantial numbers as the replacement for the An-12 as the standard workhorse freighter throughout the CIS (the rival An-70 being Ukrainian). Production is to be handled by KAPO (Kazan Aircraft Production Association), with the first off the line intended to fly in 1996. In 1995 this programme was fully funded to prototype testing. If the schedule can be kept, certificated aircraft might be available in 1998.

Span 43.5m (142ft 8⅝in); length 42.0m (137ft 9½in); wing area 195.5sq m (2,104sq ft).

Maximum payload 35t (77,160lb); maximum take-off weight 103,500kg (228,175lb).

Cruising speed 800km/h (497-528mph) at 11,000m; range 3,000km (1,864 miles) with 30t payload, 5,600km (3,480 miles) with 20t payload; required runway length 2,200m (7,217ft).

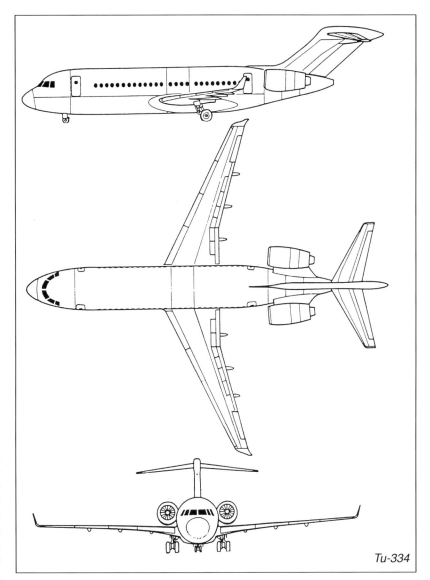

Tu-334

Tu–334

Another potentially large programme, this passenger jet was announced in 1989 as the intended replacement for the Tu-134. It was originally planned in two forms, the basic turbofan aircraft, and a stretched propfan version. The latter was shelved in 1995, partly because the whole programme was running behind schedule. The fuselage is a shortened version of that of the Tu–204, with an almost identical cockpit, and the wings are a smaller and slightly simplified version of the Tu-204 wing, with ¼-chord sweep reduced to 24.

Tu-334–100 This is the baseline version. In 1990 the Tupolev facility at

NII Zhukovskii was building a static-test airframe, and three flight articles for certification. About 20 per cent of the structure is made of composites. The flying control surfaces are fully powered, with manual reversion on the horizontal tail, signalling being fly-by-wire. Avionics can be Russian or from a choice of Western suppliers. The fuselage has doors on the left at front and rear, with service doors opposite. With the usual buffet/galley and toilet at each end, typical seating is for from seventy-two to eighty-six in mixed classes or 102 single-class (3+3 at 81cm, 32in, pitch). In 1993 it was intended to fly both the first two Tu-334s in 1994, for service entry in 1996, but this schedule slipped. As this book went to press, in mid-1995, the objective was to fly the first Tu-334, with ZMKB Progress D-436T1 engines rated at 7,650kg (16,865lb) in mid-

1995, and the second and third later in the same year, if possible with D-436T2 engines.

Span 29.77m (97ft 8in); length 31.26m (102ft 6¾in); wing area (gross) 83.226sq m (895.9sq ft).
Operating weight empty 30,050kg (66,248lb); maximum fuel 9,540kg (21,032lb); payload 11t; maximum take-off weight 46,100kg (101,631lb).
Cruising speed 800–820km/h (497–510mph) at 10,600–11,100m (34,777–36,417ft); range with maximum passenger load 2,000km (1,243 miles); required runway (balanced field) length at 30 deg C 2,200–2,300m (7,218–7,546ft).

Tu-334–100D This is the long-range version, powered by D-436T2 or, for export, BR715 engines. Tupolev hope to begin production of this version in 1997, with initial ARMAC (Russian Federation) certification a year later.

Span 32.6m (106ft 11½in); length unchanged; wing area (gross) 100sq m (1,076sq ft).
Maximum fuel 13,790kg (30,400lb); zero-fuel weight 42,920kg (94,621lb); maximum take-off weight 54,420kg (119,975lb).
Cruising speed 800km/h (497mph); range with maximum passenger load 4,100km (2,550 miles).

Tu-334–200 This is the intended production version, with 8,200kg (18,078lb) D-436T2 engines. The air-frame is slightly stretched, to seat 110–126 passengers. The main landing gears will have four-wheel bogies with tyre pressure suitable for unpaved runways. Certification is hoped for in late 1997. All-cargo and convertible versions are planned, and fuel efficiency is described as '1½ times better than the Yak-42'. For export customers an alternative engine as seen as the BMW Rolls-Royce 715-55, rated in the 18,500lb class. In each case the engines would have target-type reversers. Production is planned to include both Kiev (Ukraine) and Taganrog factories.

Span 32.61m (106ft 11⅞in); length 35.16m (115ft 4¼in); wing area 100.0sq m (1,080sq ft)
Operating weight empty 34,375kg (75,783lb); maximum fuel 13,790kg (30,401lb); maximum take-off weight 54,800kg (120,811lb).
Cruising speed 800km/h (497mph) at 11,100m (36,420ft); range with 126 passengers (11,970kg payload) 2,200km (1,367miles).

Propfan version This was expected to be powered by two ZMKB Progress D-27 engines, each rated at

13,880hp or 8-9t (17,637–19,841lb) thrust and installed as pushers to provide even better fuel economy and less cabin noise. Wing would be slightly reduced and fuselage length increased to seat 104–137 passengers, with a Class-I exit added above the wing on each side. MTO weight would be 47,000kg. In 1995 this version was shelved, at least for the time being.

Tu-336 This projected cryogenic-fuel version would have to Samara NK-112 turbofans, each rated at 8,500kg (18,739lb), fed from a long saddle tank above the fuselage housing 8t (17,637lb), of LNG. It would carry 120 passengers 1,200km (746 miles).

C-Prop

Announced in early 1994, this unconventional transport is described as a regional cryo-turboprop. Totally Westernised for export, promotional material so far is in English only, and the engines would be two pusher Pratt & Whitney Canada PT6A-67, each rated at 1,500shp. PWC has a close tie with St Petersburg's Klimov Corporation. It has not announced that any of its engines are certified on cryogenic fuels, but this should not be difficult. CAHI has been involved in developing the 'triplane' configuration, with a

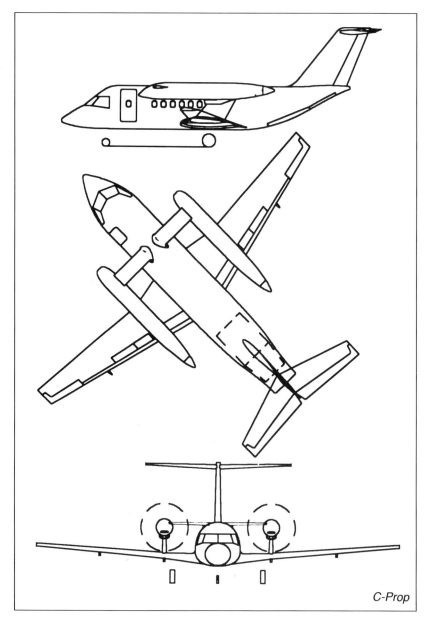

C-Prop

flapped foreplane as well as a conventional horizontal tail (the same layout has been adopted for transport projects by Molniya and Sukhoi). The insulated tanks for LNG fuel would form nacelles connecting the foreplanes and engines. The pressurised cabin would be 9.68m (31ft 7½in) long, 2.6m (102in) wide and 2.2m (86.6in) high at the centreline. Typical payloads would include thirty-two passengers (2+2) or three ISO 88in by 108in or 88in by 125in pallets. No decision to build has been announced.

Span 22.5m (73ft 9⅞in); length 21.0m (68ft 10¾in).

Liquid-gas fuel 2.4t; payload 3.4t; maximum take-off weight 13.5t (29,762lb). Cruising speed 450km/h (280mph) at 8,000m (26,250ft); range with full passenger load 1,500km (932miles); required runway length 620m (2,035ft).

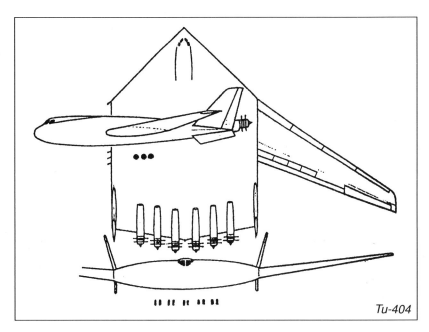

Tu-404

Tu-404 model (June 1993). (Author/ RART)

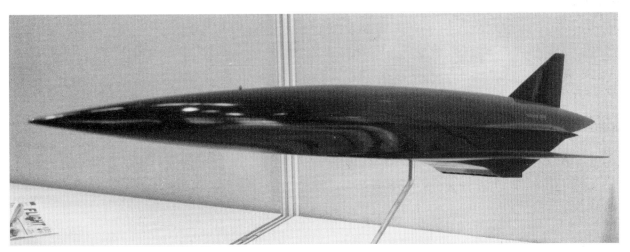

Tu-2000 model. (RART)

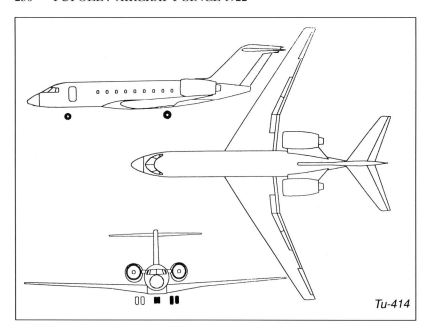

Tu-414

Tu-404

This unconventional but convincing project was Tupolev's first ultra-high-capacity 'mega-carrier'. It began with a conventional configuration resembling a double-deck 747, but the passenger cabins posed severe problems until Chief Designer Yuri V Vorobjov and team switched to a fuselage forming a very wide lifting surface, the wings being attached on each side. The cabin would be some 27m (88ft) wide, the vast floor and roof being joined by numerous ties or walls to withstand the severe pressurisation loads. In 1993 Tupolev exhibited a large model having the configuration illustrated, with a projecting cockpit near the nose, widely spaced slanting tails with underfins, and six advanced turbofans or propfans burning LNG. The ailerons would be elevons, for longitudinal as well as lateral control. Foreign partners were being sought. This seems a better configuration than

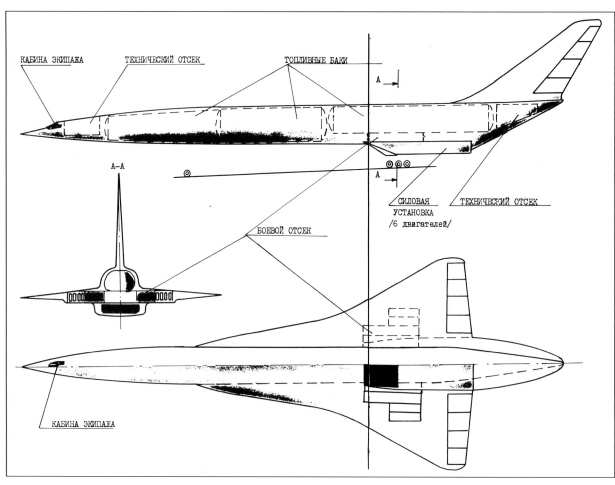

Aero-space bomber project.

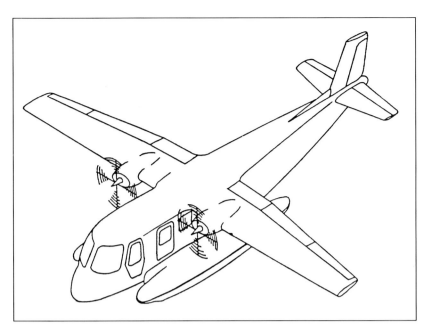

Triton (1994 sketch).

the traditional shapes the airlines appear in 1995 still to prefer.

Span about 100m (328ft); length about 52.5m (172ft); wing area about 700sq m (7,535sq ft).
Maximum take-off weight about 550,000kg (1,212,500lb).
Cruising speed 840km/h (520mph); range with maximum passenger load 13,000–15,000km (8,080–9,320 miles).

Tu-414

The sheer size of the former Soviet Union prompted ANTK Tupolev to study a superior business jet with exceptional range. The Tu-414 was announced in 1994. Like the other recent projects, the basic features can be seen in the drawing. Diameter of the highly pressurised fuselage would be 2.5m (98.4in), and the selected engine is the BMW Rolls-Royce BR 710, rated at 14,990lb. Passenger capacity could be from eight to nineteen in the executive role, or thirty as a regional transport (though the long range fits it for very long 'thin' routes). The long range results in greater

sweepback than on the 204 and 334, the wing leading edge being set at 33 deg. ANTK Tupolev hopes to receive funds from a national programme to support civil aviation for year 2000.

Span 28.8m (94ft 5⅛in); length 28.2m (92ft 6¼in).
Maximum payload 3,300kg (7,275lb); maximum take-off weight 10,000kg (88,183lb).
Cruising speed 860km/h (534mph); range 10,560km (6,562 miles) with passengers, 8,500km (5,282 miles) with 30 passengers; required runway length 2,040m (6,692ft).

Spaceplanes

Brief details have been published of two aero-spaceplanes of slender-delta form, one a research vehicle and the other a bomber.

Tu-2000 This two-seat vehicle is designed to research aerodynamics, propulsion, structures and materials, systems and computational fluid dynamics for Mach 6–8, as a means to future development of efficient air-

breathing engines for propulsion from rest to Mach 25. External-burning hypersonic turbo-ramjet engines using liquid-hydrogen fuel are being studied by CIAM.

Span about 14m (46ft); length 55–60m (180–197ft).
Take-off weight 70,000–90,000kg (154,000–200,000lb).
Speed Mach 6 (6,400km/h, 3,965mph) at 30,000m (98,500ft), with a future second stage for acceleration to Mach 15 (orbit) or Mach 25 (escape).

Bomber

This two-seat vehicle is much larger, and would have enormous liquid-hydrogen capacity in a fuselage of 8m (315in) diameter. The drawing shows the relatively tiny crew capsule in the nose, quad flight-trajectory surfaces, six underslung turbo-ramjet engines, bomb cells in the wing roots and landing gear further outboard.

Span 40.7m (133ft 6⅛in); length 100m (328ft); wing area 1,250sq m (13,455sq ft).
Equipped weight 200,000kg; maximum fuel 150,000kg maximum 20,000kg; take-off weight 350,000kg (771,600lb), thus sum of fuel/military load is 150,400kg
Cruising speed Mach 6 (6,400km/h, 3,965mph) at 30,000m (98,500ft); range 9,000–10,000km (5,590–6,200 miles).

Triton

As the last of the unrivalled succession of Tupolev designs comes a simple multi-role utility transport amphibian, being developed jointly with the KKB (Kazan Design Bureau). Powered by two 400hp DN-400 diesel engines, it would be unpressurised, carry a crew of two and bulky freight or eight passengers, and have wheels retracting into ski/floats.

Dimensions not stated.
Payload 800kg (1,764lb); maximum take-off weight 5,800kg (12,787lb).
Maximum speed 220km/h (137mph); cruising speed 180km/h (112mph) at 0–3,000m (0–9,843ft); range 800km (497 miles); ferry range 1,500km (932 miles).

Appendix 1
Engines used in Tupolev aircraft

ACh-30 This family of diesel engines was named for General Constructor (design bureau leader) Aleksei Dmitriyevich Charomskii. They stemmed from a long series of two-stroke diesels with V-12 layout, with water-cooled cylinders, which began with the AN-1 (see below). The 1,500hp ACh-30B, which until 1941 was designated M-30B, had bore and stroke of 180mm × 200mm giving a capacity of 62.34 litres (compared, for example, with 27 litres for the British Merlin). It weighed no less than 1,200kg (2,654lb), but its economy in burning fuel (which was oil, not petrol) enabled the diesel-engined Pe-8 versions to have much longer range. It had four exhaust-driven turbochargers and pressurised glycol cooling.

AL-7 General Constructor Arkhip M Lyul'ka was a pioneer of axial-compressor turbojets, and was the first to run engines with supersonic flow through the upstream stages of the compressor. The AL-7 was the first of his engines to go into series production. The basic engine was certificated in 1954 with a nine-stage compressor handling an airflow of 114kg (251lb) per second. The AL-7F was fitted with an afterburner and variable nozzle, takeoff ratings being 6,500kg (14,330lb) dry and 9,000kg (19,840lb) with maximum afterburner. The Tu-110 originally had the 5-tonne (11,020lb) basic AL-7, but later received the AL-7P rated at 7,260kg (16,005lb). Later versions for the Tu-128, with after-burner, were the AL-7F-1 with maximum afterburner

thrust of 9,900kg (21,825lb), the AL-7F-2 of 10,100kg (22,282lb) and the AL-7F-4 of 10,700kg (23,590lb).

AM-02 Dating from 1949, the second axial turbojet developed by the bureau of General Aleksandr Aleksandrovich Mikulin was a fairly simple unit rated at 4,250kg (9,370lb), though its diameter of 1,380mm was almost as large as that of the AM-3.

AM-3 This large but basically rather primitive turbojet was designed in 1947–49 by a team led by P F Zubets. They worked in the design bureau of A A Mikulin, who had previously been Tupolev's most important supplier of piston engines. The AM-3 had an eight-stage axial compressor based mainly on German aerodynamics, which in its first production form in 1952 handled an airflow of 135kg (298lb) per second with a pressure ratio of 6.4. The pre-series engines were rated at 6,750kg (14,880lb) thrust for takeoff, but this was increased in the AM-3A of 1953 to 8,700kg (19,180lb). Most production versions had an integral gas-turbine starter rated at 100hp. The service designation was RD-3, and the most important variants were the RD-3M-500 rated at 9,500kg (20,950lb) and the RD-3M-500A, cleared to a short-duration emergency rating of 10,500kg (23,150lb).

AM-34 See M-34.

The D-30-II is seen here complete with its thrust reverser. (Author's Collection)

AM-35 This V-12 liquid-cooled piston engine was derived from Mikulin's M-34 series. It had an improved cylinder head and modified supercharger, though capacity remained 46.66 litres as in the M-34FRN. The AM-35 was qualified in 1939 at 1,200hp at 2,050rpm. The AM-35A was rated in 1940 at 1,350hp.

AM-37 This derivative of the AM-35 was qualified at 1,380hp in 1940, and the AM-37F followed at 1,400hp a year later.

AM-39 A further step along the long road to higher powers, the AM-39 was a considerable family of experimental V-12 engines developed during the Second World War. The baseline engine was tested in 1942 at 1,870hp. Other versions were rated at from 1,800 to 1,900hp.

AM-43V This was yet another in the long Mikulin series of high-power liquid-cooled V-12 engines. Starting at 1,640hp in 1944, the AM-43 reached 1,950hp in the AM-43V version on 100-octane fuel, and later versions attained 2,200hp.

AM-44 Not quite the ultimate Mikulin piston engine, the AM-44 was qualified in 1945 at 1,900hp, but the Type 64 heavy bomber would have used the AM-44TK, with TK-300 turbosuperchargers maintaining 2,200hp to over 20,000ft, and 1,500hp to over 30,000ft.

AN-1 Designated from the Russian for 'aviation, oil' (as distinct from using petrol as fuel), this was the first Charomskii diesel. A two-stroke water-cooled V-12, it was first run as a complete engine in 1933, and was rated at 850hp (later at 900hp) at 2,000rpm. It was developed into the ACh-30 and ACh-40 families.

Anzani The ANT-1 was powered by an Anzani radial with six aircooled cylinders of 90mm bore × 120mm stroke. The French company advertised this model as 45hp, but Russian sources always describe it as 35hp.

ASh-2 This massive piston engine was essentially two ASh-82 units joined to make a 28-cylinder package rated at 3,600hp. It was never installed in the Tu-85 allocated to it.

ASh-21 This piston engine by Arkadiya Dmitriyevich Shvetsov was essentially half an ASh-82. It thus comprised a single row of seven cylinders with bore and stroke of 155.5mm × 155.0mm, giving a capacity of 20.6 litres. When adopted by Sukhoi for the UTB it was newly qualified in 1947 at 700hp.

ASh-73 This large two-row radial piston engine was developed by Shvetsov over many years, starting with the licensed Wright Cyclone F-series in 1933. The cylinders were redesigned with bore and stroke of 155.5mm × 174.5mm, and the ASh-73 had eighteen of these cylinders, giving a capacity of 59.6 litres. Thus, it was slightly larger than the R-3350 Duplex Cyclone fitted to the B-29. Moreover, the Tu-4's ASh-73TK had an excellent TK-19 turbosupercharger which maintained 2,400hp to

a height of 6,500m (21,325ft). The ASh-73FN and TKNV were uprated to 2,650hp.

ASh-82 This design by A D Shvetsov was one of the principal engines of the Second World War with about 71,000 produced. A compact 14-cylinder air-cooled radial, it had bore and stroke of 155.5mm × 155.00mm, giving a capacity of 41.2 litres, and the short stroke restricted overall diameter to 1,259mm (49.6in). The basic engine, as used in the Pe-8, was rated at 1,400hp. The ASh-82FN was rated at 1,630hp, or at 1,850hp on 100-octane fuel. The civil ASh-82T, rated at up to 1,900hp, was retrofitted to the Pe-8ON *Polyarnyi*.

ASh-83 This engine was derived from the ASh-82 but differed in structural detail and equipment; with direct injection it was rated at 1,900hp.

BMW Rolls-Royce BR715 This two-spool turbofan with a 1,346mm (53in) fan has a takeoff rating in the 18,500-20,000lb class. The Tu-414 was originally planned to use the 14,900lb BR710.

Bristol Jupiter This aircooled radial piston engine first ran in 1918, and became the world's most widely used aero engine in the 1920s. It had nine cylinders with bore and stroke of 146mm × 190mm, giving a capacity of 28.7 litres. Imported engines and those made in Moscow as the M-22 were usually rated at 480hp. Tupolev aircraft also used the French licensed version, the Gnome-Rhône GR9B, usually rated at 420hp, and the 570hp GR9Ak.

Bristol Lucifer This simple piston engine was designed by Roy Fedden in 1918 to use three cylinders basically similar to those of the Jupiter but with stroke reduced to 165mm. Standard ratings were 120hp or 130hp, but in the ANT-2 it is always described as 100hp.

Bristol Mercury This nine-cylinder radial was originally derived from the Jupiter by reducing the stroke to 165mm. Capacity thus became 24.9 litres. In the I-14 version the VIS2 was very conservatively rated at 580hp.

Curtiss Conqueror This high-power piston engine had two cast monobloc banks each housing six water-cooled cylinders with bore and stroke of 130mm × 159mm, giving a capacity of 25.7 litres. Normal takeoff power was 600hp.

D-20 This pioneer two-shaft turbofan could also be called a bypass turbojet, because the bypass ratio was unity; in other words, the airflow from the LP compressor that bypassed the core was the same in mass flow per unit time as the airflow through the core. Designed by Pavel Aleksandrovich Solovyev (pronounced 'Solovyov'), it weighed 1,468kg (3,236lb) and had a total airflow of 113kg (249lb) per second. Takeoff thrust rating was 5,400kg (11,905lb).

D-30 To meet the requirements of the Tu-134 Solovyev improved the compressors of the D-20 to handle 125kg (275.6lb) per second with increased pressure ratio, though he kept bypass ratio at unity. The result was a

The NK-8-2 was the original engine of the Tu-154. (Author's Collection)

series of D-30 engines, beginning with a thrust of 6,800kg (14,990lb). The D-30-II added a thrust reverser, and the D-30-III had an improved compressor enabling the same thrust to be obtained with reduced turbine temperatures.

D-30K In typically confusing Russian manner, this family of engines has nothing in common with the D-30 except the designation. To power the Tu-154 Solovyev started again with a more advanced turbofan with a bypass ratio of 2.42. Its compressors handled 269kg (593lb) per second in the first version, rated at 11 tonnes (24,250lb) thrust, but today's Tu-154s have the D-30KU-154 rated at 10,500kg (23,148lb) or the D-30KU-154-II which is flat-rated at 10,800kg (23,830lb) to 23°C.

D-436 Initially for the Tu-334, the Zaporozhye engine bureau, now led by Fyodor Mikhailovich Muravchyenko, is engaged in qualifying this high-bypass-ratio turbofan in several forms. Bypass ratio varies from 4.9 to 6.2, and takeoff thrust from 7.5 tonnes to 9.0 tonnes (16,535–19,840lb). These engines are derived from the D-36, designed under Muravchyenko's predecessor, Vladimir Lotarev.

Gnome-Rhône 14K Mistral Major In 1921 Paris-based Gnome-Rhône took a licence to make the Bristol Jupiter, but they soon found ways to avoid paying royalties. One of the more legitimate ways was to modify the engine (with stroke shortened to 165mm , the same as the Bristol Mercury), and by cutting the number of valves per cylinder from four to two they were able to produce this 14-cylinder two-row version. Its capacity was 38.65 litres, and it was rated at 800hp. Soviet licensed versions began with the M-85.

Gnome-Rhône Titan The French company also began making other Bristol aircooled radial engines without paying a royalty. One was the Titan, with five small (146mm × 165mm) cylinders, rated at 230hp.

Hispano-Suiza HS12 Though not a brilliant engine, this French V-12, with cast water-cooled blocks, was used in large numbers in many countries. It led to the Soviet licensed version designated M-100, from which stemmed the Klimov VK-100 family. Bore and stroke were 150mm × 170mm giving a capacity of 36.05 litres. In the DIP and SB Tupolev used the HS12Ybrs, rated at 760hp at 2,400rpm.

Lorraine-Dietrich In the 1920s this French company produced various high-power water-cooled piston engines. The R-3LD was powered by the 12Ed with three banks of four cylinders with bore and stroke 120mm × 180mm, giving a capacity of 24.4 litres. Takeoff power was 450hp.

M-5 This was the designation of the first high-power engine to be mass-produced in the Soviet Union. It was a copy of the US-designed Liberty of 1917. Its twelve water-cooled cylinders had bore and stroke of 127mm × 179mm, giving a capacity of 27 litres (the same as a Merlin). Standard takeoff power was 400hp.

M-14P Selected for the Tu-24 family, this is one of numerous versions of a radial piston engine with nine aircooled cylinders with bore and stroke 105mm × 130mm, giving a capacity of 10.16 litres. Takeoff rating is 355hp (360 metric hp). The original design was by A G Ivchyenko, but today's engines have been modified by I M Vedeneyev and are being produced at Voronezh under Prof A G Bakanov.

M-17 Derived from the German BMW VI of 1926, this V-12 piston engine had large water-cooled cylinders of 160mm bore × 190mm stroke, giving a capacity of 46.95 litres (compared with 27 litres for the later British Merlin). The Soviet licensed M-17 version far outnumbered the German original. It was rated at 680hp in 1929. The M-17F, refined by A A Mikulin, was by 1932 cleared to a power of 715hp or 730hp. There were several other versions, a total of 27,534 M-17 engines being produced by 1942.

M-26 This five-cylinder radial was based on the Bristol Titan, but used cylinders designed by A D Shvetsov. It failed testing at 300hp in 1929.

M-34 One of the most important engines of Tupolev's early aircraft, this water-cooled V-12 was originally assembled by A A Mikulin from parts of other engines. Its cast blocks each contained a row of six cylinders of the same size as in the M-17. The rear wheelcase was based on that of the Hispano-Suiza HS12, the supercharger was based on the Allison V-1710-A and the 'herringbone' (double helical) reduction gear was based on the Rolls-Royce Buzzard. Starting in 1931 at 660hp, it was developed to 820hp in the M-34F and 34R, 900hp in the FRN version, 1,200hp in the FRNV and 1,275hp in the RNF.

M-85 This 14-cylinder radial was the Soviet licensed version of the Gnome-Rhône K14, with cylinders 146mm × 165mm and capacity 38.65 litres. Sergei Konstantinovich Tumanskii, in the Mikulin bureau at OKB-29, developed it to 850hp by 1936.

M-86 Hardly meriting a different number, this was Tumanskii's improved M-85, rated by 1937 at up to 950hp.

M-87 A further OKB-29 development, to 930hp for the M-87A and to 950hp for the M-87B. Tupolev did not use the final M-88 versions of over 1,000hp.

M-100 Under this designation the French Hispano-Suiza HS12Y was put into licence production in 1934, and its further development was assigned to a new design bureau headed by Vladimir Yakovlyevich Klimov, who had previously been first assistant to Mikulin. Eventually over 129,000 were produced in many versions. From 1940 most were designated with a VK prefix in honour of Klimov, though Tupolev archives invariably adhered to the original M-prefixes. Later versions had two cast blocks of six water-cooled or glycol-cooled cylinders, with bore and stroke 148mm × 170mm, giving a capacity of 35.08 litres. The M-100 was qualified in 1935 at 750hp, and the M-100A gave 840hp at 2,400rpm. The M-103 with a two-speed supercharger was certificated in January 1937 at 860hp, holding power much better at high altitude. The M-103A of 1939 was rated at 960hp, rising to 1,100hp with 100-octane fuel. The M-104 and M-106 were two of numerous experimental versions. The M-105, intended for the Type 57, was a mass-produced engine in the 1,200hp class.

NK-4 This was the second turboprop to be developed at Kuibyshev under Nikolai Dmitriyevich Kuznetsov. It was tested at 4,000shp and 200 were built, but not used.

NK-6 This formidable two-shaft turbofan was developed by the design bureau of N D Kuznetsov to power supersonic bombers. The prototype went on test in 1959 at 20 tonnes (44,092lb) thrust with duct burning and core afterburning. Its applications, such as the Type 135, were cancelled.

NK-8 Based loosely on Rolls-Royce concepts for bypass turbojets (low-ratio turbofans), this two-spool engine was designed at Kuibyshev by the bureau of N D Kuznetsov. The LP compressor (fan) had two stages, and rotated with two further stages to supercharge the core. The HP spool had six stages, mainly of titanium. The NK-8-2 had a takeoff rating of 9,500kg (20,950lb), and the NK-8-2U was uprated to 10,500kg (23,150lb).

NK-12 Originally designated TV-12, this turboprop, designed by a team under Ferdinand Bradner in the Bureau of ND Kuznetzov, made the Tu-95 possible. Basically a simple engine, it has for over 40 years been the most powerful turboprop in the world. Its 14-stage axial comproessor originally handled 62kg (137lb) of air per second, running at a constant speed of 8,250rpm, giving a takeoff power of 11,995shp. Power is controlled by varying the fuel flow. The NK-12M increased speed to 8,300rpm, and the airflow to 65kg (143lb) per second, takeoff power rising to 14,995shp. The gearbox drives AV-60N coaxial propellers each with four blades and a diameter of 5.6m (18ft 4½in). In the NK-12MV provision was made for rapid autofeathering of the propellers in the event of loss of drive torque, with positive manual feathering as a backup.

NK-16 This development of the NK-12 was to have been rated at 18,100shp, rising in the NK-16ST to 16,000kW (21,450shp) but the Tu-96 was cancelled.

NK-22 In 1967 a major team at the Kuznetsov bureau qualified this two-spool augmented turbofan at its design ratings of 16,200kg (35,715lb) dry and 22 tonnes (48,500lb) with maximum augmentation. Like the NK-144, NK-25 and several other engines, it stemmed form the NK-6 and NK-8.

NK-25 This further-developed engine, produced for the Tu-22M3 and Tu-22R, has a new and highly efficient two-spool compressor handling a greater airflow with improved fuel economy. With no change in installational dimensions, and only a very modest increase in weight, maximum thrust was increased to 19 tonnes (14,900lb) dry and 25 tonnes (55,115lb) with full augmentation. It is the most powerful supersonic engine in service in the world.

NK-88 Based on the NK-8-2U, this engine has been qualified to run on cryogenic (refrigerated) liquid hydrogen, at a temperature of −255°C, a major technical achievement owing much to spaceflight rockets. Thrust remains 10,500kg (23, 148lb).

The PS-90 is the baseline engine of the Tu-204. (Author's Collection)

NK-89 Again based on the NK-8-2U, this engine is qualified to run on LNG (liquefied natural gas). This is much less cold than liquid hydrogen. Ratings are again unchanged.

NK-93 Potentially of great importance, if money can be found to complete its development, this subsonic transport engine can be regarded as a shrouded prop-fan or a high-bypass-ratio turbofan. Under development at Samara, it has a three-stage LP turbine which, via a reduction gearbox transmitting up to 31,000shp, drives contra-rotating fans with a diameter of 2.9m (114.2in). The front fan has eight blades and provides 40 per cent of the thrust, and the rear unit has ten blades and provides the remaining 60 per cent. All blades have variable pitch, the 110° range providing for reversed thrust. Takeoff rating is 18 tonnes (39,683lb), with cruise specific fuel consumption of 0.486.

NK-94 This highly efficient Samara three-spool turbofan is expected to be qualified in 1997 on LNG (liquefied natural gas). Takeoff rating will be 18 tonnes (39,683lb), the same as the NK-93 propfan.

NK-144 Yet another of the large two-shaft engines for Mach 2 aircraft, this was in effect an intermediate stage between the NK-6 and NK-22. Five versions were produced for the Tu-144, all with variable core augmentation and a convergent/divergent variable nozzle. A major problem was that, to maintain the Tu-144 at over Mach 2, it was necessary to keep augmentation burning in cruising flight, which precluded attainment of the design range. Thrusts are given in the Tu-144 section.

NK-321 This large augmented turbofan, designed by the Samara/Trud team led by Evgenii A Gritsenko, had to combine high thrust with the best possible fuel economy. It was in effect the NK-32 high-pressure core placed between a new LP turbine driving an extra three-stage fan. This increased maximum airflow to 365kg (805lb) per second, pressure ratio at rest being 28.4 and bypass ratio 1.4. A typical engine dry weight is 3,390kg (7,474lb), and maximum thrust 14 tonnes (30,864lb) dry and 24,980kg (55,077lb) with full augmentation.

The VK-1 was Klimov's improved version of the British Nene. (Author's Collection)

PS-90 Designated for General Constructor Pavel Soloviev, this modern transport engine is a turbofan with a high pressure ratio (typically 35) and high bypass ratio (4.8). The original PS-90A was qualified in 1992 at 16 tonnes (35,275lb) thrust, but did not quite reach its design target for fuel economy. Several later versions have followed, the PS-90P being rated at 16,140kg (35,580lb).

PW2240 American Pratt & Whitney high-bypass-ratio turbofan rated at 18,915kg (41,700lb).

RD-3 See AM-3.

RD-36-51A Despite its designation, this large turbojet for Mach 2 propulsion is unrelated to previous RD-36 engines, which were vertical-lift turbojets. Developed in the Rybinsk bureau under P A Kolesov, the successor to Dobrynin (see VD engines), the RD-36-51A weighed 4,125kg (9,094lb) and had a takeoff rating of 20 tonnes (44,092lb). Specific fuel consumption in Mach 2.2 cruise was 0.88.

RD-45 This was the designation for the Soviet copy of the Rolls-Royce Nene centrifugal turbojet of 1947, rated at 2,200kg (4,850lb) initially, and from 1949 as the RD-45F at 2,270kg (5,000lb). The designation came from the number of the Moscow production factory, headed by Vladimir Klimov. He developed the RD-45 into the VK-1.

RD-500 This was the designation of the Rolls-Royce Derwent V, rated at 1,588kg (3,500lb). It took its designation from the number of the factory where under Vladimir Mikhailovich Yakovlev (no relation to the aircraft designer) it was produced in small series from January 1949.

RR 535E4 British Rolls-Royce high-bypass-ratio turbofan flexibly rated at up to 18,734kg (43,100lb).

TR-3 Studied for the Type 87 bomber, this was the original designation for the Lyul'ka AL-5, an axial turbojet of 8,680kg (19,140lb) thrust.

TV-2 One of the first turboprops to be flight-cleared in the Soviet Union, the TV-2 was derived from the German Junkers 109-022 by partly German teams administered under Kuznetsov. With a 14-stage compressor driven by a three-stage turbine, it was rated in 1949 at 3,680kW (4,933hp). It was greatly developed into the TV-2M, qualified in 1955 at 5,700kW (7,650shp). To power the Type 95/I the TV-2F, of 5,050shp, was produced in a twinned version, with a coupling gearbox driving independent coaxial propellers. This engine, designated 2TV-2F, was cleared for flight in 1952 at 9,200kW (12,332shp), but was eventually abandoned.

TV7-117S Originally developed for the Il-114, this free-turbine turboprop by what is today the Klimov Corporation has five axial compressor stages followed by a centrifugal, giving a takeoff pressure ratio of 16. The dry weight is 520kg (1,146lb) and takeoff rating 1,839kW (2,466shp) flat-rated to 35°C.

TV-12 Original designation of the NK-12.

VD-4 A masterpiece of piston-engine design, by Vladimir Alekseyevich Dobrynin, initially assisted by G S Skubachyevsky, this began life at the Moscow Aeronautical Institute in 1939 as the M-250. Features included six monobloc banks each of four liquid-cooled cylinders with bore and stroke of 140mm × 138mm, giving a capacity of 50.98 litres. In 1941 this engine was qualified at 2,500hp at 3,100rpm, and it was later developed into the VD-4, named for its designer when he became a General Constructor. Coupled with blow-down exhaust turbines on each bank, and a huge exhaust-driven turbosupercharger, the VD-4K was rated at 4,300hp, and maintained 2,800hp to 36,000ft, which no other piston engine ever did.

VD-5 This turbojet was tested at 13 tonnes (28,660lb) thrust in 1956, but was not put into production.

VD-7 Dobrynin developed this powerful turbojet to meet the demands of the Myasishchev 3M ('Bison') heavy bomber. In 1960–84 the Rybinsk design bureau was headed by Pyotr Alekseyevich Kolesov, who developed the VD-7 into a supersonic version selected for the Tu-22. The VD-7M was fitted with an afterburner and variable nozzle, the maximum rating being 16 tonnes (35,275lb). The production Tu-22 was fitted with the RD-7M-2, with a different afterburner and improved nozzle, with a maximum thrust of 16,500kg (36,376lb).

VD-19 This afterburning turbojet was almost a VD-7M with diameter reduced to fit the Tu-128. Maximum thrust was 10,200kg (22,487lb).

VK-1 Named for General Constructor V Ya Klimov, this was an improved derivative of the RD-45, with a usual takeoff rating of 2,700kg (5,952lb).

VK-5 This centrifugal turbojet was the ultimate development of the British Nene, rated at 3,100kg (6,834lb) in 1949.

Wright Cyclone This famous American radial piston engine had nine aircooled cylinders with bore and stroke 156mm × 175mm, giving a capacity of 29.87 litres. Used by Tupolev at 712 or 730hp. Later licence-built as the M-25.

Wright R-3350 Also called the Cyclone 18 or Duplex Cyclone, this two-row radial had 18 cylinders of bore and stroke 156mm × 160mm, giving a capacity of 54.77 litres. In the B-29 and Tu-70 the takeoff rating was 2,200hp.

Wright Whirlwind The Whirlwind radial piston engines had nine aircooled cylinders, which in the J6 version had bore and stroke of 127mm × 139mm, increasing capacity to 15.9 litres and takeoff power to (usually) 300hp.

Appendix 2
Guns fitted to Tupolev aircraft

DA Greatest of the Soviet Union's pioneer small-arms designers, Vasilii Dyegtyaryev began to design a light machine-gun in 1921. Features included gas operation, air cooling and feed from a 47-round drum of standard 7.62mm (0.300in) rimless ammunition. By 1926 the gun was on test, and by 1929 it was the standard Red Army LMG, and also the standard air weapon for hand-aimed pivoted mounts. A typical DA weighed 7.5kg (16.5lb) and fired at 500 to 780 (usually 600) rds/min, with a muzzle velocity (m.v.) of 840m (2,755ft) per second.

PV Standing for *Pulyemet Vozdushnyi*, meaning 'machine-gun for air use', this rifle-calibre machine-gun was one of many derived from the Maxim, made by Vickers. Small-arms designer A D Nadashkevich removed the large barrel casing, converted the gun to fire the standard rimless 7.62mm ammunition and changed the feed from a fabric belt to disintegrating metal links. The resulting gun was 130cm (51in) long, fractionally longer than a DA. It weighed 14.5kg (32lb) and fired at a cyclic rate of 750rds/min, with m.v. of 870m (2,855ft) per second. It was a standard fixed gun for fighters, usually with Constantinesco synchronization.

ShKAS With this rifle-calibre machine-gun the Soviet Union strode far ahead of the rest of the world. The designation meant *Shpital'nyi/Komarnitskii Aviatsionnyi Skorostrel'nyi*, the designers' names plus 'aviation fast-firing'. Chief designers B G Shpital'nyi and I A Komarnitskii created a brilliant weapon which was lighter, simpler and more compact than (for example) a rifle-calibre Browning, and fired faster than any other gun of its day. Gas-operated, it went into production in seven forms, for fixed and pivoted mounts, all firing regular 7.62mm ammunition. A typical weight was 10.6kg (23.4lb), firing rate 1,800rds/min and m.v. 825m (2,710ft) per second. From 1937 the Ultra-ShKAS became available, with the action speeded up to a maximum of 2,700rds/min and installationally interchangeable with earlier versions.

APK, DRP Leonid Vasil'yevich Kurchyevskii was chief proponent of a family of large-calibre recoilless cannon, derived from the British Davis of 1915. The propellant charge fired the shell forwards and balanced the recoil by firing an inert mass plus the hot gas through another muzzle pointing to the rear. Tupolev aircraft mounted the APK-4 of 76.2mm (3in) calibre, the APK-11 of 45mm (1.77in) and the APK-100 of 101.6mm (4.0in).

M-1927 field gun Pavel Ignatyevich Grokhovskii was a pioneer of airborne forces, and also designed aircraft.

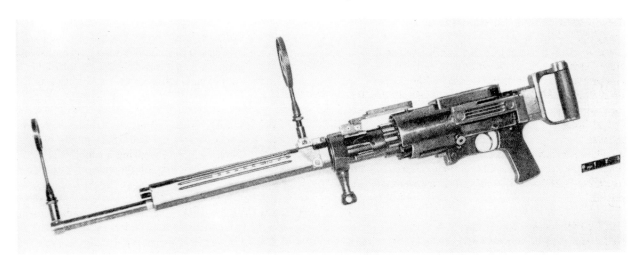

One of the ShKAS versions for pivoted mounting. (RART)

20mm ShVAK. (RART)

One of the things he did in 1934 as Director of the Airborne Forces Special Design Team was to strip down Model 1927 field guns, of 76.2mm (3in) calibre, and fit them to heavy aircraft. Details of the tests with the TB-3 are given in the section on that aircraft. Note: unlike the Kurchyevskii weapons, these were high-velocity guns with full recoil.

Oerlikon To defend some of his giant aircraft of the early 1930s Tupolev used this Swiss/German cannon, which fired 20mm ammunition of lower power than the Hispano and Soviet ShVAK. The latter gun (see below) had not then been developed. The Ikaria-made MG-FF weighed 23kg (51lb) and fired at 600rds/min with m.v. of 520m (1,700ft) per second.

ShVAK Shpital'nyi collaborated with S V Vladimirov to scale up the ShKAS to produce an equally outstanding cannon of 20mm calibre. The designation meant *Shpital'nyi/Vladimirov Aviatsionnyi Kurpokaliber*, the principal designers plus 'aviation large-calibre'. Predictably simpler and lighter than the British Hispano, it was produced from late 1936 in several versions. For turrets and hand-aimed installations the barrel length was usually 124.5cm giving an overall length of 176cm (69.3in) and weight of 40kg (88lb). Firing rate was 750–850rds/min, with an m.v. of 800m (2,625ft) per second. Fixed versions had a barrel length up to 170cm, giving an overall length of 215.5cm (84.9in), but the weight of 43kg (95lb) was still lighter than a MkII Hispano, and the m.v. of 900m (2,955ft) per second was significantly greater.

UB Meaning *Universal'nyi Beresin*, this gas-operated heavy machine-gun by Mikhail Yevgen'yevich Beresin was again the best weapon of its class in the world. As it was designed in the 1930s, and cleared for use in 1940, it could hardly fail to knock the 'fifty-calibre' Browning into the proverbial cocked hat. The chief versions were the UBK for fixed wing mounting, the synchronized UBT (not used by Tupolev) and the UBT for turrets. All weighed about 21.4kg (47lb) and fired 12.7mm (0.5in) rimless ammunition at 1,050rds/min, with m.v. of 850m (2,790ft) per second.

VYa Designated for designers A A Volkov and S Ya Yartsev, this was the first in a series of devastating aircraft guns which among other things made aircraft able to destroy heavy battle tanks. The VYa was qualified in 1940 to fire a new design of 23mm ammunition whose projectile weighed 200g (compared with 96g for the 20mm ShVAK) at an m.v. of 905m (2,970ft) per second. The gun weighed 68.5kg (151lb) and had a cyclic rate of 500rds/min.

NS-37 This 1942 gun was named for A E Nudel'man and A S Suranov, and it fired 37mm ammunition at a cyclic rate of 250rds/min. The projectiles weighed 735g, and had an m.v. of 900m (2,950ft) per second. This immense gun was 267mm (105in) long and weighed 150kg (331lb).

NS-45 In 1944 the same design team produced the same gun with a slightly shorter barrel of 45mm calibre, firing a 1,065g (2.35lb) projectile with an m.v. of 850m (2,790ft)

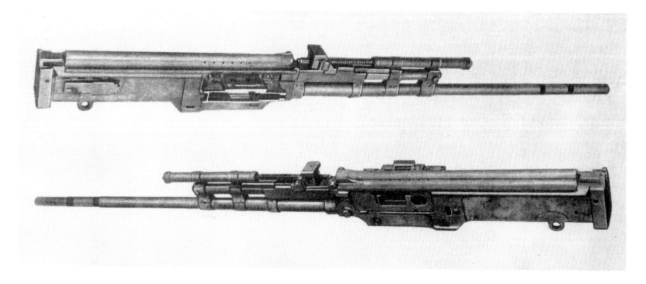

The VYa-23 cannon. (RART)

per second. The gun was 252cm (99in) long and weighed 152kg (335lb).

B-20 The replacement for the ShVAK was developed by Beresin, who did much better than merely scale up his UB to 20mm. Though it was again gas-operated, from a point well down the barrel, it was singularly neat and weighed only 25kg (55lb), compared with 42kg for the ShVAK. Length was reduced from 215cm to 138cm. It fired the same ammunition at the same cyclic rate and with the same m.v.

NS-23 In 1945 Nudel'man and Suranov replaced the VYa with this excellent gun, which fired lower-power 23mm ammunition at 550rds/min, and accordingly was a much lighter weapon. Projectile weight remained 200g, but m.v. was reduced to 690m (2,264ft) per second, enabling the gun to be shortened and to weigh only 37kg (81.6lb).

NR-23 In 1949, four years after the NS-23, this gun was produced by Nudel'man and A Rikhter. Even simpler and more reliable, and weighing just 2kg (4.4lb) more, it increased cyclic rate to 950rds/min.

AM-23 Though derived from the NR-23, this 1954 gun was strengthened and altered mechanically to achieve a cyclic rate of 1,300rds/min. Most were shorter than the NR-23, but a long-barrel version remains in use with overall length of 217.5cm (85.6in) and weight increased to 43kg (95lb).

GSh-23 Named for V P Gryasev and A G Shipunov, this twin-barrel gun was qualified in 1961. It was the first of several weapons firing a new 23mm round with projectile weight reduced in most versions to 186g whilst increasing their lethality. This results in more compact ammunition, despite an increase in m.v. to 735m (2,415ft) per second. Standard cyclic rate is 3,600rds/min, and in turret form the gun weighs 50.1kg (110.5lb). The 1976 GSh-23M, with barrels superimposed instead of side-by-side, has variable firing rate up to 4,000rds/min, and weighs 52kg (114.6lb).

Appendix 3
Missiles carried by Tupolev aircraft

KS-1 The first air-launched cruise missile to be developed in the Soviet Union, this subsonic weapon was designed in 1948–51 by OKB-155, the famous bureau headed by A I Mikoyan and M I Guryevich. It thus resembled a scaled-down version of contemporary MiG fighters. Powered by a centrifugal turbojet, it was primarily for use against surface ships and had a large armour-piercing warhead. Launched in the known direction of targets, it cruised on autopilot until its nose radar detected a sufficiently reflective target. It then switched to automatic homing. The guidance was tested on a MiG-9L dropped from a Tu-4. The production missile had span and length of 4.9m (16ft 1in) and 8.25m (27ft

1in), and launch weight of about 3t (6,614lb). Range was 150km (93 miles) cruising at 800km/h (497mph). The missile formed part of *kompleks* K87 in the Tu-16KS. The ASCC (Allied Standards Co-ordinating Committee) designation was 'AS-1 Kennel'.

KSR-2 Remarkably, this subsonic cruise missile was designed to have considerably greater range than the KS-1 even though it was powered by a rocket engine! The KSR-2 was developed from the KS-1 by the Bureau of A Ya Bereznyak, who in 1939–43 had been famed for producing the BI-1 family of rocket-engined interceptors. It was again an anti-ship weapon which, after

KS-1 displayed with its pylon launcher. (RART)

The Lavochkin K-10S, carried by a Tu-16K-10. (RART)

cruising on autopilot, switched on active radar for the terminal homing phase. Span and length were 4.75m (15ft 7in) and 8.58m (28ft 2in), body diameter remaining 1.22m (4ft). The launch weight was 3,800kg (8,377lb), and range varied with flight profile up to 330km (205 miles), with a cruising speed of nearly 1,000km/h (620mph). KSR-2P and the inertially guided KSR-2M both formed part of *compleks* K-16, carried by the Tu-16K-16 and the later dual-role Tu-16K-11-16. The ASCC called these missiles 'AS-5 Kelt'.

K-10S With a configuration reminiscent of the contemporary USAF Hound Dog, this large cruise missile could exceed the speed of sound in its dive on the target. Again it was primarily an anti-ship weapon, initially cruising at Mach 0.9 on autopilot, with turbojet propulsion, accelerating to Mach 2 whilst homing on its target in a dive. The warhead was either a large conventional armour-piercing type, or nuclear with a yield said to be 500Kt. Span and length were 4.9m (16ft 1in) and 9.95m (32ft 8in), and launch weight 4,400kg (9,700lb). A typical range was 350km (217 miles). K-10S was developed by the former aircraft-design collective of S A Lavochkin, and formed part of *kompleks* RSL-1. The ASCC name was 'AS-2 Kipper'.

Kh-20 One of the largest weapons ever carried by aircraft, this cruise missile was again a product of the MiG aircraft-design bureau. Aerodynamically it was based on the I-1 and I-3 fighters, though unlike the latter it had a centrifugal turbojet. It was again originally designed to cruise on autopilot before switching to the search mode and homing by radar on a moving ship target. Later versions had inertial guidance for use against land targets, in which case the usual warhead was

1Mt thermonuclear. Span and length were 9.2m (30ft 2in) and 14.96m (49ft 1in), launch weight 8,950kg (19,730lb) and a typical range 390km (242 miles). This awesome missile formed part of *kompleks* K-20, the only possible carrier being the Tu-95. The ASCC called it 'AS-3 Kangaroo'.

Kh-22 Recognising that the previous types of cruise missile could be shot down (indeed some were, in Egyptian service) it was decided to switch to rocket propulsion for supersonic cruise. More modern guidance systems were also developed, for missions that dodged and doglegged before homing on their target. OKB-155 (Mikoyan) again played a major role in designing the basic air vehicle, but completely new guidance systems were perfected for different missions. Three actually entered service, all part of the K-22 *kompleks* in versions of Tu-95, Tu-22 and Tu-22M. Kh-22 had inertial guidance and a 350Kt warhead, Kh-22MP had passive guidance for homing on emissions from the target, and Kh-22N had an active search/lock radar in the nose, both homing versions having various warhead options. All shared a common airframe with span and length of 2.99m (9ft 10in) and 11.67m (38ft 3in), and launch weight typically of 6,800kg (14,990lb). Range varied with trajectory profile to a maximum of 460km (286 miles). The designation (X-22 in Russian) stemmed from the Kharkov production factory. The ASCC name was 'AS-4 Kitchen'.

KSR-5 With this missile almost the same guidance, propulsion and warheads as fitted to the Kh-22 were repackaged in a smaller airframe with less than half the radar cross-section. Span and length were reduced to 2.5m (8ft 2in) and 10.9m (35ft 9in), and launch weight to 5,950kg (13,120lb). Maximum range was about 400km (250 miles). As before, suffixes N and P denoted active and passive radar guidance. Most missiles formed part of *kompleks* K-26, but a variant linked with K-11 was the KSR-11 version, carried only by the Tu-16K-11-16. A few were modified by the Raduga bureau as KSR-5NM or KSR-5MV supersonic targets flying on preprogrammed guidance. The ASCC name was 'AS-6 Kingfish'.

RKV-15B One of the first major programmes by the brilliant Raduga design collective was this air-breathing cruise missile, which was designed to set wholly new standards of range, guidance and unstoppability. New miniaturised warheads made it possible to package everything into an airframe with smaller radar cross-section than any previous Soviet strategic air/surface missile. Features include stealth overall shape, small folding wings unswept in subsonic cruise, turbofan propulsion and inertial cruise guidance followed by very accurate terminal homing using terrain comparison. Span (unfolded) and length are 3.24m (10ft 8in) and 8.09m (26ft 8in), and launch weight is 1,700kg (3,750lb). Maximum range on an all-hi profile is about 2,400km (1,490 miles). The ASCC name is 'AS-15 Kent'.

RKV-500B Part of the Kh-15 or Kh-15S *kompleks*, this highly supersonic missile was originally almost a copy by the Raduga bureau of the US AGM-69A SRAM (short-range attack missile), even to the extent of having three fins at 120deg. Wingless, it is propelled by a rocket motor with multiple-restart capability, and three versions were produced with different guidance (pure inertial or inertial plus active or passive radar homing) and either a 350Kt warhead or an armour-piercing conventional type. Length is 4.78m (15ft 8in) and launch weight typically 1,200kg (2,645lb); maximum range is usually 150km (93 miles). The ASCC name is 'AS-16 Kickback'.

R-4 The only air-to-air missile carried by a Tupolev aircraft was this large weapon developed in 1958–66. The

R-4RM air-to-air missile carried by the Tu-128. (RART)

design bureau was that of Matus Ruvimovich Biesnovat, which had much earlier produced high-speed aircraft. The missile was part of the awesomely complicated (for its day) *kompleks* 80, built into the Tu-128. Eventually the missile was qualified in two forms, the R-4R with active radar homing and the R-4T with passive infrared homing. Later both were modified into the R-4RM and R-4TM with totally solid-state electronics and a higher-performance rocket motor. Span over the cruciform wings was 1.49m (4ft 11in). The radar version had a length of 5.25m (17ft 3in), launch weight of 580kg and maximum reliable range of 60km. The IR version had a length of 5.16m (16ft 11in), launch weight of 545 kg and maximum range of 30km (19 miles). The ASCC name was 'AA-5 Ash'.

Index of Aircraft

Index of People